SOCIETY, ECONOMY AND THE MARKET
Commercialization in Rural Bengal
c. 1760-1800

SOCIETY, ECONOMY AND THE MARKET
Commercialization in Rural Bengal
c. 1760-1800

RAJAT DATTA

MANOHAR
2000

First published 2000

© Rajat Datta 2000

All rights reserved. No part of this publication may be reproduced or transmitted, in any form or by any means, without prior permission of the author and the publisher

ISBN 81-7304-341-8

Published by
Ajay Kumar Jain for
Manohar Publishers & Distributors
4753/23 Ansari Road, Daryaganj
New Delhi 110002

Typeset by
Kumud Print Service
Boileauganj, Shimla 171005

Printed by
Rajkamal Electric Press
B/35-9, G.T. Karnal Road Indl. Area
Delhi 110003

To
KARUNA DATTA (Ma)
and the memory of
AMULYA NARAYAN DATTA (Baba)

Contents

List of Tables   8
List of Figures   11
List of Maps   12
Abbreviations   13
Preface   15

Introduction: Towards a Comparative Perspective on Agrarian Commercialism   21

Chapter 1   Patterns and Processes of Agricultural Production   37

Chapter 2   The Peasantry and the Question of Peasant Stratification   88

Chapter 3   The Rural Elite: Landed Property and the Landed Gentry   133

Chapter 4   The Agrarian Economy and the Dynamics of Commercial Transactions   185

Chapter 5   Dearth and Famine: Ecological Dislocation, Subsistence and Crises   238

Chapter 6   Subsistence Crises and the Agrarian Order   285

Conclusion   324

Appendix 1   The Company's Tax Burden and the Economy   333

Appendix 2   The Problem of Bullion and the English East India Company in Bengal   342

Bibliography   365

Index   371

Tables

1. Output of Rice in Selected Parganas — 41
2. Output of Lentils — 42
3. Output of Various Grades of Oil-seeds — 43
4. Estimates of Land under Sugarcane in some Districts in the 1790s — 44
5. Labour Utilization in the Cultivation of Sugarcane in East Bengal — 45
6. Labour Utilization in the Cultivation of Sugarcane — 46
7. Output of Sugarcane, Sugar and Molasses — 47
8. Output of Tobacco — 49
9. Ratio of Cotton to Silk in Company's Investments — 51
10. Cotton in Bengal: Local Production and Amount Imported — 52
11. Patterns of Cultivation in Bengal — 57
12. *Jama* of *Khudkashta* and *Pahikashta* Cultivators in Laskarpur, 1770-1 and 1772-3 — 94
13. *Huzuri* Talluqas in Bengal, 1778 — 142
14. Sale Values of Talluqas in Bengal, 1724-93 — 146
15. Grants of *Baz-i-zamin* by some Zamindars, before and after 1765 — 149
16. *Baz-i-zamin* Holders in Muhammadshahi, 1787 — 149
17. *Baz-i-zamin* and *Raiyati* Lands in Muhammadshahi, 1787 — 150
18. Distribution of Land in Amerabad and Bhuluah, 1775 — 150
19. Distribution of Land in pargana Apole, 1777 — 150
20. Distribution of Land in Jalasore, 1792 — 151
21. *Chakeran-zamin* in some Zamindaris — 151
22. *Baz-i-zamin* in Midnapur and Jalasore, 1773 and 1787 — 152
23. *Chakeran* and *Baz-i-zamin* in some Zamindaris — 153
24. *Kharij-jama* in Eleven Districts, 1778 — 154
25. Source of Zamindari Income, 1786 — 158
26. Officials in *Khas* Talluqas of Murshidabad — 162
27. Cultivation of *Hasil* Lands in pargana Muhammadshahi — 165
28. List of Market Officials in Purnea, 1790 — 169
29. Annual Cash Expenditures of Nine Zamindaris — 171
30. Annual Cash Disbursements of the Zamindar of Mahisadal, 1773 and 1783 — 172
31. Personal Expenses of Bengal Zamindars — 172
32. Cash Salaries and Wages of Zamindari Officials — 173

33.	Expenditure Patterns of Officials Serving the 'Great Landholders' in Rangpur and Dinajpur	174
34.	Estimates of Villages in Different Districts of Bengal	187
35.	Spatial Distribution of Villages	188
36.	Professional Groups in Rangamati, 1775	189
37.	Professional Groups in Sibpur, 1791	190
38.	Professional Groups in Rangpur, c. 1807	191
39.	Consumption Patterns of the Rural Gentry in Eastern Bengal, 1789	192
40.	Patterns of Consumption of an Artisan in Comfortable Circumstances, Rangpur, c. 1807	193
41.	Patterns of Consumption of a Common Artisan, Rangpur, c. 1807	193
42.	Patterns of Food Consumption in Rangpur, c. 1807	194
43.	Price Differentials between Town and Country	199
44.	Number of Market-Places in Selected Districts	206
45.	Volume of Local Trade in Food Grains	207
46.	Food Shortages and Natural Calamities in Bengal, 1700-60	240
47.	Cropping Patterns of some Cash Crops	241
48.	Dearth and Famine in Bengal, 1761-1800	243
49.	Harvest Damage in Dearth and Famine	250
50.	Rice Prices under Dearth in the Late Eighteenth Century	255
51.	Revenue Collected During 1765-72	257
52.	Collections from Districts, 1769-70 and 1770-1	258
53.	Revenue Collected from the Worst Affected Districts in 1769-71	258
54.	*Taqavi* and *Hasil* in some Bengal Parganas, 1769-70 and 1770-1	260
55.	Mortality and Desertions in 1770: Rajshahi and Chittagong	263
56.	Extent of Cultivation in Bengal: The Evidence of the Amini Commission	265
57.	Estimates of Population in Bengal	266
58.	Balances with the *Dalals* at Dhaka, 1767-72	295
59.	Company's Investments for Bengal Piece-goods and Silk, 1766-75	296
60.	Prices of Raw Silk in Bengal, 1768-71	298
61.	Looms and Weavers in Dhaka's Manufactories	300
62.	Share of the Company and Private Traders in Production of Cotton Textiles in Dhaka, 1790-9	304
63.	Bengal's Revenue Assessment, 1700-90	333
64.	Bengal's Share of Company's Exports of Bullion for Trade with Asia	344
65.	Distribution of Company's Bullion in India	345
66.	Merchandise and Bullion Imported into Bengal by Private Traders, 1796-7 to 1806-7	346
67.	Monthly Rates of *Batta* on Arcot Rupees in the *Bazaars* of Dhaka, 1769-73	349

68.	Company's Exports from Bengal, 1802 and 1803	351
69.	Bullion into and from China, 1768-9 to 1770-1	352
70.	Balance of the Company's China Trade, 1761-2 to 1770-1	353
71.	Transfer of Bullion to other Presidencies, 1761-2 to 1770-1	354
72.	Estimates of the Drain from Bengal, 1757-8 to 1793-4	356
73.	The Drain as a Portion of the Gross Produce (Selected Years)	357

Figures

1. Agricultural Output, *c.* 1794: A Sectoral Comparison — 76
2. Index Numbers of Rice and Sugar Prices, 1700-1800 — 77
3. Index Numbers of Rice Prices in Calcutta, 1754-1800 — 78
4. Index of Rice and Jaggery Prices in Calcutta, 1754-1800 — 79
5. Food Grain Prices in Lower Bengal, 1700-1800 — 221
6. Comparative Rice Prices in Birbhum, 1784-1813: Coarse Rice — 222
7. Comparative Rice Prices in Bengal, 1700-1800: Common Rice — 223
8. Bengal Rice: Prices, 1700-1800, Price Line, Trend and Moving Averages of Common Rice — 224
9. Lower Bengal Rice: Prices, 1700-1800, Price Line, Trend and Moving Averages of Common Rice — 225
10. Prices in Lower Bengal: Oil and *Ghee*, 1700-1800 — 226
11. Calcutta: Mustard Oil and *Ghee* Prices, 1754-1800 — 227
12. Rice Prices in Bengal Districts: Ordinary Rice — 228
13. *Aman* Rice Prices: Burdwan and Birbhum, 1784-1813 — 229
14. 1770 Famine Prices: Ordinary Rice — 272
15. 1788 Famine Prices: Ordinary Rice — 273
16. Monthly Rice Prices in Midnapur, October 1769 to September 1770 — 274
17. Rice Prices in 1788 Famine: Ordinary Rice — 275
18. Bengal's Revenue, 1766-84 — 339
19. Revenue and Harvests as per cent of *Jama* — 340
20. Company's Goods and Bullion Exports to Asia, 1733-1805 — 359
21. Gold and Silver in Bengal's Revenue, 1796-7 to 1808-9 — 360

Maps

1. Bengal in the Eighteenth Century: The Region and its Major Places   14
2. Agricultural Profile of Bengal: Zones of High Activity and Moribundity   69
3. The Spatial Sweep of Famine: 1769-70 and 1788   249

Abbreviations

Add. Mss.	Additional Manuscripts
BDR	*Bengal District Records*
BPC	Bengal Public Consultations
BRC	Bengal Revenue Consultations
BRFW	Proceedings of the Board of Revenue at Fort William
BM	British Museum
BRP	Bengal Board of Revenue Proceedings
BR Misc	Proceedings of the Board of Revenue (Miscellaneous)
CEHI 1	*The Cambridge Economic History of India, Volume 1, c.1200-c.1750*, edited by Irfan Habib and Tapan Raychaudhuri, Indian reprint, Delhi, 1984.
CEHI 2	*The Cambridge Economic History of India, Volume 2, c.1757-c.1970*, edited by Dharma Kumar with Meghnad Desai, Indian Reprint, Delhi, 1984.
CCR	Proceedings of the Calcutta Committee of Revenue
CCRM	Proceedings of the Controlling Council of Revenue at Murshidabad
Circuit	Proceedings of the Committees of Circuit
Coast & Bay	Letters from Coast and Bay
Committee	Proceedings of the Controlling Committee of Revenue
CPC	*Calendar of Persian Correspondence*, Calcutta, various dates
FR	Factory Records
FR 2	*The Fifth Report of the Affairs of the East India Company, Volume II*, edited by W.K. Firminger, Calcutta, 1917
Grain	Proceedings of the Board of Revenue, Grain Branch
JPS	*Journal of Peasant Studies*
HM	Home Miscellaneous
IESHR	*The Indian Economic and Social History Review*
IOL/IOR	India Office Library/Records
Khalsa	Proceedings of the Board of Revenue, Khalsa Branch
LCB	Letter Copy Book of the Resident at the Durbar
Ms/Mss	Manuscripts
PCR	Proceedings of the Provincial Councils of Revenue
Reports	*Reports from the Committees of the House of Commons*
WBDR, ns	*West Bengal District Records, New Series*, Calcutta, various dates
WBSA	West Bengal State Archives

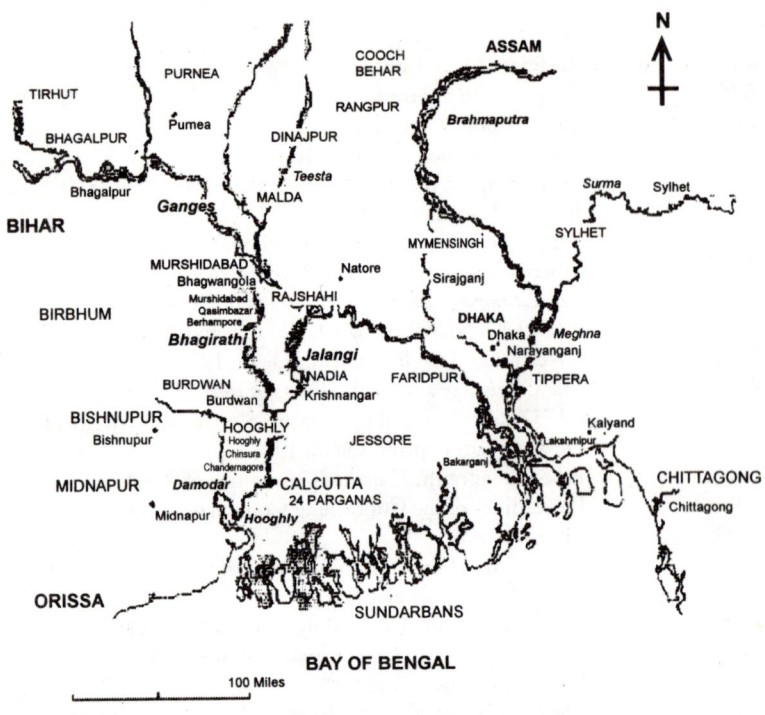

Sources: Based on James Rennell, *Bengal Atlas*, London, 1781; P. J. Marshall, *Bengal: The British Bridgehead, Eastern India, 1740-1828*, London, 1987.

MAP 1. BENGAL IN THE EIGHTEENTH CENTURY: THE REGION AND ITS MAJOR PLACES.

Preface

SOME OF THE most innovative historiographical revisions in recent years have occurred in the study of the eighteenth century. From being considered an age of decline, fresh researches on various regions have revealed the innate dynamism of this century. Yet its full face is still far from complete. One major gap is the second half of the century.[1] Nowhere is this more evident than in the case of Bengal. For far too long existing historiography has tended to view the late eighteenth century in Bengal as a period of crisis,[2] when the supposed economic prosperity of the early eighteenth century[3] was irrevocably halted. This was achieved by a combination of state, i.e. the English East India Company's, exploitation and the disastrous famine of 1769-70, an event, which according to consensual wisdom, decimated 10 million people or one-third of the province's population and unleashed a prolonged agricultural recession. The essential focus of such studies hinges around the quantum of revenue imposed and collected by the state and not on the actual operation of the agrarian economy. Therefore what is characterized as the economic history of the province is actually its revenue history, or at best a history of changes in the structure of landed property between 1765 and 1793.[4] While there is no denying the important role performed by the revenue squeeze in the provincial economy, there is a pressing need to contextualize that role in the overall framework of economic, mainly agricultural, production. Revenue is a form in which the surplus is appropriated by the state, but it is the mode in which that surplus gets created which provides the crucial insight into the domain of production.

In this context, this work is about agricultural production, consumption and the dynamics of agrarian commercialism in rural Bengal in the late eighteenth century. It is also about revising some of the established views of India's first brush with British rule. The basis of the study are the revenue records and other papers of the English East India Company and its officials. The Company's archives provide the largest single body of evidence on the economic history of Bengal for this period. They contain enormous amount of hitherto unutilized information on the essential elements which shaped the face of Bengal's rural society and economy: the peasantry, landed property, agricultural production, markets, local trade as well as provide invaluable details regarding subsistence crises and coping. These issues are the central concerns of this book.

The main purpose of the ensuing discussion is to demonstrate that Bengal underwent an unprecedented degree of commercialization in this period. There is no doubt that in previous decades the province had become pervaded by intricate commercial networks linked to the trade in textiles. Later (that is, in the nineteenth and twentieth centuries) Bengal went through a significantly new commercial phase under the impact of indigo, jute and tea. Nonetheless, there seems to be a strong case to argue that while the early eighteenth century textile-based commerce set the tone of a long-wave of commercialization, it was the late eighteenth century which provided the distinctive departure by bringing about the commercialization of the province's *rice producing economy and of the social relations which were embedded in it*. The clearest indications of these developments are the rapid proliferation of a sharecropping peasantry in this period (Chapter 2) and the increasing intervention of the grain-merchant (*byapari*) in agricultural production (Chapter 4).

There are also many areas and problems which this book does not address, or addresses only partially. There is very little in the subsequent pages on land-revenue administration. The role of the English East India Company and the problem of the drain of wealth from Bengal are either not dealt with or touched upon only to the extent that they actually influenced the working of rural commercial transactions. The issues associated with these problems are briefly discussed in the appendix, the reasons for which are twofold. First the major contours of these issues are fairly well established. Second, there is pressing necessity of going beyond these issues in order to reconstruct the structures on which these administrative and exploitative edifices were based. This is not to deny the central purpose of the Company's rule after 1765, which was to use Bengal's resources for complex commercial, military and political purposes; but to study that exploitation (undoubtedly the dominant preoccupation of existing historiography) divorced from the material milieu of production seems a largely pointless exercise.

There is also very little discussion of the so-called cash crops such as mulberry, opium or indigo in the following pages. These were commercial products par excellence, and their vicissitudes have been quite exhaustively documented by historians.⁵ Additionally, these were closely associated with the *international* movements of Asian trade. My concern in contrast is only with those commodities which had the most direct bearing on the daily lives of people. Chapter 1, therefore, is an attempt to reconstruct the patterns of production of those crops most closely associated with the *internal market* and for which there was a palpable demand on a day to day basis. Cotton figures in this chapter because it was essentially designed for internal consumption despite the high international demand for piece-goods. Ninety per cent of Bengal's cotton cloth was manufactured for supply to the Indian markets. By contrast, silk had become an export commodity in the service of European trade by this period. This is clearly evident from the declining trend in Bengal's silk exports to Asia between 1748 and 1789 and the rising trend of Company investments for that commodity during that period, and especially after the grant of the Diwani in 1765.⁶

The central purpose of this book is to suggest that late eighteenth century Bengal (*c.* AD 1765 to *c.* AD 1800) witnessed a number of structural changes in its agricultural economy and rural society. These changes were caused by a conjuncture, in this period, between four critical elements. The changing linkages between the state and the market brought about by the political transition in the province was one element of that conjuncture. Other elements were the growing demand, especially in rural areas, for food in the province and a secular upward swing in the prices of agricultural produce. Finally, the apparently fortuitous instances of dearth and famine between 1769 and 1793 contributed significantly to the conjuncture. The order in which these various elements have been listed does not reflect each one's hierarchical importance in the conjuncture, but they are intended to delineate the political and the economic elements in that configuration of forces and events which shaped Bengal's economic life in this period. The net result of this conjuncture, I propose, was to lead the province into one of its most intense phases of agrarian commercialism, the uniqueness of which was the commercialization of its *rice producing small-peasant economy*.

I am grateful to many scholars for shaping my ideas over the years. Peter Marshall encouraged my initial forays in this direction when he supervised my Ph.D. thesis on the economy of rural Bengal in the late eighteenth century at Kings College, London. To him I will always remain beholden for the many insights he shared with me, for his mastery over the records pertaining to this period and for his personal warmth and affection. If he sees many of his ideas reflected in this work the resemblance is more than coincidental. C.A. Bayly and David Arnold gave me the benefit of their critical comments. Peter Robb took the trouble of writing a detailed critique of the thesis which helped me in making substantial modifications to bridge wide gaps in my analysis. To Harbans Mukhia goes my gratitude for nurturing my initial attempts at historical research and for sustaining it with his constant interest and encouragement. Sabhyasachi Bhattacharya painstakingly read and commented on various portions as this book was taking shape and discussions with Dilbagh Singh helped in pinpointing some central issues concerning the agrarian history of India in the eighteenth century. In many respects Binay Bhushan Chaudhuri has been a pioneer in the economic history of early-colonial Bengal, and a lot of my concerns are derived from engaging with his arguments. Burton Stein encouraged me to look outside obvious arenas and to search for commercial networks in hidden corners of society. I have tried to do that in my limited capacity. That he will not be able to read this book is only one of the many tragedies that his death has caused. Let me, however, hasten to add that all these people are in no way responsible for the shortcomings that abound. The culpability for those is entirely mine.

I also take this opportunity of saying a special thank you to Ratna for patiently coaxing a very lazy person into taking this step and to Rhea for having found all this so boring. I hope I haven't let either of them down.

NOTES AND REFERENCES

1. The notable exception is C.A. Bayly, *Rulers, Townsmen and Bazaars: North Indian Society in the Age of British Expansion, 1770-1870*, Cambridge, 1983. For Bayly too the second half of the eighteenth century is a period where 'the trail runs cold' (p. 4).
2. For a recent reiteration of this old view see A.K. Bagchi, 'Markets, Market Failures and the Transformation of Authority, Property and Bondage in Colonial India', in Burton Stein and S. Subrahmanyam (eds.), *Institutions and Economic Change in South Asia*, Delhi, 1996, pp. 50-2.
3. For a recent example of such an argument see Sushil Chaudhuri, *From Prosperity to Decline: Eighteenth Century Bengal*, Delhi, 1995, and for a critique of Chaudhuri see my review of the book in *Studies in History, New Series*, vol. 13, no. 1, January-June 1997.
4. The classic example of such an approach is N.K. Sinha, *The Economic History of Bengal: From Plassey to Permanent Settlement*, vol. 2, Calcutta, 1968; also see B.B. Chaudhuri, 'Agrarian Relations: Eastern India', in *CEHI 2*, pp. 86ff.

5. See for instance, B.B. Chaudhuri, *Growth of Commercial Agriculture in Bengal, 1757-1905*, Calcutta, 1964; for a discussion of the growing constraints on commercial farming in the late eighteenth century see Bhaskar Mukhopadhyay, 'Forced Commercialisation in Early Colonial Bengal: A Model and Beyond', *Calcutta Historical Journal*, vol. xv, nos.1-2, July 1990-June 1992; also Bhaskar Mukhopadhyay, 'Orientalism, genealogy and the writing of history: The idea of resistance to silk filature in eighteenth century Bengal', *Studies in History, New Series*, vol. 11, no. 2, July-December 1995.
6. See Rajat Datta, 'Markets, Bullion and Bengal's Commercial Economy: An Eighteenth Century Perspective', in Om Prakash and Denys Lombard (eds.) *Commerce and Culture in the Bay of Bengal, 1500-1800*, Delhi, 1999, p. 332.

INTRODUCTION

Towards a Comparative Perspective on Agrarian Commercialism

Historically, peasant production has been much more than the cultivation of a variety of crops. It has been the central point of an agrarian economy around which various other social relations are articulated. Thus, the state of the entire agrarian economy, defined as the intermeshing of natural conditions, social relations, and institutional networks in the countryside, were determined by the fluctuations in the domain of agricultural production and peasant enterprise.[1] The primacy of agriculture was all the more vital in such economies because even non-agricultural, or craft, production was undertaken largely as an adjunct to agricultural production. In many cases, Bengal's artisans were basically peasants (that is, they cultivated land primarily with their own or their domestic labour) who entered non-agricultural production as an extension of their main enterprise as cultivators. A definition of peasant activity, made in the context of South-East Asia as the 'effort devoted to the production and cultivation of crops and the time spent on a multitude of home or artisan activities'[2] is, in general outline, applicable in the context of pre-industrial/pre-colonial societies. Thus, in many parts of Bengal salt was manufactured by the 'tillers of the land' who used the slack agricultural season to acquire supplementary incomes by working in the salt-pans (called *khalaries*).[3] Spinning of thread was definitely done by women in peasant-households, and in some cases even weavers were said to have the 'double resource of tilling the lands'.[4]

The following pages examine the dynamics of peasant production and agrarian commercialism in eighteenth century Bengal, a region deeply involved in the circuits of interregional and international commercial exchanges of growing complexity between the seventeenth and eighteenth centuries.[5] These circuits made the peasantry of Bengal operate in a complex matrix of production designed to meet a whole range of requirements generated in the tiny village market (the *haat*) to the burgeoning demand of some of their produce in the international market, and all this happened within the overarching primacy of a rice producing economy. It is therefore crucial that one specifies the end-purpose of peasant production to make any sense of the nature of interaction, if any, between commercial processes and commercialization of agricultural production.

It is being increasingly realized that there was a clear commercial context of agricultural production in medieval Bengal. Richard Eaton has argued that the ability of the province to achieve a remarkably successful high agrarian, mainly rice, surplus underpinned 'an agricultural and manufacturing boom' in later medieval Bengal and created favourable conditions for an expanding 'overland and maritime trade that linked Bengal even more tightly to the world economy'.[6] Nor was Bengal alone in this respect. The intricate connections between peasant production and an integrated produce market created by the formation of intermediate and large marketing centres (*bazaars* and *ganjs*) was a remarkable development over much of northern India in the late eighteenth and early nineteenth centuries.[7] South India underwent a transformation in its traditional peasant production into 'commercialized peasant agriculture'—that is, production for a market on an increasing scale— during the course of the eighteenth and nineteenth centuries by a close nexus between peasants with assets in land and merchants in towns.[8] There is, therefore, substantial and growing evidence to show that much of the South Asian economy in the late pre-colonial epoch was organized in similar 'hierarchic interdependency'[9] between peasant production and a tiered network of markets in ascending importance.[10]

The exact fit between agriculture and commercialism in late precolonial period is certainly problematic. Obviously, much depended on the overall milieu in which producers functioned. P.J. Marshall suggests that the balance of advantages and disadvantages in the cultivation of various crops, particularly cash crops, in Bengal's agriculture was more complex than the cultivator's response to an underlying profit motive. More crucial were the considerations of the relative pitch of revenue and the availability of ready cash. Bayly argues that the cultivation of cotton by the early nineteenth century Bundelkhand peasantry was determined by a complex combination of the available labour surplus in a peasant family, quality of land, the revenue regime, the terms of the available cash advance and the prevailing prices on the world market; the mere existence of high prices was not a sufficient inducement for the 'middle or poor peasants' to take to the cultivation of cotton. Ludden views peasant production in pre-colonial south India as a function primarily of a given ecology, viz., soil, water and drainage, which imparted its particular blend of crops (that is, agricultural specialization) to communities in their respective agricultural zones.[11] Additionally, producers often combined two occupations, agriculture with artisanal production, in their respective households.

Yet it would be difficult to argue, as Marx did, that the combination of agriculture with artisanal production in a peasant household symbolizes the working of a pre-capitalist 'natural' economy; nor would it be acceptable to see this phenomenon in a Chayanovian perspective of a labour-consumer balance in a peasant household.[12] Also, the fact that peasants often consciously resisted expansion of the domain of 'commercial' farming (e.g. cotton in Bengal) and sought to intensify the cultivation of rice should not be treated as an irrational or perverse economic choice.

Pre-colonial Asian agriculture was far from homeostatic. In fact, it is the nexus between atomistic peasant production (in personal holdings) servicing, and in turn being stimulated by, a hierarchy of demand-requirements (including that of revenue/rent) in an ascending chain of internal and global markets which are the key features in the typological construction of the late pre-colonial peasant economy of Asia.

Thus Caglar Keydar finds peasant household-production in late Ottoman Turkey 'immersed' in a series of 'interlocked' factor and product markets.[13] Studies in the economy of Ming and Ch'ing China have conclusively demonstrated the articulation of similar markets in China with overseas trade, flows of bullion and taxation in cash. These studies have also documented their effects on the creation of regional specialization of different provinces into grain and cash crop producing areas in the seventeenth and eighteenth centuries.[14] The peasant economy of Tokugawa Japan, which was relatively 'closed', nevertheless underwent, a phase of intensive commercialization of agriculture which was achieved largely by the proliferation of rural periodic 'peasant' markets, linked into a network of markets of ascending importance, leading to the large urban centres.[15] Towns meant a substantial urban presence which demanded bulk rural-urban trade in rice.[16]

Japan's economic experience of the eighteenth century is important for any analysis of peasant production in Asia because the entire commercial edifice during the Tokugawa period was constructed mainly on the basis of rice and by the *commercialization of its rice-producing economy.*[17] Peasant initiative in expanding the acreage devoted to rice for subsistence and for the market was the prime mover and the hallmark of this commercialization, and this was accompanied by the transition from rent in kind to rent in cash in the Japanese countryside.[18] There is no reason to assume the uniqueness of Japan in the eighteenth century, for, as Clifford Geertz shows, it was not before late in the nineteenth century that one can effectively speak of the divergence of the Japanese 'path' from that treaded by other countries of Asia.[19]

The example of rice production in China also demonstrates links between commercial impulses and the so-called subsistence sector in agriculture. Here, state initiative in the early medieval phase encouraged the cultivation of better strains of rice which fitted well with the enterprise of the primary producers. The result was a steady expansion in agricultural production up to the end of the eighteenth century[20] which was characterized by a high degree of commercialization in which most farmers 'produced in part for the market and some, probably along the coast, produced primarily for the market'.[21] 'Market expansion' and 'market conditions' have been seen as the important variables in stimulating agricultural production in southern China in the eighteenth century.[22]

The cases of Tokugawa Japan and Ch'ing China amply demonstrate the intricate connections between cereal production, peasant economies and commercialized agriculture in late pre-colonial Asia; and at the typological level they also show the futility of (i) adhering to any rigidly antithetical notions of 'subsistence' *versus* 'commercialized' agriculture, and (ii) limiting oneself to the growth of the so-called 'cash' crops while studying the workings of commercialized peasant production. Unfortunately, there have been few attempts to break these rigid, and largely incorrect, boundaries in the context of agricultural production in eighteenth century South Asia: the reasons for which are the total absence of a comparative *Asian* perspective and an excessive preoccupation with the impact of early-colonial regime on the economy. The latter preoccupation is most pervasive in the historiography of the province of Bengal. Here, the second half of the century is interpreted as a period of agricultural recession brought about by the revenue burden of the English East India Company and by the severe famine of 1770, and commercial production is located exclusively in those sectors (like opium and indigo) which were structurally in the service of the Company's commerce.[23] This argument needs to be substantially modified as there is now fairly substantial evidence to show the existence and strengthening of an integrated provincial economy in this period, operating in the main through a hierarchy of markets and trading networks in primary staples like rice and other items of internal consumption.[24] In fact in a recent re-examination of the evidence regarding the nature of opium and indigo cultivation in Bengal, Chaudhuri indeed places a greater emphasis on 'autonomous' market forces, in relation to the use of force, in determining the pace and direction of the cultivation of these crops. He also points towards peasant resistance and the availability of alternative sources of credit as some of the other factors mediating in favour of the cultivators.[25]

Significantly, the central indicator used to measure the health of this economy was the price of rice and not of any other commodity. It was also generally accepted that the prevailing price of rice influenced the state of other agricultural and non-agricultural prices;[26] and that it was access to rice, and not to any other 'cash' crop, which was crucial in ensuring physical survival during a crisis of subsistence.[27] This was perhaps natural. The product value of the different varieties and qualities of rice and paddy accounted for more than 45 per cent of the gross agricultural output. It must, however, be borne in mind that the dominance of the food grain sector in total agricultural production, and the influence of cereal prices on the prices of other farm and non-farm products was not limited to Asian agriculture.[28] Rice in Bengal was far from being a subsistence crop.[29] Large portions of the land under rice (sometimes as high as 70 per cent in some districts) were cultivated by peasants under a complex system of cash advances made by the grain merchants (see Chapter 4). These loans were aimed at denying the peasantry any direct access to local markets, and to ensure the advance hypothecation of their product *prior to the completion of the production cycle*. This feature, namely, the extent of mercantile penetration in primary agriculture (discussed in Chapter 4), perhaps imparted a distinctive and unique thrust to the processes involved in the commercialization of Bengal's rice production, thereby making it substantially more commercialized than most contemporary economies in eighteenth century Asia.

In Tokugawa Japan, cash cropping was often the monopoly of the big feudal *hans*, while rice was produced in peasant farms. The system of cash loans for rice was practically undeveloped, but an extremely elaborate system of mercantile credit and investment of capital did exist for other products and commodities.[30] The merchants function was to purchase grain from the lords' or the farmers' markets, or to commute the peasants rice revenue into cash, and to move their stock for sale to the larger cities and to the capital.[31] In Ch'ing China, the marketing of rice was apparently the main item of local trade (so much so that rice was often used as the standard to reckon the price of other goods along with existing metallic currencies), but the process of moving grains and other commodities from the villages of origin was accomplished 'largely through inequalities in the form of rents and surpluses for marketing', a system which favoured the landlords the most.[32] Merchants did invest in the cultivation of tea, sugarcane and tobacco,[33] but remained largely dissociated from the production of rice.

The extent of mercantile involvement in the production of rice in Bengal appears to have been unique also within the Indian subcontinent. Two studies of regional grain trading in eastern Rajasthan and Bihar[34] demonstrate the existence of an elaborate hierarchy of markets and specialist grain merchants, but provide no evidence to show whether these networks had any influence on peasant-merchant relationships. In the case of eastern Rajasthan such developments were probably stymied by the extremely intrusive role of the state in the marketing of grain which, as Bajekal argues, forced these traders to operate more in the nature of state agents than as autonomous merchants. In Bihar, grain trading was virtually free of state control, but the grain-merchant had very little to do in the processes of agricultural production. Production loans were apparently provided, not by specialist traders, but by a coterie of peasant-cum-small-traders, the *grihastha-byaparis* (literally, the householder cum trader), who are described by Banerjee as the 'dominant peasants' who used their position in the village society to branch out into trade.[35]

Yet, situations like Bihar and Rajasthan only go to show that there was, in the eighteenth century, an increasing incidence of agrarian commercialism, and that this process was caused by the substantive integration of production with the market. Taking advantage of the producers vulnerabilities, merchants often succeeded in imposing oligopsonic controls over local resources. The burdens imposed on the cultivators by the pressure of revenue naturally aided these merchants. Since revenue was overwhelmingly collected in cash (at least this was so in Bengal), the produce had to be sold, and the quantity of cash which would accrue to the producer was directly related to the price line. Witness, for instance, the following description in this connection given by Colebrooke: 'when the crops of corn are very abundant, it is not only cheap, but wants a ready market. As the payments of rents are regulated by the season of the harvest, the revenue is due and must be paid, whether there be or not be a vend for the produce. . . . Thus the eagerness of the vendors [further] reduces the price.'[36] Merchants would buy these stocks at almost distress prices, store in their *golahs* (granaries) and then sell when prices were again favourable from their points of view.[37] The price situation was therefore of critical importance in all market transactions; but it would be unreasonable to argue that they (both prices and transactions) were determined by an untrammelled operation of the laws of supply and demand, or that everyone benefited equally. The village was at a relative disadvantage in such a system of trade. Since each of these markets was connected with the next rung in an increasing scale of commodity transactions, the primary focus of trade naturally tended to shift away from the village. Given this situation no single peasant-household could have had an appreciable impact on the formation of prices even at the level of the village *haat*.[38] Moreover, the oligopsonic power of the local merchants over trade in agricultural produce, and the monopsonic control of the East India Company in sectors like cotton piece-goods and silk (later to be supplemented by opium and indigo) made for a situation where the formation of prices depended on command and coercion to an extent equal to the influence exerted on them by the interplay of demand and supply.

It is precisely because of such imperfections that these markets became more than areas of contact between potential buyers and sellers. They reflected prevailing social inequalities in the sphere of exchange which were crucial in the processes of commercial accumulation. In fact, as I have argued elsewhere, the various devices adopted by merchants to restrict the peasants' independent access to such markets was part of an elaborate system which (i) ensured the former's pivotal role in the mechanisms by which the surplus in kind (extracted by the revenue-tax-rent combine) was converted into surplus into cash; and (ii) facilitated the intrusion of merchant-capital into the circuits of peasant production through the advance contracts-debt servicing channel.[39] Thus, instead of the peasants going to the market, the market seems to have come in a fairly big way to the peasants of eighteenth century Bengal.[40] A symptom of this was the rapid mushrooming of intermediaries between the merchant and the cultivator in rural Bengal in this century.[41]

A proliferation of mercantile intermediaries was necessary in order to minimize the degree of risks involved in operating in an economy where production and procurement centres (the villages) were often widely dispersed. Crucial also was the role these intermediaries performed in keeping a close check on prices in order to forestall, or at least minimize, the effects of variations in price, especially those massive fluctuations which accompanied both bad and plentiful harvests.[42]

Situating Eighteenth Century Bengal

The earlier analysis has tried to indicate that there were perhaps certain unique features in processes of agrarian commercialism in eighteenth century Bengal. The complexities involved in the entire process must have involved a number of variables, of which three seem to have been fundamental.

First, there were the anterior developments which occurred in the seventeenth century. These were (*a*) a shift from spices to textiles by the European trading companies; (*b*) the transfer of imperial tribute from Bengal to Delhi in the form, among others, of high value textiles; (*c*) the export of food grains to other provinces on a regular basis both through sea and land.[43] The significant development of the eighteenth century was the emergence of an integrated provincial market in food grains, knitted together by a dense network of surface and river communication,[44] which extended and broadened the range of commercial production and commercialized transactions initially set into motion by items like silk and cotton. A substantial part of the discussion in this book is concerned with the development of this integrated market.

Second, there was the integration of Bengal in a global economic network by the expanding world-economy of the eighteenth century. The political-economy of this integration had the following dimensions:

To keep in motion the great machine of the Company commerce, [for which] it is absolutely necessary that the provision of Cargoes for the return of ships be ensured. Hence the origin of previously investing Funds at the Aurangs [manufactories];[45] hence also the necessity of being secured in due realization of goods for which money is advanced. . . . The accession of the Company to the Government of the Country [i.e. the Diwani in 1765] did not change these [earlier] principles and usages, tho' [there was] the immense increase of their Funds and of the wealth of their Servants, all to be centr'd finally in Europe and to be conveyed thither only in goods. . . .

The impact of such an integration was quite fundamental, as it '*exceedingly enlarged the demands for the Market of Bengal*'.[46]

Third, there was the political dimension. Bengal was the first province in India to undergo *colonial* conquest whereby its revenue resources were now at the disposal of the Company to invest in trade and empire-building. Profits accruing from monopsonic controls over silk and piece-goods were, in 1772, buttressed with a declared official monopoly over salt. Monopsonic controls were subsequently extended over opium and indigo. On the other hand, the Company's declared intention was to ensure 'fair trade' in those commodities, like that in agricultural produce, which were not directly connected with the international flow of commerce. Thus, after 1757 state intervention assumed apparently contradictory forms of rigorous control in the marketing of certain commodities, and a relatively striking non-interference in the movement of others; and the latter's design was to ensure that the local merchant was at 'liberty to carry his merchandize where ever he thinks proper for sale'.[47]

The other aspect of the Company's rule was revenue, mainly its assessment and timely collection in cash. In fact, this was the culmination of a process already set into motion by the previous regime, the Nizamat; but there was a major difference: the Company's state persistently refused all suggestions to revert to revenue in kind during periods of price slumps in order to reduce the pressures on the peasantry, which was a major departure from one of the abiding principles of Mughal revenue policy.[48]

The British conquest of Bengal had some fundamental effects on the province's economy. Between 40 and 45 per cent of the agricultural output taken as revenue[49] was annually converted into cash. Obviously, a greater degree of appropriation in money depended on the facilities available for the circulation of cash and commodities. There were, henceforth, significant changes in the structures of local markets, and in the relationship between the state and these markets. Crucial in this respect was the policy followed by the Company, between 1773 and 1790, of dissociating landed proprietors (especially the zamindars) from collecting local taxes on trade (*sa'ir jihat*) in their respective territories.[50] This removed, to a large extent, one of the principal problems hampering the easy movement of goods under the Nizamat. The fact that the Company tried to break the existing bottlenecks in local marketing systems is an example of a major feature of early colonial policy, viz., 'to create markets where they [did] not exist before, or to regulate existing markets'.[51] One immediate effect of this (essentially political) decision was the remarkable geographical proliferation of markets in Bengal during the second half of the eighteenth century. Sources are replete with descriptions of the 'extension of hauts and bazars' all over Bengal in this period (see Chapter 4). Therefore the assertion that 'British rule greatly disrupted [the pre-British] network of markets, most of which were oriented towards domestic consumption'[52] is not supported by the available evidence, at least not for this period.

There was a triadic integration—that of the peasant-household, agriculture and trade—with production for the market in Bengal. Of this the integration of the peasant-household, both as sellers and as producers, was of vital importance because it determined (despite the many burdens and pressures on it) the amount of agricultural produce which could *actually* be marketed. It also shaped the various tenurial and sub-tenurial arrangements, particularly that of sharecropping, which apparently mushroomed in Bengal during this period, and made the growth of a sharecropping peasantry in the eighteenth century an instrument of commercialized agriculture rather than an appendage of a certain (and largely undefined) subsistence security system of a 'pre-capitalist' variety, or a 'feudal' exploitation of a nebulous sort of 'feudalism'. A large part of the subsequent discussion will be concerned with analysing the specifics of this triadic integration.

NOTES AND REFERENCES

1. For an excellent overview of this centrality in the Chinese and Japanese context, see Francesca Bray, *The Rice Economies: Technology & Development in Asian Societies*, Oxford, 1989; also Francesca Bray, 'Patterns of Evolution in Rice-Growing Societies', *JPS*, vol. 11, no. 1; October 1983.
2. S.A. Resnick, 'The Decline of Rural Industry under Export Expansion: A Comparison among Burma, Philippines and Thailand', *Journal of Economic History*, vol. 30, no. 1, March 1990, p. 51.
3. See IOR, BRC, P/49/52, 29 May 1775; and ibid., P/50/3, 12 August 1777; ibid., P/51/45, 5 August 1789 for statements regarding the combination of agriculture with artisanal production in the salt sector.
4. WBSA, CCRM, vol. 8, 5 December 1771.
5. This growing complexity in commercial involvement can be seen from the case of silk, the most prized of Bengal's exports during this period. In the seventeenth century the market value of this commodity, and the returns to Bengal by its sale, were determined by interregional demand. Witness, for instance, the following assessment of Bengal's silk market in 1661: 'silk sells in Agra as the price of it in Cossimbuzar riseth or falleth. The exchange of money from Cossimbuzar to Pattana and Agra riseth or falleth as the silk findeth a vent for it in Pattana or Agra' (vide, 'Register of Papers Relating to the English and Dutch East Indies', BM, Add. Ms., 34123, fol. 42). However, in the eighteenth century we see international demand establishing its primacy in place of interregional demand in Bengal's trade in silk. This

is clearly demonstrated by the fact that there was a sharp fall in the exports of silk from Bengal to other parts of Asia throughout the second half of the century. Simultaneously, there was an equally spectacular increase in the East India Company's investment for silk (Rajat Datta, 'Markets, Bullion and Bengal's Commercial Economy', p. 332).
6. Richard M. Eaton, *The Rise of Islam and the Bengal Frontier, 1204-1760*, Delhi, 1994, pp. 203, 210. See also Rajat Datta, 'Peasant Production and Agrarian Commercialism in a Rice Growing Economy: Some Notes on a Comparative Perspective and the Case of Bengal in the Eighteenth Century', in Peter Robb (ed.), *Meanings of Agriculture: South Asian History and Economics*, Delhi, 1996. Eaton's, *The Bengal Frontier* (esp. pp. 194-227) is an excellent overview of the long-term expansive nature of such a process. Though the nature and role of markets as symptoms of this great agrarian expansion are not discussed by Eaton in a major fashion, his findings provide a much-needed long-term perspective which significantly corroborates some of my own formulations made in 'Rural Bengal: Social Structure and Agrarian Economy in the Late Eighteenth Century' (Ph.D. thesis, University of London, 1990), and 'Peasant Production and Agrarian Commercialism', both of which were written much before Eaton's extremely valuable contribution became available.
7. Bayly, *Rulers, Townsmen and Bazaars*, pp. 74ff.
8. David Ludden, *Peasant History in South India*, Delhi, 1989, pp. 9, 162 and *passim*.
9. The term is from Frank Perlin, 'Proto-Industrialisation and Pre-Colonial South Asia', *Past & Present*, no. 98, February 1983.
10. For a recent conspectus on issues relating to agriculture and agrarian commercial growth see, Peter Robb (ed.), *Meanings of Agriculture*; also see the various discussions in S. Subrahmanyam (ed.), *Merchants, Markets and State in Early Modern India*, Delhi, 1989; S. Bose, *South Asia and World Capitalism*, Delhi, 1990. Such commercial processes occurred also beyond the frontiers of South Asia. For instance, the formation of a 'national market' in bulk commodities in eighteenth century China has been seen as the outcome of similar linkages between various local and regional production and marketing centres (Susan Naquin and Eleanor Rawski, *Chinese Society in the Eighteenth Century*, New Haven, 1987, pp. 98-100), and its impact is easily discernible in the spread of cotton cultivation in the provinces of Fukien, Kwangtung and Kiangsu in the Ming and Ch'ing periods (cf. Kang Chao, *The Development of Cotton Textile Production in China*, Cambridge Mass., 1977, pp. 13-24). Also, the eighteenth century upturn in European grain prices apparently led to the rapid spread of maize in the Ottoman peasant agriculture, which also underwent a mini 'cotton boom' under the influence of French and American Revolutions [R. Kasaba, 'Incorporation of the Ottoman Empire, 1750-1820, *Review*, vol. 10, nos. 5 and 6 (supplement), Summer/Fall 1987, pp. 807-28; also M. Cizacka, 'The

Incorporation of the Middle East in the World Economy', *Review*, vol. 7, no. 3, Winter 1985, pp. 353-77].
11. P. J. Marshall, *Bengal: The British Bridgehead, Eastern India, 1740-1828*, Cambridge, 1987, p. 170; Bayly, *Rulers, Townsmen and Bazaars*, pp. 245-6; Ludden, *Peasant History*, pp. 52ff.
12. Karl Marx, *Pre-Capitalist Economic Formations* (ed. E.J. Hobsbawm), London 1978; also Karl Marx, *Capital: A Critique of Political Economy, vol. 3*, Harmondsworth, 1982, p. 931. For the Chayanovian perspective, Basil Kerblay, 'Chayanov and the Theory of Peasant Economies', in T. Shanin (ed.), *Peasant and Peasant Societies, Selected Readings: Second Edition*, Penguin, 1988, p. 178.
13. 'Small Peasant Ownership in Turkey: Historical Formation and Present Structure', *Review*, vol. VII, no. 1, Summer 1983, p. 5; Halil Inalcik, 'Capital Formation in the Ottoman Empire', *The Journal of Economic History*, vol. 29, no. 1, March 1969.
14. D.W. Perkins, *Agricultural Development in China, 1368-1968*, Chicago, 1967, pp. 144-8; W.S. Atwell, 'Notes on Silver, Foreign Trade and the Late Ming Economy', *Ch'ing-shih wen-t'i*, vol. 3, no. 8, December 1977; R.M. Myers, 'The "Sprouts of Capitalism" in Agricultural Development during the mid-Ch'ing Period', ibid., vol. 3, no. 6, December 1976; Naquin and Rawsky, *Chinese Society in the Eighteenth Century*, Chapter 4.
15. F.W. Moulder, *Japan, China and the World Economy: Toward a Reinterpretation of East Asian Development*, Cambridge, 1977, pp. 36-7; Tsuneo Sato, 'Tokugawa Villages and Agriculture', in Chie Nakane and Shinzaburo Oishi, *Tokugawa Japan: The Social and Economic Antecedents of Modern Japan*, (trans. Conrad Totman), Tokyo, 1991, pp. 70-1.
16. Gilbert Rozman, 'Castle Towns in Transition', in M.B. Jansen and G. Rozman (eds.), *Japan in Transition: From Tokugawa to Meiji*, Princeton, 1988, pp. 274-5, 318.
17. See Yasukazu Takenaka, 'Endogenous Formation and Development of Capitalism in Japan', *The Journal of Economic History*, vol. 29, no. 1, March 1969, pp. 141-62; also F. Bray, *The Rice Economies*, pp. 210-17.
18. F. Bray, 'Patterns of Evolution', pp. 18, 21.
19. Geertz shows on the basis of Javanese agriculture that the main reason of this divergence was the colonization of other Asian countries and Japan's escape from this experience (Clifford Geertz, *Agricultural Involution: The Process of Ecological Change in Indonesia*, Berkeley, 1963, pp. 130-43). Another argument advanced is the concurrence of the growths in agricultural and non-farm productivity in the nineteenth century which did not occur elsewhere in Asia [cf. T.C. Smith, 'Farm Family By-Employments in Preindustrial Japan', *The Journal of Economic History*, vol. 29, no. 4, October 1969; B.F. Johnston, 'The Japanese "Model" of Agricultural Development: Its Relevance to Developing Nations', in C.R. Wharton (ed.), *Subsistence Agriculture and Economic Development*, London, 1970]. It must however be

emphasized that 'by-employment' was an adjunct—a 'side occupation'—of the peasantry which did not reduce the overarching primacy of rice cultivation: 'they were auxiliary and supplementary to the farm household' [O. Saito, 'The Rural Economy: Commercial Agriculture, By-Employment and Wage Work', in Jansen and Rozman (eds.), *Japan in Transition*, pp. 405-6].

20. F. Bray, 'Patterns of Evolution', p. 16; also F. Bray, *The Rice Economies*, p. 207.
21. D.W. Perkins, 'Persistence of the Past', in Perkins (ed.), *China's Modern Economy in Historical Perspective*, California, 1975, p. 4.
22. Evelyn Sakakida Rawski, *Agricultural Change and Peasant Economy in South China*, Cambridge Mass., 1972.
23. See B.B. Chaudhuri, 'Agricultural Growth in Bengal and Bihar, 1770-1860: Growth of Cultivation since the famine of 1770', *Bengal Past and Present*, January-June 1976; B.B. Chaudhuri, 'Eastern India', in *CEHI 2*; B.B. Chaudhuri, *Growth of Commercial Agriculture in Bengal, 1757-1860*, Calcutta, 1964.
24. See Chapters 4 and 6; also Rajat Datta, 'Merchants and Peasants: A Study of the Structure of Local Trade in Grain in Late Eighteenth Century Bengal', *IESHR*, vol. 23, no. 4, October-December 1986; Marshall, *Bridgehead*, pp. 13-14; D.H. Curly, 'Rulers and Merchants in Late Eighteenth Century Bengal', unpublished D. Phil. thesis, University of Chicago, 1980.
25. B.B. Chaudhuri, 'The Process of Agricultural Commercialisation in Eastern India during British Rule: A Reconsideration of the Notions of "Forced Commercialisation" and "Dependent Peasantry"', in Peter Robb (ed.), *Meanings of Agriculture*, pp. 35-70.
26. Colebrooke, *Remarks*, pp. 67f.
27. See Chapter 5; also Rajat Datta, 'Subsistence Crises, Markets and Merchants in Late Eighteenth Century Bengal', *Studies in History* (new series) vol. 10, no. 1, 1994. That Bengal was not alone in attaching such a degree of importance to rice in determining the prices of agricultural products is borne out from other similar developments in China and Japan (cf. W.S. Atwell, 'Some Aspects on the "Seventeenth Century Crisis" in China and Japan', *Journal of Asian Studies*, vol. 45, no. 2, February 1986, pp. 230-3). For a discussion of the economic significance of rice prices in China see Yeh-Chien Wang, 'The Secular Trend of Prices during the Ch'ing Period (1644-1911)', *Journal of the Institute of Chinese Studies (Hong Kong)*, vol. 5, no. 2, 1972.
28. Fluctuations in the price of wheat exerted a most critical pressure on the prices of other farm produce in England throughout the eighteenth century; in fact the Corn Laws were a device adopted by the state to reduce the massive oscillations in the price of wheat in that period (see J.D. Chambers and G.E. Mingay, *The Agricultural Revolution, 1750-1880*, London, 1966, pp. 107-9). The main trend of French agriculture during the seventeenth and eighteenth centuries was to increase the cultivation of land under the

cultivation of cereals; and the causal connections between cereal and non-cereal prices were most sharply brought into focus during the severe harvest failures in the 1770s and 1780s which exacerbated the crisis of the *ancien regime* [compare Jean Jacquart, 'French Agriculture in the Seventeenth Century', in P. Earle (ed.), *Essays in European Economic History*, Oxford, pp. 166-82; and Peter Jones, 'The French Peasantry of France on the Eve of the French Revolution', *History of European Ideas* (special issue *The European Peasantry on the Eve of the French Revolution*, ed., Peter Jones), vol. 12, no. 3, 1990, pp. 335-61].

29. Subsistence production is defined here, following Wharton, 'as a situation where the fruits of an individual or group productive efforts are dictated more toward meeting immediate consumption needs . . . without any or few intermediaries or exchange . . .' [C. Wharton, 'Subsistence Agriculture: Concepts and Scope', in Wharton (ed.), *Subsistence Agriculture and Economic Development*, p. 133].

30. Takenaka, 'Endogenous formation', pp. 144f. Satoru Nakamura, 'The Development of Rural Industry', in Nakane and Oishi (eds.), *Tokugawa Japan*, p. 91. It is possible that the Japanese peasants had access to these sources of credit which may have reduced their dependence on loans which (as in Bengal) were often usurious.

31. G. Rozman, *Urban Networks in Ch'ing China and Tokugawa Japan*, Princeton, 1973, p. 134.

32. Ibid., pp. 85-7.

33. Naquin and Rawski, *Chinese Society*, p. 99.

34. Madhavi Bajekal, 'The State and the Rural Grain Market in Eighteenth Century Rajasthan'; and Kumkum Banerjee, 'Grain Traders and the East India Company: Patna and its Hinterland in the Late Eighteenth and early Nineteenth Centuries', in S. Subrahmanyam (ed.), *Merchants, Markets and the State*, pp. 90-121, 163-89.

35. Kumkum Banerjee, 'Grain Traders', p. 166. This was perhaps (and here one can only speculate) due to the overarching influence of caste in the social hierarchy of rural Bihar (Marshall, *Bridgehead*, p. 37). The existence of relatively comfortable peasants, connected to their counterparts by ties of caste, may have prevented the grain-merchants from intruding into agricultural production in Bihar. By contrast, in Bengal, where the caste factor in production was much weaker, and where relative affluence was based on control over resources and labour and not on the caste-based exploitation of a destitute peasantry (see Chapter 2 for a discussion) it was perhaps easier to exploit production in the context of an expanding trade in grain.

36. Colebrooke, *Remarks*, p. 67.

37. Rajat Datta, 'Merchants and Peasants', pp. 29-33.

38. This is a modified version of Caglar Keyder's argument regarding late Ottoman Turkey ('Small Peasant Ownership in Turkey', pp. 53-5).

39. See Chapter 4.
40. Somewhat analogous developments may have occurred in Bihar in the same period for three reasons. First, Bihar apparently had a more developed town life than Bengal in the eighteenth century. Second, there seems to have been a substantial degree of integration of peasant and artisanal production with the local markets along with an interregional and overseas flow of goods. Third, there was a robust trade in grain and peasant involvement in the market and a network of trading intermediaries (Marshall, *Bridgehead*, pp. 38-9; Banerjee, 'Grain Traders', op. cit.). The effects these had on the structure of peasant production is unfortunately not clear; but this lack of clarity may not necessarily symbolize economic 'backwardness', it may equally reflect the pressing need to do more detailed research on late pre-colonial Bihar.
41. Rajat Datta, 'Merchants, Markets and Subsistence Crises', pp. 85-8.
42. See Chapter 4.
43. For a comprehensive overview of these changing patterns of European trade, see Om Prakash, *The New Cambridge History of India, II, 5: European Commercial Enterprise in Pre-Colonial India*, Cambridge, 1998; Om Prakash, 'European Trade and Asian Economies: Some Regional Contrasts, 1600-1800', in L. Blusse and F. Gaastra, *Companies and Trade*, The Hague, 1981, pp. 189ff.; for the other aspects of the seventeenth century developments cited here, see British Museum, Add. Ms. 34123; IOR, HM, vol. 456 F.
44. Tilottama Mukherjee, 'Of Roads and Rivers: Aspects of Travel and Transport in Eighteenth Century Bengal', M. Phil. dissertation, Jawaharlal Nehru University, 1997.
45. The *dadni* or the advance contract system.
46. Report of Commercial Occurrences, 6 August 1789, IOR, HM, vol. 393, pp. 279-80; emphasis added.
47. The term 'fair trade', and the policies to be followed in connection were established by Warren Hastings in a Minute dated 9 March 1773, IOR, HM, vol. 217, pp. 444-9; also see IOR, BRC, P/49/38, 23 March 1773.
48. James Steuart, 'Memoirs of the Coinage of Bengal', IOR, HM, vol. 62; John Shore's Minute in IOR, BRC, P/51/50, 28 October 1789; this aspect is discussed in Appendix 1.
49. Minute of John Shore, 18 June 1789, in *FR 2*.
50. See Chapter 4.
51. Cyril S. Belshaw, *Traditional Exchange and Modern Markets*, Delhi, 1965, p. 74.
52. Amiya Bagchi, 'Markets, Market Failures and Transformation', p. 52.

CHAPTER 1

Patterns and Processes of Agricultural Production

TO ALL APPEARANCES the agrarian economy of Bengal operated under a paradox. On the one hand, as all contemporary observers and officials remarked, the natural fertility of the soil was remarkably high. For Warren Hastings, the wealth of Bengal was due to the soil's 'fertility and the number and industry of its inhabitants'.[1] James Rennell saw the province as possessing some 'of the most fertile lands in the Universe'.[2] William Tennant observed, in 1797, that Bengal had 'an excellent soil and climate, and possessed of almost every variety of cultivated grains, and competent number of hands to raise them. . . '.[3] One of the most revealing descriptions of such intrinsic fertility of land comes from the district of 24-Parganas in 1762, where 'the fertility and goodness of the soil is proved by its producing everything, almost spontaneously'. Here:

You have the Cocoanut, Date, Palmeira, Serepul, Tamarind, Plantain and several other trees not known to me. We passed through fields of Mustard, Melons, Gram, Flax, Pease, Cotton & ca. and yet the Inhabitants are far from being either industrious or skilful. *They scarcely scratch the Superfices of the Earth to make it yield* this increase in many things; and in others they sow without any previous preparation at all. . . . The country every where abounds with cattle.[4]

In Chittagong:

Ploughing is indeed by no means intended to turn the Ground. It merely scratches the Surface, divides the Sward; and gives access to the teeth of the Harrow, or rather the Bullock rake to lay hold of the weeds.

On the other hand, most producers actually worked under the most resource constrained circumstances. It was generally recognized that almost without exception these poor peasants were unable *even* to *commence* agricultural operations without outside financial aid.[6] 'The lower Ruatts [*raiyats*] live from hand to mouth and without tacavi or advances cannot afford to cultivate a piece of ground.'[7] 'The want of capital employed in agriculture and manufactures, cripples every enterprise' was how Reverend William Tennant described the state of production in the province.[8] James Rennell believed that 'the labourers suffer[ed] such great oppressions' that they were rendered unable to work their lands to full capacity.[9] Later estimates made by Buchanan-Hamilton in the districts of Rangpur and Dinajpur (*c.* 1807) revealed that nearly 52 per cent of the agricultural population in Rangpur[10] and 77 per cent in Dinajpur[11] were sharecroppers, agricultural labourers or 'needy farmers'.

With regard to its organization, Bengal's agriculture was overwhelmingly undertaken on small peasant farms operated by individual peasant families with the use of domestic labour. It was, therefore, a classic example of petty production: household centred and labour intensive, where the majority of producers were perennially constrained by a shortage of productive resources. The other major aspect of Bengal's agricultural production was the apparent predominance of the so-called 'subsistence' sector, rice and paddy, in overall production. The so-called cash crop sector was subsidiary to this major productive enterprise.

Agricultural Pròduction: Land, Labour and Output

Agricultural production—the utilization of land to procure various crops—was organized around three harvest seasons: *aman* (winter), *aus* (spring) and *boro* (summer). Of the three, *aman* and *aus* were considered the most important, both in terms of output and their share in the payment of revenue since all the major cash crops were grown in these two seasons. The winter rice was commercially the most valuable as it was grown principally for sale.

The evidence available (Table 11) certainly suggests that the grand purpose of agricultural production was the cultivation of paddy (*dhan*) for rice. For Tennant:

The culture of almost every plant, and particularly of the *gramina*, in proportion as it has long been diffused, induces numerous varieties. The several seasons of cultivation, added to the influence of soil and climate, have multiplied the different species of rice, into endless variety. From the awned and unawned, from that growing on the mountains to that produced in humid situations, there are various diversities adapted to every circumstance of soil, climate and season. . . .'[12]

Land-use for the cultivation of rice was a complex process. The first priority was obviously access to water, and here the central concern was to grow the maximum permissible on lands within the range of the seasonal overflow of the various rivers and streams. 'The annual inundation, to which the soil is principally indebted for its fertility, regulates the sites of cultivation, and in some measure the time of sowing and reaping' was how Taylor described the principal considerations behind the cultivation of rice in Dhaka.[13] In Midnapur, rice was grown on *jala* (low-lying) lands for reasons of their close access to water: the *jala* was further sub-divided into eight different types depending on each one's elevation, and different types of rice were grown on each.[14] Kyd's survey of cultivation on the western banks of the Hughli river shows that the broadcast summer rice was sown on the lands of the lowest elevation (called *dowrah*), while the transplanted winter rice was planted on lands called *caddurie* which were 'a little elevated, not more than five or six inches, above the common level of the [*dowrah*]'.[15]

Under normal circumstances, the entire agricultural output, as well as the payment of revenue *kists* (instalments), was concentrated in the nine months between *Baisakh* (April-May) and *Pous* (December-January), 'no crops of any consequence are cut after the month of Poose [*Pous*], and the intervening months of Maug, Phagun and Choite [January-April] are taken up in cultivating those lands from which the resources of the ensuing year are derived'.[16] The preparation of land for rice was as follows:

Pous to *Phalgun*	(January to March)	for *aman* rice
Pous and *Magh*	(December to February)	for *aus* rice
Pous	(December-January)	for *boro* rice

The number of times the land had to be ploughed obviously depended on topography. In the western parts of Dhaka where the soil 'consisted of red *kunkur* [gravel] and of different strata of clay', the *aman* rice land required to be ploughed sixteen times,[17] whereas the soil on the western banks of the Hughli was rendered fit for this harvest by ploughing it twice.[18] In the 24-Parganas, the rice land was prepared by barely 'scratching the superfices of the earth'.[19] The *boro* harvest did not require the land to be ploughed more than once. In eastern Bengal it was grown 'on churs, or low marshy ground after the waters have subsided, consequently it [was] not sown till late in the season'.[20] In western Bengal, *boro* rice land was 'the lowest admitting of cultivation', which 'require no manure, produces a constant succession of rice *ad infinitum*, without any discernible decrease of vegetative power, requiring only [to be] annually sown with successive varieties of grain'.[21]

In the largely flat alluvial plains of southern and south-eastern Bengal, the same land was used to procure multiple crops of rice in succession. Thus in Midnapur, *jala* land of the '1st kind . . . produces two crops of rice at the same time, the first called ause . . . and the other aumeen'.[22] In eastern Bengal:

> The aumun crop being cut in Aughun [November-December], the Reapers scatter Khassarie [vetch] amongst the stubble, while the soil is yet moist and soft from the inundation, which springs up without any trouble, and is gathered in Phagun [February-March]. The stubble and stalk of the khassarie are then set on fire and the ashes ploughed in with the soil. The boro dhan [paddy] mixed with aumun is then sown; this [*boro*] being of very quick growth is ripe in Jeyte or Assar [May to July]; it is then cut and the aumun which is of slower growth rises with the water [i.e. the seasonal inundation] and is cut in Aughun [November-December].[23]

Naturally, the cultivation of rice was labour intensive, though in the small peasant-holdings the entire exercise was perhaps undertaken by using the labour available in individual households. Apart from ploughing, labour was required most in '1st for clearing the grounds from roots of grass, 2nd for weeding and 3rd for reaping'.[24] Transplanted winter rice entailed the further task of preparing the nursery for rearing the seedlings, which were then re-planted in the main field once they had attained a height of about 18 inches.[25] The costs of cultivating a *bigha* of land with rice of this kind was Rs. 3.87 during Kyd's survey of the western Banks of the Hughli in 1791;[26] in Commilla (eastern Bengal) the comparative costs were Rs. 2.30 in 1789.[27] The seed yield ratio appears to have ranged from 1:24 in parts of western Bengal[28] to 1:20 in Dhaka.[29] Data available from parganas in Commilla, Rangpur, Midnapur and Burdwan show the following output of the *aman* harvest per *bigha* (see Table 1).

Various leguminous plants were grown in close association with rice, often on the same land. As discussed earlier, the rotation of *khassari* on rice lands, and then using the stubble to fertilize the land for the next rice harvest seems to have been well developed in eastern Bengal. While vetches would grow in flooded soil, other lentils, especially those called *kalai, moog* and *mussoor*, 'much eaten by the natives, and particularly

PATTERNS AND PROCESSES OF AGRICULTURAL PRODUCTION   41

TABLE 1. OUTPUT OF RICE IN SELECTED PARGANAS
(maunds per bigha)

Parganas in	Average produce (maunds)
Commilla	8
Rangpur	10.72
Midnapur	10.5
Burdwan	10.71

Sources: IOR, BRC, P/51/40, 15 July 1789; ibid., P/52/42, 9 March 1792; ibid., P/52/50, 19 October 1790, and IOR, BRP, P/71/22, 20 March 1790.

when split, is called dall' required just the right amount of moisture for 'if sown on [land] too wet, [lentils] will not vegetate, if on too dry, will never thrive, nor make any adequate return for labour'.[30] Therefore lentils and vetches depended on proper timing, and this is why *kalai* in western Bengal was planted on the rice ground 'in the intervals unoccupied by the tufts of rice ... in the intermediate month of Cautic (October-November) on the taking off of the rains, before cutting down the 2nd crop [*aman*] of rice, [the land now] being no longer overflowed but drenched in moisture'; once harvested in *Phalgun* (February-March), 'the ground is then prepared three or four times in readiness for the Baisak [*aus*] sowing of the ensuing season'.[31] In Midnapur, *kalai* was considered valuable enough to share the land alternately with sugarcane; it was grown on *cola* (elevated) land of the '4th kind [which] yields Ook, sugarcane, one year and Kalai the next'.[32] In eastern Bengal, jute (*paat*) lands, after the harvest, were made 'ready to receive ... pease and pulses of various descriptions [whose] harvest is reaped a short time before it is expedient to recommence the culture of the [jute], and the land bears a succession of [such] crops for many years'.[33]

The output of these leguminous plants from different districts is not known, but the evidence from Commilla and Rangpur indicates that the produce from one *bigha* of land, at least in east Bengal, was as shown in Table 2.

It is necessary to bear in mind that these lentils and vetches had an important position in the provincial demand for food, as they were crucial ingredients in the daily diet of the people. Thus, this crop was essentially designed for sale, and a substantial part of the intra-regional trade in food products, particularly from western in to eastern Bengal, was in split pulses or *dal*.[34]

Tennant considered the 'universal use and vast consumption of vegetable oils' in Bengal as 'prejudicial to agriculture' because 'much

TABLE 2. OUTPUT OF LENTILS (*maunds per bigha*)

Region	Lentil	Produce
Commilla	Gram	5 maunds
	Peas	7 maunds
	Lentil, *moog*	2 maunds
Rangpur	*Kalai*	1.98 maunds
	Gram	1.5 maunds
	Mussoor	1.98 maunds
Western Bengal	*Mussoor*	2 maunds

Sources: IOR, BRP, P/71/22, 20 March 1790; ibid., P/71/22, 20 March 1790; IOL, Ms. Eur. F. 95, fol. 35.

land and a great proportion of the cultivated land' was used by such cultivation, which Tennant thought 'trench deeply upon the productive ground for human sustenance'.[35] Obviously his reservations were not well founded for, as Figure 1 (p. 76) shows, oil seeds constituted a mere 4 per cent of the gross agricultural output in 1794. Nevertheless, its cultivation was important for local production and trade. In Jessore, for instance, *til* (sesame) was considered a crop of the 'first class' which bore 'a proportion of four annas [i.e. 25 per cent] to the jumma'.[36] Among the main oil producing plants were those of sesame, mustard (*sarson*) and *teesy* (linseed);[37] some oil was also extracted from poppy seeds: in Rangpur, 'the poppy seed is used solely for making oil [and] the Ryotts sell the poppy seed to the Pikars [traders' middlemen] and Oilmen in the Mofussil'.[38]

Among the vegetable oil plants most coveted for its culinary and commercial importance was mustard. In the alluvial plains of Gangetic west Bengal, mustard was often sown on the banks of the river after the monsoonal overflow of its water had subsided. Here the chief resource was the sediments deposited by the seasonal inundation which Tennant considered 'the richest of all soils'. The mustard seed was sown broadcast 'and gently covered with a harrow, or by scratching the mould with a branch of a tree', after which 'the crop makes its appearance seemingly in great abundance, and one of the most beautiful to the eye which the country affords'.[39] In his survey of the western banks of the Hughli, Robert Kyd noticed that both mustard and sesame were grown on a category of land called 'dangeah', this being the most elevated land where production was principally dependent on 'artificial labour', manuring and water-retention 'by means of artificial ledges of about 8 or 10 inches high'; the 'dangeah' was primarily meant for the cultivation of sugarcane, but it simultaneously yielded two crops of 'sersong and teel', i.e. mustard and sesame.[40] In general, the mustard plant was sown on relatively higher ground in order to prevent an excessive penetration of water in the month of *Kartic* (October-November) and reaped in *Magh* or *Phagun* (January or March).[41]

Teesy was grown on rice lands, preferably upon those which 'have lain fallow for 3 or 4 years'. The ground was ploughed three times and the seed sown in *Agrahan* (November-December). The crop was harvested 'four months after the seed is sown or about the month of Faugun and Choite [February-April] . . . & oxen employed to tread the seed from which the oil is extracted'; 'the treading of oxen [also] destroys the plants and bark adhering to them [the seeds]'.[42] The output of these various oilseeds per *bigha* in different areas was as shown in Table 3.

There was a general consensus among contemporaries that the cultivation of sugarcane and the manufacture of refined sugar had declined substantially during the course of the late eighteenth century. 'Formerly sugar was one of the staple articles of Bengal', but by 1776 'the sugar trade of Bengal [had] in fact [become] annihilated' because of 'the increase in the price of raw material & of labour';[43] but by the time of the Permanent Settlement the exports of sugar from Bengal had once again become 'considerable' in relation to those from rice and other food grains.[44] What these descriptions indicate is perhaps a relative decline in the *export* of sugar from Bengal which should not be taken to imply a reduction in the importance of sugarcane as a valuable cash-crop or a comparable reduction in the importance of sugar in *domestic*

TABLE 3. OUTPUT OF VARIOUS GRADES OF OIL-SEEDS
(*maunds per bigha*)

District	Oilseed	Output
Sylhet	Teesy	4
Commilla	Til	2
Commilla	Mustard	3
Rangpur	Mustard	1.9
Muhammadshahi	Mustard	1.5
Hughli	Mustard	3
Hughli	Til	1.27

Sources: IOR, BRC, P/51/40, 15 July 1789; ibid., P/71/22, 20 March 1790 and 22 March 1790; IOR, HM, vol. 375, p. 305; IOL, Ms. Eur. F. 95, fol. 35.

consumption or internal trade. Peasants in Chittagong cultivated 'a good deal of Sugar Cane about [their] houses', and during his journey through this district, Buchanan 'heard the creaking of mills employed in expressing the juice'.[45] As will be discussed subsequently, the cultivation of sugarcane was based on investments made by the local merchants who, in turn, directed the produce, especially of refined sugar, to feed the large and intermediate towns. The history of sugar in our period offers a good example of the *re-direction* of agricultural commodities from export to internal consumption under the influence of a rising demand for food in the province, whose main features have been discussed in Chapter 4.

By nature, the cultivation of sugarcane was an expensive business, and the shortage of resources among the majority of peasants meant that the land actually under its cultivation at any point in time, or in any particular district, was bound to be relatively small. This can be seen from the district-wise output of sugar in Table 4.

Nevertheless, this was a valuable crop, and great care was taken to ensure its success wherever it was sown. In Birbhum, 'lands appropriated to the cultivation of sugarcane [were] of various qualities distinguished in to first, second and third agreeable to the quantity & quality they produce[d]'.[47] The cane was universally planted in high (*colla*) of the best kind in *Baisakh* (April-May) and cut over the months of *Magh* (January-February) and *Phalgun* (February-March).[48] The sugarcane ground:

Being previously manured with rich soil taken from low bottoms at the rate of one basket load p[er] square fathom, mixed with the expressed old seed of Sersong

TABLE 4. ESTIMATES OF LAND UNDER SUGARCANE
IN SOME DISTRICTS IN THE 1790s

District	Sugarcane land (*bighas*)
Burdwan	25,000
Birbhum	14,663
Purnea	26,500
24-Parganas	765.40
Jessore	760.80
Dinajpur	13,959.50
Dhaka	10,000

Sources: IOR, BRP, P/72/8, 29 October 1792; ibid., P/72/10, 26 December 1792.

[mustard] ... in the proportion of two and a half maunds per biggah and trenched one span deep forming ridges one and a half cubits [30 inches] broad; the sugarcane is planted in Bysaac [*Baisakh*] by slips taken from the preceding years stock in two lines or rows on each ridge at the distance of one cubit. One biggah contains 6500 slips. . . .[49]

In Chittagong:

The leaves of the preceding crop . . . are burned, and the ashes with cowdung serve for manure. The field is then well laboured, and levelled, and afterwards divided into beds, 20 feet long and 2 broad. These beds are separated by ridges, about half a foot high, and as much broad, which serve to confine the water bestowed on the young plants. In these beds, the joints of the sugarcane are placed obliquely with one end projecting about three inches above the surface. The joints are placed in the bed sometime in one Row, sometime in two, and one distant from another about one foot. The joints are carefully watered, till they push out young shoots. The earth is then gathered up about them in ridges and this operation serves for weeding.[50]

Sugarcane also required extensive manuring in three stages. The first manuring was done while ploughing with 'a compost made of soil or mud taken from the bottom of tanks, rotten vegetables and cows dung'; the second stage was when 'oil cake broken to dust' was mixed in the soil 'when the plants are put in the ground'; and finally 'when they [the cane] attain a height of one cubit [the cultivators] strew the ground over with the dust of oil cake in the proportion of 2 maunds per Begah'.[51] Sugarcane also involved a fairly high intensive use of labour in its production. Table 5 provides the figures for the number of labourers to cultivate a *bigha* of sugarcane in east Bengal.

The amount of labour-time socially necessary for cultivating 5 *bighas*

TABLE 5. LABOUR UTILIZATION IN THE CULTIVATION OF SUGARCANE IN EAST BENGAL

Process involved	Labour and duration
Ploughing	4 labourers for a month
Enclosing the beds	2 labourers for a month
Planting	4 labourers for a month
Making beds and rows	2 labourers per month
Cutting and pressing the juice	10 labourers for 8 days
Boiling the juice	3 labourers for 8 days

Source: IOR, BRC, P/51/40, 15 July 1789.

TABLE 6. LABOUR UTILIZATION IN THE CULTIVATION OF SUGARCANE

Process involved	Duration
Ploughing	5 ploughs for 10 days
Original nurseries for the cane	45 days
Transplanting shoots to main nursery	20 days
Watering the nursery beds	10 days
Transplanting shoots in the main field	50 days
Watering shoots in the fields	40 days
Clearing roots and trenching between rows	50 days
Stripping dead leaves from plants and weeding	60 days
Binding the cane together	100 days
Cutting the cane	240 days
Total necessary labour time	615 days

Source: IOR, BRP, P/72/8, 29 October 1792.

in the 'environs of Calcutta' was estimated (in 1792) in the fashion as shown in Table 6.

The costs involved in an operation of this magnitude were certainly beyond the capacity of ordinary cultivators. For instance, the district of Purnea, where there was apparently no 'deficiency of land fit for the cultivation of sugarcane', very little of the cultivated land was actually utilized for its production because: 'The produce is so small in proportion to the first expense added to that of cultivation, that the cultivator is not reimbursed in the first year & does not gain his profit until the end of the second year. The Ryots therefore prefer the cultivation of articles which yield an early profit.'[52]

Additionally, sugarcane sapped the vitality of the soil in three years, after which it needed to be kept fallow for more than four years 'before it recovers its vegetative powers'.[53] Peasants tried to replenish the soil by planting legumes (*kalai* and *mussoor*) in the intervening spaces between the cane-stems,[54] but it was still an expensive proposition both in terms of labour and capital costs. Thus, in Mymensing, 'the ground prepared for planting the [sugar] cane must be raised and manured at a great expence, and the labour of cultivating it is not compensated in the same proportion with the cultivation of other productions better suited to the species of Land it occupies'.[55]

The connection between scarce resources and choice of crops by the peasants will be discussed below, but it was precisely this constraint which allowed the sugar-merchant to become instrumental in maintaining and furthering its cultivation. Advance contracts taken by the cultivators from the merchants for the cultivation of sugarcane were widely prevalent in west Bengal; it was certainly the major form of cultivation in Birbhum.[56] Prinsep's testimony regarding the cultivation of cane and sugar (in 1793) shows that the main financial source for the peasant cultivating sugarcane was the 'mahagen', who levied an interest of 'one anna monthly for every rupees advanced, which is equal to 75 per cent per annum, and never less than 6 pice' or 37.5 per cent.[57]

The available evidence for the output of sugarcane from a *bigha* of land, and the amount of refined sugar or molasses (*gur*) which could be procured from a maund of sugarcane juice is outlined in Table 7.

Though the cultivation of betel-leaf (*paan*) did not find adequate emphasis in Colebrooke's *Remarks on the Husbandry and Internal Commerce of Bengal*, it was nevertheless an important item of agricultural production. The leaf, grown entirely for commercial purposes was reared in specially constructed gardens called *voroj*, and these were completely distinct, and required a different kind of cultivation, than the leaves which most villagers grew in their own kitchen-gardens. The latter was called *gachhoya paan* (literally touching trees) because they were grown as creepers planted to the roots of trees. The betel-vine was then naturally allowed to climb along the trunk in order to produce an inferior quality leaf for domestic consumption, and none of it was sold.[58] In contrast were the leaves grown in special sites whose entire output was designed for sale. The vine-slips were planted in *Kartic* (October-November) but there were no specific reaping seasons as these leaves, once matured, would bear leaves for 9 to 10 years in west Bengal, and up to 6 or 7 years

TABLE 7. OUTPUT OF SUGARCANE, SUGAR AND MOLASSES (*in maunds*)

Region	Output of cane (per bigha)	Sugar	Molasses
Hughli	8	0.37	0.62
Burdwan	15 to 18	0.25	-
Dinajpur	14	0.25	0.62
Purnea	10.2 to 13	0.01	0.15
Commilla	12	-	0.33
Jessore	11.5	-	-
Birbhum	18.2	0.13	00.56

Sources: IOL, Ms. Eur. F. 95, fol. 34; IOR, BRP, P/72/10; Prinsep, *Bengal Sugar*, p. 26.

in the east, and the leaves were 'taken away from time to time as they ripen'.[59] The betel-vine: 'The leaf of which forms part of the composition of the article of luxury for the natives . . . is raised on artificial eminences from 3 to 5 feet above the common level under a kind of shed composed of the stems of the Dooncha [?], the roof and walls loosely compacted so as to be pervious to the air and rain, and to afford a cover only against the violence of solar rays.'[60]

Buchanan gives the following description of betel-leaf cultivation in Chittagong:

The betel-leaf is a very delicate plant, and requires a vast attention in the cultivation. The ground on which it is raised must be high. After it has been well wrought, and cleared of weed, in order to completely carry off the water it is, by narrow trenches, divided into beds two cubits wide. Upon these the earth from the trenches is thrown, and in the middle of each row a row of betel slips is planted, at a distance of two spans. Sticks about 8 feet long are then stuck up in form of the St. Andrew's Cross, in order to support the plant which is a creeper. Other sticks are laid, horizontally over the crosses, and on them is spread straw or grass to keep off the Sun. . . . The wind must also be excluded from this delicate vegetable by covering the sides of the Garden with palm leaves or straw.[61]

To construct such gardens, the earth had first to be dug to the depth of 10 inches, then filled with soft mud mixed with a manure of mustard oil-cakes which remained after the oil had been extracted. The entire garden had to be enclosed by a bamboo wall and the roof was made of bamboos mixed with mud and straw. The outer wall was tied with rattan canes. The vines were planted in straight beds and rows, manured with the same mixture of mud and oil-cakes and then constantly watered and weeded and watched for worms. The vine-stems were never allowed to rise to the roof-level. The stem was constantly shortened, not by cutting from the top but by holding the stalk from near the root, drawing it out and then making a fold in the stalk which was covered with mud and manure. This process was repeated each time the stalk grew beyond the maximum permissible height.[62]

Such activities were, undoubtedly, both labour and capital intensive. In west Bengal the average cost of planting a *bigha* with *paan* was estimated at Rs. 30 in 1791,[63] but in the east it was higher: in Commilla the costs were Rs. 81.94 a *bigha* in 1789,[64] while in Dinajpur they were Rs. 168 in 1807.[65] Returns also differed between the west and the east: in Hughli a *bigha* of betel-leaf would give a return of Rs. 475 a year, in Commilla the return was Rs. 200.[66]

The importnce of betel in the commercial world of rural Bengal can hardly be understated. The high costs involved would require certain command over resources, a necessary pre-condition for its cultvation, thereby limiting it to the relatively af-fluent and provides an important example of a highly specialized form of market-gardening. However, the case of the more inferior varieties, the *gacchoya paan* for instance, which was seldom sold, would indicate kitchen-gardening within a peasant's domestic holding in order to acquire an inferior version of a commodity which had become a vital ingredient in the patterns of consumption in the province.

Local merchants had established near-monopolies in the internal trade in betel-leaf during the Nizamat (Chapter 4) and this seems to have continued in our period too. For instance, the sole trader in betel-leaf for the whole of Dhaka city in 1789 was Ganesh Das Tiwari.[67] The chief reasons behind these localized monopolies were (*a*) that trade in betel-leaf was extremely profitable despite the perishable nature of this commodity, (*b*) there was an extensive and growing demand for this leaf. As Buchanan noted, its internal demand for consumption was 'enormous',[68] and (*c*) the cultivation of betel-leaf was expensive and peasants were extensively funded by betel-merchants who made 'advances to the Ryots many months before they [the leaves] were fit for reaping'.[69] Thus betel-leaf production provides another good example of the inter-relationship between peasant enterprise and commercialization of agricultural production in the late eighteenth century.

According to Colebrooke, 'tobacco [is] a very profitable culture [and] it is eagerly pursued' in Bengal.[70] We have some evidence regarding the estimated produce of this crop from different districts (Table 8).

TABLE 8. OUTPUT OF TOBACCO (*maunds per bigha*)

District	Output in maunds
Purnea	50,000
Nadia	38,000
24-Parganas	4,000
Midnapur and Jalasor	3,237
Bishnupur	1,500
Jessore	5,000
Muhammadshahi	2,400
Sarfarazpur	1,825
Rangpur	3,00,000

Source: IOR, BRP, P/72/10, 15 June 1789.

These figures show that the cultivation of tobacco was unevenly disseminated in the province. While Rangpur produced the largest quantities of tobacco, its neighbouring district of Dinajpur produced practically none, its land being considered 'not favourable for the culture of it'.[71] Nevertheless, tobacco was one of the principal items of non-food consumption and its internal demand was substantial: in most districts the local leaf was 'not equal to the consumption'. There was apparently a fairly well-marked social difference in the use of tobacco: the 'lower class of natives' stuck to the local, and often insipid, product, while the 'more opulent natives' tended to consume finer quality tobaccos regularly imported from Chunar, Patna and Rajmahal.[72] Boatloads of tobacco left from northern Bengal, especially Rangpur, bound for Dhaka, Murshidabad and Calcutta and the government often found it difficult to regulate this trade: 'it is indeed a difficult task to stop all boats as the navigable passages are almost innumerable', especially during the 'wet season'.[73]

'The culture of tobacco is laborious, as it requires the ground to be broken up by repeated ploughings. The tobacco, though transplanted, needs one or two weedings and a hand-hoeing. It is frequently visited by the labourer to nip the heads of young plants, and afterwards to pick off the decayed leaves.'[74] Tobacco:

Succeeds best in strong soil, is sowed in Bhadun [August], transplanted in Assin [September-October] at the distance of one-fourth cubits and is ripe and cut down in Poose and Maug [December to February]. The plants as they rise to maturity (when about a cubit in height) are stript [sic] of their menus shoots and top allowing only from 10 to 15 leaves to remain as the state of the soil admits. The leaves are dried in the sun and then packed up for use.[75]

By nature, tobacco required the richest, as well as the highest, soil in the village. In Rangpur, tobacco was invariably grown: 'contiguous to the Ryotts houses where they can avail themselves of the assistance of their families, in paying constant attention to the preparation of the Ground and the cleaning of the plants which is indispensably necessary'.[76] Additionally 'the goodness or badness of tobacco depends entirely on the care and attention paid to the cultivation of it & the time & labour & the expence of manure requisite to bring the ground to a proper state of producing the greatest quantity of the best tobacco are very great'.[77]

This meant that many peasants did not undertake its cultivation. Thus, in the 24-Parganas, 'the ryott finds greater advantage by the culture of grain [as] tobacco grounds require much pains to be taken with them'.[78] The problems here were those of resources and returns, and (like the cultivation of sugarcane and betel-leaf) merchants had a strong influence in production. Almost the entire tobacco output from Nadia was 'produced by means of money supplied to the Ryotts by the Mahajans'.[79] Rangpur's tobacco was either directly financed, or was contracted for in advance, by the merchants of 'Dacca, Moorshedabad, Jungypoor, Cutwah, Bogwongola, Chandernagore [and] Calcutta through their Gomasthas'.[80] The output of tobacco leaves per *bigha* was as follows:

In Hughli	4 maunds
In Jessore	3 maunds
In Rangpur	3.41 maunds

While detailed historical studies exist on the production of cotton textiles in our period,[81] very little is known about the state of the cultivation of *copass*, raw cotton, by the cultivators. According to Colebrooke, cotton was cultivated throughout Bengal, but the quantity produced was not enough to sustain the demand for cloth which had 'given rise to a very large importation from the banks of the Jamuna and the Dakhin', and the reason was '[either] the increase of manufactures or decline of cultivation'.[82] There can be no doubt that there was an immense domestic and foreign demand for Bengal's cotton piece-goods in our period. More than 61 per cent of the Company's investments in 1788 were for cotton piece-goods.[83] The ratio of the Company's investments for raw silk and cotton is given in Table 9.

There was furthermore the *internal* demand for cotton textiles. Colebrooke computed an annual production of cotton textiles worth Rs. 60 million in *c*. 1794 which was 10.21 times greater than the Company's investment for cotton piece-goods in 1788 and 9.15 times

TABLE 9. RATIO OF COTTON TO SILK IN COMPANY'S INVESTMENTS

Year	Ratio
1787	1:3.3
1788	1: 2.7
1795	1:2.8
1799	1:3.5

Sources: I am grateful to P. J. Marshall for providing the figures of investments in 1787. For other years see IOR, BRP, P/70/49, 12 December 1788; N.K. Sinha, *The Economic History of Bengal, Volume 3, 1793-1848*, Calcutta, 1970, pp. 1-3.

greater than the investment made in 1793.[84] These influences provide the reason for explaining the increasing imports of raw cotton from north and western India. In a survey of the cotton industry undertaken by the Company in 1789, the proportion of local production to imports are shown in Table 10. The imports were principally *via* Patna, Mirzapur, and Surat.

Such data must also indicate the unwillingness of the cultivators to increase the production of *copass* in order to meet the rising provincial demand. In fact, some areas (like the 24-Parganas, Tamluk, Sylhet) practically grew no cotton at all,[85] in others (like Purnea) cotton production had actually declined,[86] while elsewhere (Dinajpur and Rangpur) the quality of cotton was at best of an 'indifferent' kind.[87] Two sets of considerations appear to have been of decisive importance to the peasants. First, there were the major considerations of resources and returns, as well of local soil conditions which influenced peasants not to extend the cultivated area for cotton despite the existing demand. Second, the prices of locally grown cotton were apparently much higher than of cotton which was imported. Cotton in the Deccan and northern India was 'raised more cheaply than in Bengal, that it [supported] a successful competition, notwithstanding the heavy expenses of distant transport by land and water. . .'.[88] In fact one of the major reasons for the decline of cotton cultivation in Purnea was the relative cheapness of the cotton imported from Mirzapur and Patna 'which undersells the cotton in Purnea'.[89]

The cultivation of *copass* shows distinct intra-regional differences. Cotton produced in Dhaka was primarily of a superior variety called *nurmah* in order to feed the production of fine muslins there,[90] other districts tended to grow relatively inferior varieties, variously called

TABLE 10. COTTON IN BENGAL: LOCAL PRODUCTION AND AMOUNT IMPORTED

District	Annual requirement of *copass*	Local (maunds)	Imported (maunds)
Birbhum and Bishnupur	37,000	20,000	17,000
Jessore	6,000	2,400	6,000
Rangpur	36,500	1,500	35,000
Nadia	16,800	4,000	12,000

Source: IOR, BRP, P/71/11, 1 June 1789.

bhoga and *muhree*, in order to meet the requirements of a largely indigenous demand.⁹¹ The pull of production centres also influenced the cultivation of cotton in other districts. Thus, in Sylhet, where cotton was only grown in small quantities 'on the small hills', the peasants tried to cultivate *nurmah* cotton in order to supply Dhaka,⁹² while in Burdwan, which supplied raw cotton to Birbhum, Rangpur, Nadia and Murshidabad, the emphasis was on the cultivation of *bhoga* and *muhree* cotton for manufacturing coarser cloths.⁹³

Despite its commercial importance, cotton occupied only a tiny fraction of individual peasant plots. A common expectation that richer peasants would tend to give a greater percentage of their lands to an important cash crop like cotton is not fulfilled from Bengal's case. In their surveys, the Company officials, to their surprise, found that peasants with relatively larger resources would devote at most one *bigha* of their land to it while the so-called 'poorer class of ryotts' cultivated half or a quarter *bigha*.⁹⁴ Moreover, cotton was seldom grown alone. Since the usual distance between each cotton plant was about 2 feet, the cultivator would plant the intervening spaces with 'different kinds of grain, which being of a quicker growth is always reaped before'.⁹⁵ The cotton ground was also sown with turmeric or vegetables.⁹⁶ In Rangpur cotton and ginger were simultaneously grown on the same piece of land.⁹⁷

At first sight such behaviour appears inexplicable. Here was a crop with an extensive, and increasing, demand, but there was no matching effort to increase the area under its cultivation. The reasons must therefore lie in reasons other than a purely commercial consideration. Soil was obviously a very important consideration. By nature cotton required a thick loamy soil 'in a situation high and dry to admit of the water running easily off'.⁹⁸ The initial costs were high since: 'The care required in its cultivation, the number of people employed in clearing the ground [and] keeping it continually free from weeds, destroying the insects which some quality in the plant breeds makes it not only expensive but equally laborious.'⁹⁹ Additionally, cotton was a precarious crop:

The plant itself being liable to many accidents, the culture of it is attended with much risk. A superfluity of rain rots & decays it & a drought kills it. And of so very delicate and tender nature it is, that even a few days of unseasonable weather so much hurts the growth & reduces the quantity of cotton, that the same number of plants which yield in one year 10 maunds will at another produce only 4.¹⁰⁰

Very little is known about the output of *copass* from a *bigha* of land. Fragmentary information suggests that a *bigha* of land could yield crops ranging from 0.5 maund (in Jessore), 2.5 maunds (in Midnapur) and about 5 maunds (in Rajshahi).[101]

The So-called 'Inefficiency' of Bengal's Agriculture

For western observers agricultural production in our period appeared extremely inefficient. Bengal's peasants did not use manure in adequate quantities to raise productivity from land, they did not practice fallowing in order to restore the natural vitality of the soil, and had no knowledge of crop rotation of the type (fodder–cereal–fodder) which was practiced by the peasants in eighteenth century Europe. These were listed as some of the major 'imperfections of husbandry'[102] in order to explain, what was considered by these observers to be the essential backwardness of Bengal's (and Asian) agriculture.

The fact that the greatest portion of cultivated land, under rice, was not extensively manured,[103] except naturally, was very much in keeping with the intrinsic nature of rice. Rice cultivation depends almost exclusively on adequate water supplies for most of its nutrients, and the area under wet-rice, therefore, is not dependent on large supplies of manure. Additionally, the fertilizing power of water enables the fields to be cultivated continuously without fallowing.[104] Deposits of silt and rotting vegetation left by the seasonal inundation was enough to give successive rice harvests without the need for any additional, or artificial, manure.

Nevertheless, manuring was practiced for those crops which needed to be manured. Unfortunately, very little is known about manuring practices in different districts but a comparison of Robert Kyd's survey of the western side of the Hughli river with Buchanan's survey of Rangpur does show certain common features. Leaving the roots of a previous harvest to rot in the ground was a commonly used natural manure for the cultivation of the *aman* rice. Burning the rooted stubble of one crop and then ploughing in its ashes for the next was another widespread form of manuring used for the cultivation of rice and lentils. As shown earlier, more elaborate, and costly, forms of manure were used for cultivating sugarcane, cotton, tobacco and betel-leaf. Here expressed oil seed (especially mustard) cakes were mixed with the top soil and then evenly distributed over the beds and rows in order to provide additional nutrients to the plant and to replenish the soil subsequently.[105] Marling or dressing lighter soils with clay or loamy earth in order to enable them to retain water and manure near the surface was not in vogue ('the use of the earth from marshes and ditches is very little, if at all, practiced'),[106] probably because of the costs involved, which could be prohibitive;[107] and this can perhaps be ascribed as one likely cause for preventing the spread of cotton cultivation in our period.

'The Indian allows [the land] a lea, but never a fallow' was Tennant's way of describing the absence of fallowing by the peasant of Bengal.[108] Apart from showing that lands in Bengal were intensively cultivated, Tennant's statement also demonstrates the essential divergence between Bengal's (or South Asia's) agriculture from that of contemporary Europe. In Europe, fallowing was essential to restore the fertility of the heavy and wet clay soil after two or three years of successive cropping. Here, fallowing was more than giving some respite to the land; it also entailed successive ploughings to break the surface thoroughly and get rid of weeds.[109] Obviously this process tended to raise costs of production and by the end of the seventeenth century, the rigid rotation of two crops and a fallow had come to be recognized as one of the major problems in the way of raising the productivity of land in Europe.[110] In contrast, fallowing was not an important feature of Bengal's agriculture because the soil here was more amenable, i.e. it could be broken up more easily by 'scratching the superfices of the earth to make it yield'.[111] In such situations, fallowing would mean an unnecessary loss of output and unduly high average costs of production.[112]

Nevertheless, the awareness that lands tend to get exhausted by repeated cultivation, and that it needs some rest to recuperate, was well developed in Bengal. Crops like sugarcane, cotton and mulberry sapped the land after three successive harvests, and lands bearing these crops were rejuvenated in a four-yearly cycle. In lands bordering upon the Hughli river, such lands were fallowed for about a year during which time 'the ground is restored with manure and fresh soil . . . and exposed once a month by the plough to the influence of the Atmosphere'.[113] In parts of Rangpur: '4 years *fallow* are usually allowed after 2 crops. The first year the ploughing is rather difficult, and the grass roots require to be collected and burnt; but on the whole the trouble of cultivation on such lands is exceedingly small, and the crops are good. The land is fallow for two or three years, and produces abundance of grass, but it is of a wretched quality.'[114] However, compared to Europe, the practice of fallowing was undertaken on a much smaller scale in Bengal. Here only a tiny fraction of lands producing selected crops were actually fallowed, whereas in Europe fallowing meant that from a quarter to a third of the arable land was made to lie unused for up to three or even four years.[115]

The reasons why some European observers (for instance Colebrooke and Tennant) noted an absence of crop-rotation in Bengal are unclear.

Contemporary European agricultural practices appear to have been conducted basically on a three-course—cereal, fodder, cereal—rotation of crops, though in some parts, for instance in Flanders, twelve-course rotations had been introduced by 1800; the so-called Norfolk system of rotation (turnips, barley, clover and wheat) was being gradually extended in to the rest of England along with an increasing tendency to cultivate part of the fallow with fodder crops.[116] Evidence from Bengal does suggest a fairly widespread existence of crop rotation. The simplest form was the rotation of rice and lentils on relatively lower lands, on higher lands rice and mustard were sown in regular succession. *Sih-fasli* (three harvest) lands, usually of the best intrinsic quality, produced rice with either two crops of oil-seeds or lentils, or they were used for producing successive crops of mustard, cotton and lentils in a regular annual rotation.[117] In fact, Buchanan listed 'alternate cropping' as one principal agricultural practice to prevent soil-exhaustion in parts of Rangpur.[118] Interestingly, alternate cropping, or the sequential use of fodder crops and corn crops to obviate fallowing (also called convertible husbandry) had only just started spreading to the enclosed and fairly large farms of England in the late eighteenth century.[119]

The Rationality of Peasant Production

Figure 1 shows the sector-wise contribution of different crops in the total agriculture of the province in 1794. Food grains (rice, paddy and millet) comprised 45.5 per cent of the output, whereas cash crops (cotton, mulberry and tobacco) contributed 20.5 per cent. Pulses and oil seeds, comprising about 20 per cent of total production, were certainly designed for sale, but they were not at par, in terms of exportable and commercial importance to crops like cotton and mulberry, and therefore may be seen as part of the overall production of food. The features of agricultural production, as revealed by Colebrooke's estimates, are also confirmed by some other contemporary observers[120] and from scattered evidence available from specific districts.

Fragmentary data available for this seem to suggest that more than 60 per cent of the cultivated land in central Bengal in the late 1770s was given over to the cultivation of the primary staples, rice and paddy.[121] Pargana Lashkarpur, in eastern Bengal, had a total agricultural output estimated at 3,24,000 maunds in 1773, of this paddy constituted 2,72,000 maunds.[122] In 1787, Birbhum's gross agricultural output was

TABLE 11. PATTERNS OF CULTIVATION IN BENGAL
(selected areas, in bighas)

Year	Village/District	Total land[a]	Food crops[b]	Non-food crops[c]	Unknown
1774	Gopalpur/Burdwan	129.16	87.65	12.5	29.01
1776	Rangamati/Murshidabad	1,272.9	688.5	269.5	314.55
1789	Swaruppur/Rangpur	44,731	37,176.08	6,941.12	613.80
1790	Pratappur/Yusufpur	725.80	687.95	34.15	3.70
1790	Bheerche/Yusufpur	452.1	398.4	43.95	9.75
1790	Prasadpur/Yusufpur	275.86	224.4	46.15	5.31
1790	Odurpara/Jessore	229.6	186.35	28.2	15.05
1790	Subankarati/Jessore	168.5	112	45.55	10.55

Note: [a]Land actually under cultivation; [b]Paddy, wheat and millet; [c]Mulberry, tobacco, cotton, lentils, opium, orchards.

Source: WBSA, PCR, Burdwan, vol. 2, 26 September 1776; ibid., Murshidabad, vol. 8, 15 February 1776; IOR, BRC, P/51/50, 4 September 1789; ibid., P/52/3, 10 February 1794.

valued at Rs. 32,78,156. Of this Rs. 7,00,000 were considered the value of crops for 'maufactured produce', while the rest, Rs. 25,78,156, belonged to the out-turn from the lands producing rice and other grains.[123] Buchanan-Hamilton's surveys (1807) suggest that nearly 90 per cent of the arable land, yielding about 93 per cent of the agricultural output, was given over to the primary sector in Dinajpur, whereas in Rangpur, the turnover from this sector was 90.8 per cent of the value of the gross produce.[124]

Unfortunately, the absence of measured area statistics for this period does not allow us to study the actual distribution of different crops on the cultivated land. Some data are nevertheless available. These are outlined in Table 11.

Such data certainly suggest that the agricultural landscape in eighteenth century Bengal was one where oceans of paddy fields were dotted with a few islands of commercial crops. This situation did not differ substantially even in those areas which constituted the core suppliers for Bengal's textiles industry. Thus, in the major silk producing district of Rajshahi 'there is no Ryott that does not hold grain lands & in greater proportion than Mulberry ones'.[125] Peasant production near Qasimbazar was similarly structured as the following figures detailing the patterns of cultivation in the lands of eight peasants near that town show:[126]

Total lands held by eight *raiyats*: 67.26 *bighas* of which:

Food grains	30.01 *bighas*
Mulberry	9.65 *bighas*
Lentils	24.7 *bighas*
Vegetables	2.9 *bighas*

The following evidence pertains to the land-use of Hari Das, a *raiyat* of *mauza* Srirampur, pargana Muragatcha in the 24-Parganas in 1786:[127]

Total land	8.53 *bighas*
Rice land	5.5 *bighas*
Cotton land	1.00 *bighas*
Lentils	2.01 *bighas*
Bamboo land	0.01 *bigha*
Pond and tank	0.01 *bigha*

Such evidence from a province which supplied more than 80 per cent of the Company's annual trade in cotton piece-goods and silk[128] must appear paradoxical. It certainly appeared so to contemporary observers. The almost monotonous complaints regarding the 'indolence' of the 'native' peasant as a barrier to raising the productive capacity of the land and introducing 'new' (that is, European) methods of cultivation were essentially designed to show the intrinsic agricultural backwardness and technological deficiency of the average Indian peasant and to counterpose the supposedly superior European (specifically British) methods of cultivation and land-use. Central to this entire exercise was the assumption that the British cultivator was a rational economic agent who responded to the stimulus of the market, whereas the Indian (or Bengal's) peasant was not.[129]

Robert Kyd, the superintendent of the Company's Botanical Gardens on the outskirts of Calcutta (Sibpur) had the following to say about the possibilities of introducing new horticultural techniques in Bengal:

I can give the most positive assurances of this climate [of Bengal] admitting the resource derivable from the cultivation of Peach, Plumb, Apricot, Cherry; with the Chinese fruits . . . which succeed in perfection [in Bengal] and may in the process of a few years be disseminated throughout the Company's possessions if the Factory at Canton execute the Commission, lately sent, in furnishing a proper supply. . . . I may further add that the culture of the English Melon and grapes of every kind, may be introduced with great success, altho' the latter will succeed better within the precincts of the Town [Calcutta], than in the neighbourhood of Calcutta, which form the lowness of the soil, intense moisture and heat of the atmosphere in gendering innumerable insects which destroy the foliage in such a degree as to prevent their bearing. Add to this, the defect of the soil, which bears testimony of the element from which it has apparently emerged, apparently at no distant period of time, being strongly impregnated with salt.[130]

Kyd's remarks reveal that there were no deeply structured 'cultural' barriers to the introduction of new products among the cultivators; after all the principal vehicle of Kyd's 'positive assurances' of disseminating the new horticultural products 'throughout the Company's possessions' would have been the peasant. Kyd additionally demonstrated that the receptiveness, or otherwise, to new products depended mainly on climate and soil, two vital elements in the rationale of peasant production all over the world.

To the European eye, the average peasant in Bengal appeared extremely cautious in choosing the crops to be sown and the extent of the land which was to be under a particular crop at any point in time. Cultivators 'seldom engage in extraordinary speculations' was how their initial response to indigo cultivation was described in 1790.[131] From the cultivator's point of view this cautious approach to indigo was entirely justified for, apart from the coercion which its cultivation entailed, the production and marketing of this commodity was far from efficient, at least in our period. Capital invested in indigo was often precarious, the interest rates were too high to be attractive, there was an absence of a ready market for indigo *within* Bengal and the sales of this commodity in European markets often suffered from excessive delays.[132] Such cautiousness originated from the peasants perception of the risks involved in production, and their response was either an extremely slow acceptance[133] or a conscious antipathy, and the latter was misconstrued as their 'indolence' in official circles.

An antipathy of this type was demonstrated by the peasants to the Company's efforts of finding an Irish solution to problem of famine and dearth in Bengal, viz., by the cultivation of potatoes on an extensive scale. Despite the apparently strenuous efforts made by officials to ecourage the peasants to take to its culture by 'offering seeds to any who would engage to cultivate them', they clearly failed 'to prevail upon the Ryots to assist Government in its kind intentions [*sic*] in their favour'. The fact that the peasants did not take to potatoes with the speed the Company desired immediately brought forth the charge of 'native ignorance' and their 'natural' unwillingness 'to lend their aid to the cultivation of it'. The real reasons of peasant antipathy were entirely different.

First, the potato seeds were extremely expensive: a maund of seed potatoes was priced at Rs. 4 in October 1798. Second, the cultivators were not convinced of the suitability of potatoes to local soils. Those of Birbhum clearly stated that the soil there was 'unfavourable of the plant thriving, and being brought to maturity in this district; the soil in general being formed of a hard substance, and mixed with small stones or conker [*kankar* or gravel] renders it uncongenial for the culture of potatoes'. The peasants of Dhaka objected to this item on grounds that 'the soil is in general unfit for it and that in Bowal the only part of the district where good potatoes might be raised, the number of wild hogs is so great that their potato fields would be continually liable to the devastation of those destructive animals'. Third, Bengal abounded in a number of: 'Esculent roots of a farinaceous and nutritive quantity which are a common food among the lower class of people [such] as yams, the Spanish potatoes and the numerous varieties of Caychoo [local tubers] . . . all extremely productive and . . . capable of being stored and preserved for a greater length of time by due attention to prevent the access of humidity.'

These local tubers were naturally considered 'more congenial and better adapted to the soil, and climate, and [therefore] less precarious in point of produce than potatoes'. Yet some peasants did express their willingness to cultivate potatoes. In Cooch Behar, the *raiyat* appeared 'extremely willing to extend the cultivation of potatoes, provided they are supplied with a sufficient quantity of seed in the first year'.[134]

Thus, one of the key components of peasant production was their perception of risks. This attitude was entirely reasonable in the context of the late eighteenth century because of the almost continuous threats from famine and dearth to their subsistence and production. Risk avoidance was certainly one important reason behind the preponderance of rice and paddy in cultivated lands. Corn (i.e. rice and paddy), noted Colebrooke:

> Though not equally profitable with dearer articles [i.e. cash crops] serves to alleviate the risk attending the cultivation of them; for they [the 'dearer' articles] seem precarious in exact proportion to the greatness of the profit which they are expected to afford. On the failure of his mulberry or his sugar cane, the peasant, had he no corn, must suffer the extremities of want; but raising in that and other grain a sufficiency for mere subsistence, he can wait the supply of his other wants from the success of other cultures; or, he can reserve a hoard from the crop of a successful year to meet the difficulties of one that is calamitous.[135]

Nevertheless, it would be wrong to assume that Bengal's agriculture was entirely a risk-prevention enterprise. Under conditions of relative normality, average peasants could be seen attempting to try their hands at cultivating the 'dearer' and 'precarious articles' of production. In Jangipur, for instance, cultivators often strove to strike a workable balance, on their lands, between cash crops and the cultivation of essential staples. Here jute and paddy were sown on the same lands 'adjoining to each other' because '*paat* [jute] will bear a length of Drought which would destroy the Grain, nor does it suffer from excessive rains', so that 'in case of the failure of one, from casualties of weather, the whole together may yield him a profit'. Interestingly, there was very little mulberry grown in Jangipur despite the soil being 'well adapted to the Mulberry plant'; and the reasons for this were that mulberry required relatively larger amounts of capital, and that its cultivation was susceptible to frequent losses caused by weather or by 'disasters to the silk worms'.[136]

Peasants realized the commercial potential of different crops and, circumstances permitting, to attempt to cultivate for a market. An enquiry into the agricultural conditions of the district of Baldakhal in eastern Bengal had the following to say about the peasant's choice of crops: 'Cotton in this district is an article of universal cultivation, for the Ryot whose land is not favourable to the produce, provides himself with cotton land in some other place more proper for its growth.'[137]

In 1788-9, the Company made enquiries about the cultivation of plants yielding a yellow dye 'possessing the Quality of Madder'. The description of this dye from Birbhum is revealing: 'From the small fibre of the root of the tree cultivated in this district, called Auch, a yellow dye is extracted; and of which should it be found to answer the purpose of Madder, large quantities may be obtained, as on its becoming an Article of Demand, the *Ryotts will not fail to increase the cultivation of it.*'[138]

In September 1792 the Board of Revenue ordered the Company officials to make the peasants and merchants more aware of the commercial advantages of sugarcane since 'in future the demand for this commodity will be such as to yield to them an ample profit on any quantity that may be brought to the market'. A positive response to the stimulus of a market was anticipated: '*If the profit* accruing from this increased price is *in the first instance secured to the Cultivators*, there can be no doubt of their extending their plantations to such an extent as in the course of few years will enable this country to export large quantities of sugar not only to England, but to different parts of India where the commodity is in demand.'[139]

Though the exact details of governmental efforts in the cultivation of sugarcane are absent, we are reasonably certain about the expansion in the cultivation of another cash crop, jute (*sona paat*), in Rangpur in our period. The first inquiries regarding the state of jute cultivation in Rangpur were made in 1794, which were rather imprecise both about output and lands under cultivation.[140] For the greater part of our period at least *paat* was cultivated almost entirely by the fishermen on small spots of land for making nets,[141] but by the first decade of the nineteenth century the cultivation of jute had doubled in Rangpur because of the 'demand from Europe and the advances given by the Company'.[142]

The case of betel-nut cultivation in Chittagong is revealing in this regard. Here, and despite being a highly taxed crop, its cultivation was spreading rapidly in the last decades of the century: 'Owing probably to the great number of Burma boats that now come here and supply themselves ... in place of importing the betel from Sumatra [and] every man is planting the [betel-nut] tree around his house.'[143]

Thus, the average peasant of our period blended an attitude of extreme caution with a healthy regard for commercial profit. But these peasants, like all producers, functioned in a material milieu which was usually beyond their power to control, and the influence of this environment was critical in determining the specific features of agricultural production and its social content.

Other factors were the capital costs involved in the cultivation of cash crops and the critical shortage of productive resources in the hands of the small-peasants. We are told that the estimated cost of cultivating mulberry in Murshidabad varied from Rs. 5 to Rs. 6 per *bigha* under normal circumstances in 1771; during unfavourable seasons, the cost rose to Rs. 10 and even Rs. 15. Cultivating grain was much cheaper: under favourable conditions it varied from Re. 1 to Rs. 2 and rose at most to Rs. 3 per *bigha* when conditions were not so conducive.[144] By 1780 the costs of cultivating a *bigha* of land with mulberry around Murshidabad had gone up to Rs. 21 in the first year, Rs. 9 in the second and a further Rs. 17 were required in the third year as recurring expenses to maintain production. After three successive years of cultivation the soil would become exhausted and 'the whole process is renewed on fresh ground'.[145]

In 1789, the comparative costs of cultivation in pargana Baldakhal were as follows:

Crop	Cost of cultivation (Rs. per bigha)
Two successive rice harvests (do-shari)	2.30
One crop of rice (ek-shari)	1.75
Millet	1.69
Gram	1.06
Lentils	1.50
Peas	0.44
Oil seed (til)	1.75
Mustard seed	0.37
Cotton	4.62
Sugarcane	12.62

In Dinajpur, the average cost of cultivating food grain was Rs. 2 a *bigha*, that of mulberry was Rs. 7.44, whereas the capital stock required for the cultivation of a betel-leaf garden (usually half a *bigha* in size) was Rs. 168 plus a maintenance cost of Rs. 73.62 per year.[147] From the figures given by Colebrooke, the all-Bengal average cost of cultivating a *bigha* of land with different crops works out as follows: grain, Rs. 2; tobacco, Rs. 4; and mulberry, Rs. 15.25.[148]

These cost differentials were almost prohibitive for peasants already suffering from scanty resources and heavy tax burdens. In Murshidabad, 'the extreme poverty of the Natives induces them to forego the prospect of greater advantages involving . . . the cultivation of cotton than of rice or other sorts of grain'.[149] In Rajshahi, 'the lands employed in Mulberry require much labour time and expence to fit them for the purpose that which the poor Ryott who lives from hand to mouth can seldom afford'.[150] In Rangpur, 'the Mulberry Plant is not cultivated in every Mehal, for every Ryott does not engage in it. To prepare and bring into cultivation, land proper for the Mulberry is attended with an increase of expence to the Ryotts.'[151] 'The poverty of the people is the principal impediment to the culture of cotton' in Sylhet,[152] and in Dinajpur, 'the gain resulting to the Ryott in rearing this article [cotton], can only be judged by comparison & the proportion this culture bears to that of Grain, leads to a conclusion that the Ryott finds considerably more advantage in cultivating the latter than the former'.[153] Finally, in Jessore:

The greatest disadvantages under which the culture of cotton plant labour are the general unfavourableness of the sale [i.e marketing] & the want of means in the Ryots to defray the charges of cultivation which it appears are great. . . . There is land enough to push the cultivation to a considerable extent if the Ryots had the necessary means; but poverty reduces them to the necessity of raising crops upon them, less expensive in the process of culture & [which] afford a more speedy return.[154]

In Nadia 'the cultivation of sugarcane is very expensive and requires a capital which the generality of Ryots do not possess'.[155] The major barriers to the extension of sugarcane cultivation were described by James Prinsep in the following words: 'Nothing seems to oppose an immediate and great increase of sugars here [in Bengal], but the disinclination of the Ryots to speculate upon future contingencies, which they cannot comprehend [sic], and their individual poverty, which forbids them to undertake what they cannot accomplish.'[156]

These descriptions sum up the constraints upon the bulk of the peasantry which were of fundamental importance to the *economics* of small-peasant production in our period.

The other important factor in determining the kind of crops peasants sowed on their lands was revenue. The fact that lands devoted to the cultivation of cash crops were subjected to a greater tax squeeze than those sown with food grains is a commonly known fact. In Midnapur lands producing three cash crops (mulberry, cotton and tobacco) a year were assessed at Rs. 10.19 per *bigha* whereas rice lands paid only Rs. 3.37 per *bigha* in 1774.[157] The average rate of revenue for cash crops in Birbhum was Rs. 1.35 per *bigha*, while rice and paddy lands paid no more than Re. 0.375 per *bigha* in 1787.[158] In 24-Parganas, the difference in the rates of assessment between rice and lands growing cash crops could range from Rs. 2.4 to Re. 0.50 depending on the crop being assessed.[159] In Muhammadshahi, the revenue of the *aman* (winter) rice land was Rs. 2 and the lands producing tobacco paid Rs. 3 per *bigha* in 1789.[160] Obviously, higher rates of revenue reflected higher returns which could be anticipated from cash crop, but such rates, nevertheless, had a definite impact on patterns of cultivation. The prime consideration here was not merely the pitch of the revenue demand (though that was important) but the apparent tendency for the demand to sometimes fall more heavily on cash crops. Thus, the Board of Revenue, while observing the potential of extending the cultivation of sugarcane in the province was equally keen to prevent: 'The landholders . . . from deriving an advantage by raising the rates of pottahs of the sugar cane lands instead of looking to the extension of the sugar plantations for an increase of the rents of their Estates [which] would not only be unjust as well as repugnant . . . but entirely counteract the beneficial effects.' The Board's logic clearly was that if taxes were raised on the cultivation of cash crops, the 'cultivators will derive no advantage, the quantity of land [under sugarcane] will diminish, and its price will be exorbitant as it will include the amount of additional tax on land'.[161]

These fears were quite justified as is shown in the case of some peasants in the 24-Parganas who, in 1780, attempted to plant tobacco on what were previously rice growing lands. The zamindars were quick to seize upon the financial opportunity opened up by such a move and refused to 'receive the Malguzarry according to the established Jummah [i.e. for rice land] but insisting on paying our malguzarry *increased* account in tobacco lands'.[162]

The zamindars were not alone in such matters. Though the Company tended to blame these zamindars for such taxation, its own attitudes towards revenue did not differ substantially. In fact, the important reasons for the decline in production of mulberry in Rajshahi in the 1780s were the high costs of cultivation and the financial predicament of the 'poor Ryott who is *allowed no remission in his rents*', to which were added the problems created by the famine of 1788. In Burdwan, 'the cultivation of the mulberry plant has of the late years much decreased . . . & one principal cause of this decrease has been owing to the very exorbitant rent demanded from the farmer some of [whom] pay so high as 14, 16 or even 18 rupees per Beggah'.[164] A similar position was faced by the *chassars* of Midnapur. The revenue squeeze was listed as an important cause in the decline in mulberry cultivation in the parganas of Cossijura, Narajol and Midnapur because: 'The annual amount of the revenue of their [*chassars*] Mulberry lands ought properly to be paid in instalments in no less time than twelve months, but they are compelled to pay it in nine.'[165]

The influence of revenue on production is also made clear from the case of cotton cultivation in Chittagong during the eighteenth century. Under the Nizamat, the cultivators of Chittagong had enjoyed concessional terms of revenue to the extent that 'no distinction exist[ed] in regard to Publick [*sic*] Revenue between cotton [growing] lands and such as [were] appropriated to the cultivation of rice'; the cultivators paid revenue 'thro the channel of their zemindarries an annual revenue agreeable to the established rates of assessment throughout the province', and they were 'not subject to any enhanced demands of revenue in case they [appropriated] their lands to the cultivation of cotton in lieu of rice'. Given such concessional terms of revenue, the cultivators of Chittagong shifted from cotton to rice and *vice versa* '*ad libitum* as the prospect of an advantageous sale, or more frequently, their own private wants and emergencies induced them'. The situation seems to have changed during the course of the second half of the century, when despite 'the profits arising from the cultivation of this article [cotton] being greater than those arising from the cultivation of Rice . . . *[its] cultivation has not become more general'*.

The Movement of Agricultural Prices and the Dynamics of Agricultural Production

Influencing production decisions with equal, if not greater, effect as considerations of capital costs was the purely market consideration of prices. Figure 2 provides the unweighted index of the average prices of common rice and sugar in Bengal between 1700 and 1800 (1713 = 100) and is based on the data compiled and interpolated from the ledger and account books of the Company. The prices of 1769-70 and 1787-8 have been excluded from the present exercise as they were years of famine. There is a fairly clear long-term upward trend in price. The other significant feature of this figure is the much pronounced increase in the price of rice in comparison to that of sugar.

The significant increase in the price of rice, particularly during the second half of the century, can also be seen from Figure 3 showing the unweighted index of common rice prices in Calcutta between 1754 and 1800 (1713 Bengal averages = 100) computed from the series of agricultural prices in Calcutta provided by W.B. Bailey in the *Asiatick Researches (1816)*. Common rice in the city seems to have undergone an average increase of about 219.02 per cent in the 46 years between 1754 and 1800; and once again the sharp inflationary swing between the 1760s and 1790s is clearly visible. Moreover, the provincial tendency of rice prices to escalate at a faster rate than sugar is also visible from Figure 4 (1754 = 100, based on data provided by W.B. Bailey) which shows the comparative price indexes of common rice and jaggery in the city of Calcutta between 1754 and 1800.

The fact that the internal rate of increase in rice prices was faster than the relative increase in the price of sugar over most of the eighteenth century can be taken as an indication of the state of relative prices in the cash crop (sugar) and the food crop (the so-called subsistence) sectors; and this would (at least logically) have two related outcomes: (*a*) that the inter-sectoral terms of trade in agriculture would tend to move in favour of rice, and (*b*) a greater effort would now go into the production of rice and its marketing.[168] How this was achieved would differ from region to region as well from household to household, but one can broadly identify two trends. The first is the substitution trend, and the second, I call, the spreading-out trend.

Dharm Narain's analysis[169] of the impact of price movements on twentieth century Indian agriculture shows that a perfectly rational productive choice for a peasant (operating in the context of a differential increase in the prices of agricultural produce) would be to endeavour to make the most of the available resources by switching to crops which were relatively the most profitable, even it meant abandoning (for varying degrees of time) a relatively high-value but high-cost crop and a fall in income. In the context of the two crops being discussed, the switch from sugarcane to rice could profitably be made as rice was easily cultivable, and at a much lower cost, on land meant for the cultivation of sugarcane while the reverse was not always feasible because of the high initial and operating costs of the latter. Therefore, it would not be unreasonable to see peasants prefer to switch from sugarcane to paddy cultivation on their lands, at least in some localities of the province. This indeed happened in Purnea and Mymensing.[170] However, the fact of substitution, even on a limited scale, cannot be taken to be a reversion to subsistence agriculture (de-commercialization) or a perversity of economic choice. Substitution on an extensive scale could have emerged as the general trend only in the context of an actual decline in the price of sugar. Given the fact that prices of the two actually rose, but at different speeds, the efforts to extend the cultivation of rice could be sustained only under the spreading-out trend.

It is not often realized that in a favourable land-labour environment, that is a situation of excess land to labour (like that of Bengal in the eighteenth century), rising agricultural prices do not necessarily encourage efforts to raise the intrinsic productivity of an existing peasant holding. Equally plausible responses are 'the greater use of land through bringing more land under cultivation and the use of more capital in the form of land improvement'.[171] There is overwhelming evidence to show that the eighteenth century was characterized, among other things, by some new areas of agricultural reclamation in Bengal. Though the pace of reclamation was faster in some areas (for instance in the eastern and south-eastern parts of the province), it was significant for it was predominantly financed and activated by local landed proprietors, the holders of charitable grants and by the grain merchants by an elaborate system which comprised the direct provision of agricultural loans and the grant of leases of cultivation (*pattas*) on time-bound concessional terms of revenue. While the latter mode of encouraging the extension of land was limited to landed proprietors, the use of advance loans to needy cultivators was a device adopted principally by the grain traders, who were also the people who seem to have profited most in the bargain.

The Modalities of Agricultural Reclamation

Long-term changes in the course of Bengal's river systems had already resulted in shifting the core areas of the agricultural frontier towards eastern Bengal, especially towards the delta (Map 2). Consequently, there was a shift in the balance in the production of food between western and eastern Bengal.[172] Contemporary evidence seems to indicate that the process was still continuing in the late eighteenth century when it was recognized that the economic advantages had shifted to 'the countries in the lower part of Bengal'.[173] The contemporary awareness of this shift also becomes apparent, for instance from Robert Kyd's survey of the western bank of the Hughli river in 1791 during which he noticed that the 'granaries' of Bengal were now located in areas like Dhaka and Bakarganj in the east, when previously these had been situated in Burdwan, Hughli and Rajshahi.[174] Bakarganj functioned as the leading supplier of the inter-regional trade in foodgrains. There was apparently also a flourishing illlegal channel of this trade. Thus, 'Sloops and other small craft frequently come to Backergunge for the avowed purpose of purchasing grain, and tho' the native commanders and supercargoes assert it is destined for Calcutta, Chittagong or other places within the Province; yet there is every reason to believe under this pretext, that they proceed thro' the outlets in the Sunderbunds to sea, and by [that] evade the orders of Government.'[175]

On the other hand, these ecological changes forced certain areas in western Bengal to emerge as unstable and even deficit areas. The regressive nature of the Company's early revenue experiments plus the devastation caused by the famine of 1769-70 accentuated the problems. It is no secret that the burden of the Company's taxes, and the rigours of its revenue experiments between 1760 and 1771 were borne primarily by western Bengal which also bore the brunt of the famine whereas eastern Bengal escaped largely unscathed.[176] Burdwan's agriculture and manufactures had declined considerably by the mid-1770s.[177] By 1775, Purnea was an 'impoverished country' with a 'lowered value of lands and their produce',[178] as was Rajshahi where the decline of agriculture, population and commerce was 'too evident a melancholy truth', even to

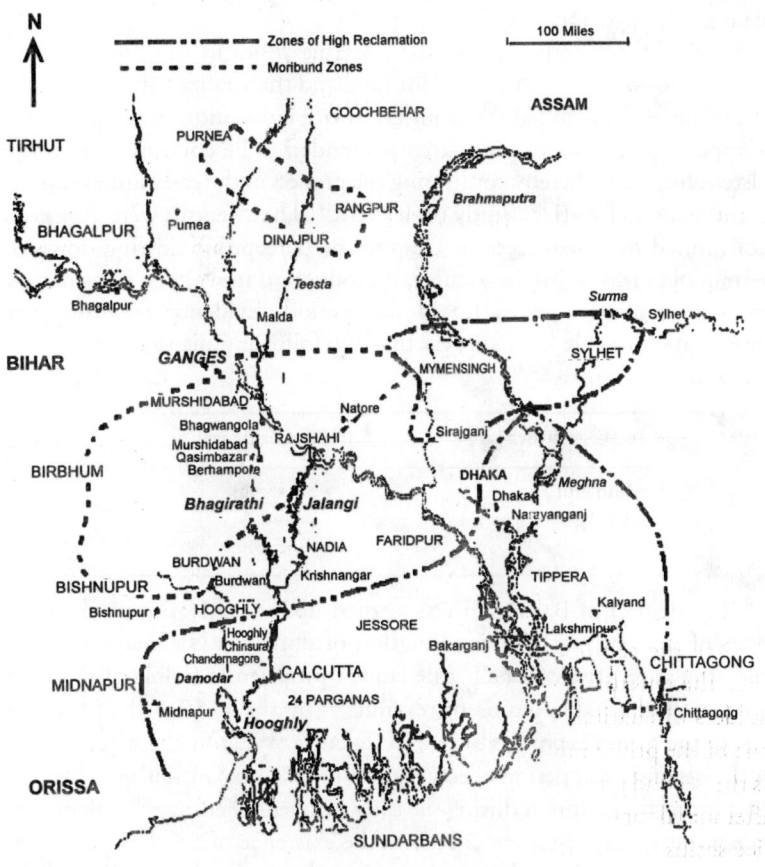

Source: Map adapted from Rennell, *Bengal Atlas*.

MAP 2. AGRICULTURAL PROFILE OF BENGAL: ZONES OF HIGH ACTIVITY AND MORIBUNDITY.

the officials of the East India Company.[179] The overall results were observable shortages in these deficit areas, which in turn made them depend heavily on the imports of essential items from relatively surplus areas of the province.

Despite the emergence of such unstable zones in the province, the overall buoyancy in the demand for food and the vitalization of networks of exchange[180] expanded the frontiers of rice cultivation in the province. It appears that most of such attempts tended to be concentrated in the eastern frontiers, thereby continuing the process of the eastward extension of the agrarian frontier already under way.[181] The attempts were, however, not limited to that direction. Despite the perceptible slowing down of certain old areas of high agricultural production in western Bengal, new areas were being opened up in this period. Evidence regarding the spread of cultivable land suggest that the following districts underwent rapid reclamation:

In Western Bengal	In Eastern Bengal
Midnapur	Bakarganj
24-Parganas	Jessore
	Chittagong

The district of Bakarganj was opened up for cultivation in the early years of the century by a combination of state (the Nizamat) initiative and the enterprise of small-scale landed proprietors (talluqdars)[182] and holders of charitable grants of revenue;[183] by the 1750s it had become one of the prime exporters of rice to Calcutta.[184] Eaton characterizes this as the 'second great period of economic and social expansion' of Bakarganj after initial forays into it during the early thirteenth century.[185] Bakarganj's rice seems to have become crucial to the existence of another important centre of production, the town of Qasimbazar, in western Bengal. In 1754, when the Nawab threatened to blockade the imports of rice by the English from Bakarganj, there was an immediate panic 'of the misery the place must be reduced to' by any such act.[186] Even during the famine of 1769-70, Calcutta continued to receive a major portion of its supplies from this district,[187] and by 1791 it was recognized as one of the major granaries of Bengal. Thus, it is clear that the pace of reclamation in Bakarganj continued unabated throughout the eighteenth century.

The integration of Bakarganj's rice growing economy with the regional demand in food grains led to three social developments. First, there was the formation of a hierarchy of trading interests ranging from the wholesaler (*aratadar* and *goldar*) and middleman (*paikar*) to the peddlar-cum-agent (*faria*) in the villages. Second, there appears to have been a rapid reclamation of the Padma estuary under the combined initiative of the talluqdar and the merchant. Between the talluqdar and the peasant there developed an intermediate social stratum, 'the Muslim religious gentry' who played an active role in imparting a great degree of stability to this expanding frontier.[189] Third, we see a proliferation in tenurial and sub-tenurial arrangements, engineered by the cultivators themselves in order to keep the best spots of land under their own control and to attract labour on a permanent basis on lands which were initially often inhospitable.[190]

The pull of regional demand saw the opening of another vast area in eastern Bengal—the Sundarbans,[191] or the estuarine forests in Jessore, where, according to an estimate made in 1784, there were more than 6,00,000 *bighas* (2,00,000 acres) of reclaimable land available.[192] Reclamation was undertaken by the zamindars who granted waste lands on concessional rates of revenue to settlers (*abadkar*). Apparently, *abdkari* was seen as an attractive proposition as zamindars, talluqdars, principal salt contractors whose territories 'are situated upon the borders of the Sunderbuns' were reported to be displaying remarkable 'alacrity' in accepting such offers.[194] Of the landed gentry involved in the process of *abadkari*, the lead was taken by the talluqdars who, 'have in proportion to their abilities advanced very considerable sums of money to the Ryotts, to enable them to clear away and bring the lands into a state of cultivation, purchase implements of husbandry & ca. . . .'.[195]

These settlers provided the initial investment and when the clearing had proceeded to a certain point, they settled the 'ryotts upon the land thus partially cleared to bring it into cultivation'.[196] The other social group closely involved in the process of agricultural expansion were the grain merchants interested in controlling the rice trade from here to the cities of western Bengal and to Dhaka. This strategy was realized by a system of advance loans to the cultivators. The annual practice of these merchants 'to advance to the poorer class of ryotts a sufficient quantity of grain to sow their lands, to be repaid in kind at the time of cutting their crops' had become the most widely prevalent method of agricultural production by the late 1780s.[197] The process of reclamation begun in this period was slowly coming to a close by the 1870s when the great forests of estuarine Jessore had almost entirely been converted into immense rice tracts.[198]

Francis Buchanan's descriptions of reclamation in Chittagong and Tippera, in 1798 suggest a rapid expansion of land under cultivation in areas like Lakshmipur and Noakhali. The rapidity is evident from a report from Lakshmipur in December 1758 which anticipates that '700 Connys of Jungul will be entirely cleared and settled by the next season . . . at the expence of the Riotts themselves'.[199] Another village, Ameegaon, was 'lately a chaar [dry bed of a river] covered with long grass which . . . when cut [was] sold to the salt boilers. The district is now cultivated, and the Fenny river is encroaching on the opposite side.'[200]

In fact the newness of reclamation in some of these areas was apparent to Buchanan 'by the stumps of the trees still remaining in the rice grounds'.[201] From Chittagong it was reported in 1761 that 'We have the pleasure to inform you [that] our tenants daily increase and many have already begun to clear away such lands as are uncultivated, and as the expence of doing it is entirely their own, we allow them the Indulgence of such lands from three to four years free of rent.'[202]

In Chittagong, such areas were called *no-abad* (newly settled). Jaynarain Ghosal, the principal salt-trader and ijaradar appears to have been at the helm of such activities. By 1770 'upwards of 200 Niabad Puttahs' had been granted to him by many 'of the Capital Zemindars . . . & in consequence large tracts of land have been cleared'.[203] Areas opened for cultivation as recently as the 1780s in Chittagong were Choonooty, Totacolly and Edgaon. The modes of reclamation were described as follows:

A man of some consequence, a Dewan, a Phousdar or the like, gets a grant of some uncultivated district. Different persons, who have a little stock, apply to him for Pottahs or leases of certain portions, and in clearing their portions these men are often assisted by the Zemundar, or possessor of the original grant, with a little money, as a temporary support. But this money becomes a debt, which they are obliged to repay, when they are able.

A reclaimer required a declaration of authenticity from the *no-abad jarib daroga*, an official appointed to measure these newly cultivated areas, upon which they were given the privilege of paying revenue at concessional rates for varying lengths of time, and entered in the revenue records as such.[204] Eaton argues that these men of 'consequence' were overwhelmingly 'associated with formal or informal Islam',[205] but Buchanan's evidence tells us that to a large extent, the new settlers were the 'Bengalese', people from the 'western parts of the province, the

Arakanese and Maghs',[206] thereby suggesting a more complex social participation in the process. In fact, the most important person involved in the process of reclamation was Gocul Ghoshal, a Brahmin salt trader-cum-contractor, who was the established *no-abadi* zamindar of Joynagar and was considered the defacto owner of the salt producing island of Sandwip among many other places.[207] Reclamation was also accompanied by the construction of tanks and wells to retain water. These tanks were apparently being constructed on a large scale as 'at present every man, who gets a few rupees wishes to perpetuate his name by digging a tank'.[208]

In the 24-Parganas, waste lands were classified under two heads, *khas pateet* and *mahsulat pateet,* the former being land never previously cultivated, while the latter referred to lands 'only fallen waste from the desertion of the Ryotts.[209] There existed a category called *abadi pattadars* who were entrusted with the task of reclaiming waste land. Cultivators were attracted by them on concessional terms as 'the Ryotts at the end of the year, when their balances are demanded of them run away and enter into the service of Abady Pottadars'.[210] In 1758, out of a measured area of 8,16,416 *bighas,* 'zemindars collect Rents on only 4,54,804 *bighas,* the rest being either barren or untenanted or assigned over to servants, idols & ca'.[211] In 1779, 433 *abadi pattadars* were instrumental in reclaiming 97,950 *bighas* of land, and by 1791 the ground cleared had increased by another 1,06,219.3 *bighas.*[212]

The pull of Calcutta was the major stimulus behind agricultural reclamation in this district. It was also fortunate in escaping the ravages of the famine of 1769-70 and its aftermath.[213] One major indication of reclamation is the data of villages that are available. In 1778 there were 3,124 villages in this district (Table 14). Less than a century later, in 1872, these had increased to 4,978.[214] Another feature described by Hunter was that the cultivation of rice was undertaken by peasants, barring a few, by taking advances from grain merchants who sent their agents to different villages in August or September to make such advances which were to be repaid, with interest, after the harvest. Zamindars also made such advances, though in their case repayment had to be in cash. Given the substantial evidence of such loans in our period,[215] one can safely assume that the genesis of such loans in the 24-Parganas occurred in the late eighteenth century.

With regard to Midnapur, George Vansittart wrote in November 1767 that:

When in April and May last [1766], I visited the several Pergunnahs belonging to the Midnapore and Jallesore Chucklas, I found upon enquiry that there were near 80000 Begas of land uncultivated, exclusive of what was purposely left waste for roads & ca., and what was deemed fit for cultivation. Of this quantity I then provided for the cultivation of 34000 Begas and I flatter myself that the whole will be cultivated in this and another year. . . .[216]

Of this, 24,900 *bighas* of land had been brought under cultivation by 1768, and a further 24,200 *bighas* were to be opened up by 1769.[217] The alluvial tracts bordering the Roopnarain river was being vigorously opened up for cultivation in this period. The *jalpai* (land liable to be under water) lands in the south of the district were crucial suppliers of firewood to Midnapur's salt industry, and as such were only marginally under cultivation. Though we hear of a proposal of clearing 10,000 *bighas* of such land in Mahisadal 'on the south side of the Roopnarain' in 1788,[218] it was only in the late nineteenth century that these lands were finally brought under the plough.[219]

The connection between extension of agricultural land and the extension the rice growing frontier in Midnapur is strongly established by the papers appended to the Amini Commission report of 1778. The assessed revenue of Midnapur was Rs. 8,84,388 in 1771-2. This had increased to Rs.10,43,985 in 1776-7. For Hijli, the *jama* had increased from Rs. 2,78,536 to Rs. 2,94,945 in the same period. The components of this increase were detailed as follows:[220]

Source of Increase	In Midnapur (Rs.)	In Hijli (Rs.)
From new lands brought under cultivation	75,696	7123
From lands cultivating high grade crops	6,668	Nil
From rice lands paying revenue in kind	52,644	1414
From new taxes, imposts and assessment of secreted lands	24,639	7922

It appears that in this district, 'Lands which yield rice have been gradually gained from jungles by adventurers who have obtained grants from Government and risked money to bring them under cultivation.'[221]

Who were these 'adventurers'? First, there were the 'ryotts', who were involved in the cultivation of a type of waste land called 'Pooroah' in Tamluk. This was 'common waste land which has not been cultivated' and which these peasants brought under cultivation 'on the term of three years', that is revenue free for the first three years, which was 'esteemed on easy footing'.[222] Since the minimum cost of bringing a *bigha* of such land into cultivation was Rs. 1.52,[223] we may presume that this would be possible only for peasants possessing some material resources. For larger projects of reclamation or those which involved the clearance of *shikast-pateet* ('broken waste ground'),[224] the initiative was taken by the landed gentry, especially the talluqdars, a form of landed property which apparently mushroomed in this district in the eighteenth century.[225]

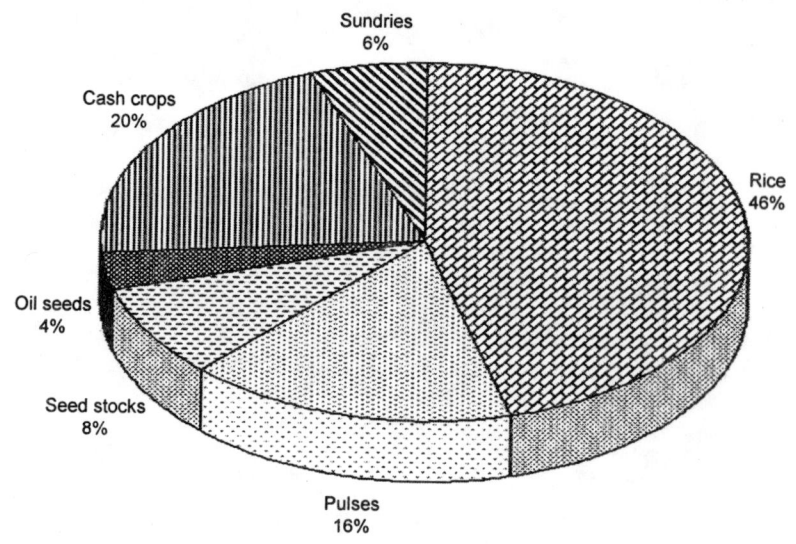

Note: Gross Output: Rs. 33,00,00,000.

Source: Colebrooke, *Remarks*, pp. 15-16.

FIGURE 1. AGRICULTURAL OUTPUT, c. 1794: A SECTORAL COMPARISON.

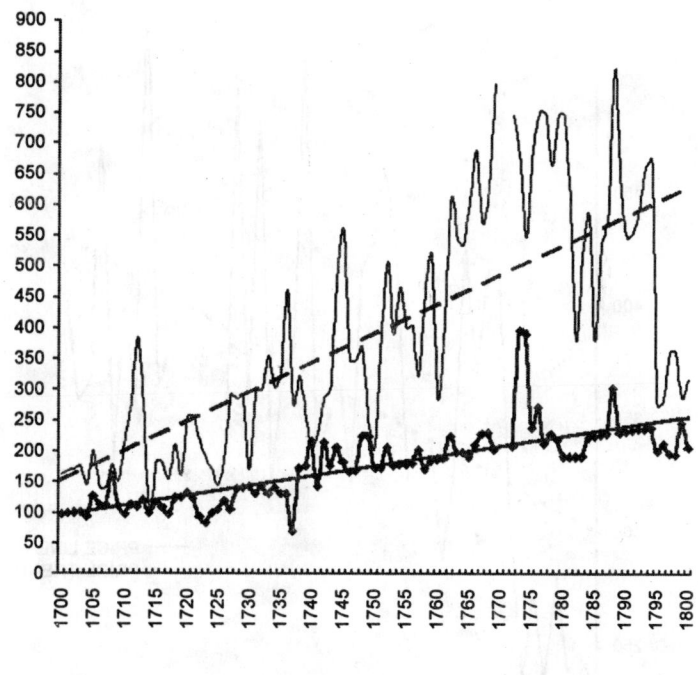

—— INDEX RICE —●— INDEX SUGAR —— TREND RICE —— TREND SUGAR

Note: Bengal Averages, 1713 = 100, Famine Prices Excluded.

Source: A.S.M. Akhtar Hussain, 'A Quantitative Study of Price Movements in Bengal during Eighteenth and Nineteenth Centuries', Ph.D. thesis, University of London, 1977, pp. 277-8.

FIGURE 2. INDEX NUMBERS OF RICE AND SUGAR PRICES, 1700-1800.

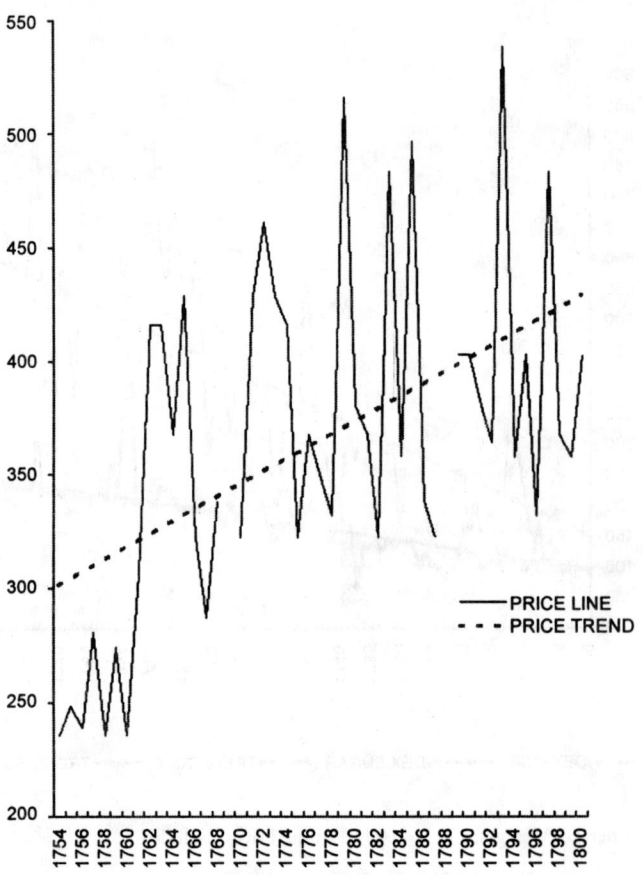

Note: 1713 = 100, Famine Prices Excluded.

Source: W.B. Bailey, 'Statistical View of the Population and ca. of Burdwan', *Asiatic Researches*, vol. 12, 1816.

FIGURE 3. INDEX NUMBERS OF RICE PRICES IN CALCUTTA, 1754-1800.

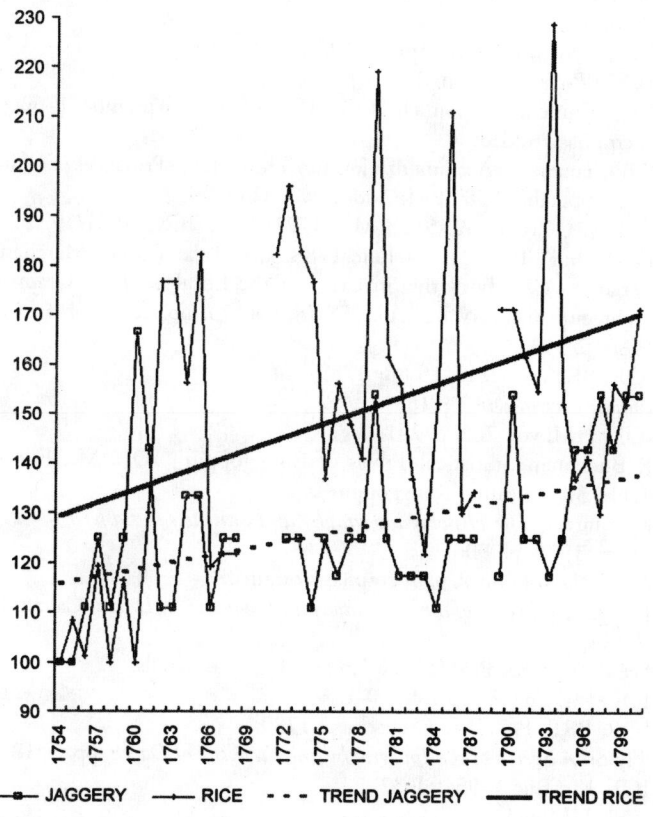

Note: 1754 = 100.

Source: W.B. Bailey, 'Statistical View of the Population and ca. of Burdwan', *Asiatic Researches*, vol. 12, 1816.

FIGURE 4. INDEX OF RICE AND JAGGERY PRICES IN CALCUTTA, 1754-1800.

NOTES AND REFERENCES

1. 'State of Bengal in 1784', in G.W. Forrest (ed.), *Selections from the State-Papers of the Governors-General in India, volume 2: Warren Hastings*, London, 1910, p. 21.
2. IOR, HM, vol. 765, 6 April 1772, p. 241.
3. *Indian Recreations 2*, p. 8.
4. Hugh Cameron to Council, IOR, HM, vol. 47, 5 September 1762, pp. 43-4; emphasis added.
5. F. Buchanan, 'An Account of a Journey Through the Provinces of Chittagong and Tipperah, 1798', BM, Add. Ms., 19286, fol. 2.
6. WBSA, CCRM, vol. 5, 23 May 1771; IOR, BRC, P/51/19, 21 March 1788; ibid., P/51/21, 14 June 1788; also Rajat Datta, 'Merchants and Peasants: A Study of the Structure of the Local Trade in Grain in Late Eighteenth Century Bengal', *IESHR*, vol. 23, no. 4, October-December 1986, p. 393.
7. IOR, BRC, P/51/40, 15 July 1789.
8. *Indian Recreations 2*, p.18.
9. IOR, HM, vol. 765, p. 241.
10. F. Buchanan, 'Statistical Tables of Ronggoppur', IOL, Ms. Eur. G. 11, Table 30 (hereafter 'Ronggoppur').
11. M. Martin, *The History, Topography and Statistics of Eastern India*, vol. 3, Delhi, 1976, p. 906.
12. *Indian Recreations 2*, p. 9; emphasis original.
13. J. Taylor, *A Sketch of the Topography and Statistics of Dacca*, Calcutta, 1840, p. 124.
14. See IOR, BRC, P/51/15, 28 January 1787 for details.
15. IOL, Ms. Eur. F. 95, fols. 30-1.
16. IOR, BRP, P/71/14, 7 September 1789.
17. Taylor, *A Sketch of the Topography and Statistics of Dacca*, pp. 3, 125.
18. IOL, Ms. Eur. F. 95, fol. 32.
19. IOR, HM, vol. 47, p. 44.
20. IOR, BRP, P/70/38, 4 February 1788.
21. IOL, Ms. Eur. F. 95, fol. 30; emphasis original.
22. IOR, BRC, P/51/15, 28 January 1787.
23. Ibid., P/51/40, 15 July 1789.
24. IOR, BRC, P/51/40, 15 July 1789; also Colebrooke, *Remarks*, pp. 61-3.
25. IOL, Ms. Eur. F. 95, fol. 32.
26. Ibid.
27. IOR, BRC, P/51/40, 15 July 1789.
28. IOL, Ms. Eur. F. 95, fol. 31.
29. Taylor, *A Sketch of the Topography and Statistics of Dacca*, p. 125. The provincial average according to Colebrooke was 1:15 in 1794 (*Remarks*, p. 63).

30. IOR, BRC, P/51/22, 23 July 1788.
31. IOL, Ms. Eur. D. 95, fol. 32.
32. IOR, BRC, P/51/15, 28 January 1787.
33. IOR, HM, vol. 375, p. 318.
34. Taylor, *A Sketch of the Topography and Statistics of Dacca*, p. 129.
35. *Indian Recreations, 2*, p. 11.
36. IOR, BRC, P/51/22, 23 July 1788.
37. IOR, BRP, P/70/40, 3 April 1788.
38. Ibid., P/72/15, 19 April 1793.
39. *Indian Recreations, 2*, pp. 136-7. Tennant's description of the mustard plant was from Chandernagar in 1797.
40. IOL, Ms. Eur. F. 95, fols. 33-5.
41. IOR, BRC, P/51/15, 28 January 1787.
42. IOR, HM, vol. 375, pp. 305-7.
43. IOR, BRC, P/49/62, 5 June 1776. Also see James Prinsep, *Bengal Sugar*, London, 1794 (IOL, *Tracts*, vol. 436 a).
44. Minute of the Board of Trade, 23 October 1793, IOR, HM, vol. 375, p. 41.
45. BM, Add. Ms., 19286, fol. 8.
46. See IOR, BRP, P/72/10, 26 December 1792 for such merchants in Birbhum; also see IOL, Ms. Eur. D. 75, vol. 2, book 4, fols. 70-1 for merchants in Rangpur.
47. IOR, BRP, P/72/10, 26 December 1792.
48. Ibid., P/70/40, Appendix to Proceedings of April 1788; IOR, BRC, P/51/15, 28 January 1787.
49. IOL, Ms. Eur. F. 95, fols. 33-4.
50. BM, Add. Ms., 19286, fol. 13.
51. 'Substance of a Conversation with Godadur Mundel, of the Village of Deca, in the Purgunnah of Renalty, and Province of Burdwan, distance 24 Coss (50 Miles) from Calcutta', in Prinsep, *Bengal Sugar*, pp. 77-8; also F. Buchanan, 'Survey of Ronggoppur' (IOL, Ms. Eur. D. 75, vol. 2, book 4, fol. 82) for the use of oil cake for manure in lands producing sugar, tobacco and betel-leaf.
52. IOR, BRP, P/72/8, 29 October 1792.
53. Ibid., P/72/10, 26 December 1792.
54. IOR, BRC, P/51/15, 28 January 1787; also F. Buchanan,'Survey of Ronggoppur', IOL, Ms. Eur. D. 75, fol. 46.
55. IOR, BRP, P/72/10, 26 November 1792.
56. BRP, P/72/10, 26 December 1792.
57. *Bengal Sugar*, pp. 54-5.
58. IOL, Ms. Eur. G. 75, vol. 2, book 4, fol. 41.
59. IOR, BRC, P/51/15, 28 January 1787; ibid., P/51/40, 15 July 1789.
60. IOL, Ms. Eur. F. 95, f. 42.
61. BM, Add. Ms. 19286, fol. 22.

62. IOR, BRC, P/51/40, 15 July 1789.
63. IOL, Ms. Eur. F. 95, fol. 42.
64. IOR, BRC, P/81/40, 15 July 1789.
65. Martin, *Eastern India*, vol. 3, pp. 864-5.
66. Compare, IOL, Ms. Eur. F. 95, fol. 42 and BRC, IOR, P/51/40, 15 July 1789.
67. IOR, BRC, P/51/50, 24 October 1789.
68. IOL, Ms. Eur. D. 75, vol. 2, book 4, fol. 48.
69. IOR, BRC, P/51/17, 17 March 1788.
70. *Remarks*, p. 76.
71. Ibid., 26 May 1789.
72. Ibid., 30 May 1789.
73. James Rennell, 'On the Severall Avenues by which Tobacco is or may be bought to Moorshedabad, Dacca, Calcutta & Ca from the Northern Provinces, 25 August 1773', in IOR, HM, vol. 217, p. 192.
74. Colebrooke, *Remarks*, p. 75.
75. IOL, Ms. Eur. F. 95, fol. 37.
76. IOR, BRP, P/71/10, 1 June 1789.
77. Ibid.
78. Ibid., 29 May 1789.
79. Ibid., 1 June 1789.
80. Ibid.
81. For instance, D.B. Mitra, *The Cotton Weavers of Bengal*, Calcutta, 1979; H. Hossain, *The Company Weavers of Bengal: The East India Company and the Organization of Textile Production in Bengal, 1750-1813*, Delhi, 1988.
82. *Remarks*, p. 84.
83. Total investment Rs. 95,96,698. Investment for piece-goods, Rs. 58,75,567 (IOR, BRP, P/70/49, 12 December 1788). N.B. 1788 was a famine year.
84. IOR, BRP, P/71/11, 1 June 1789.
85. Ibid., P/71/10, 20 May 1789, 29 May 1789, 31 May 1789.
86. Ibid., 26 May 1789.
87. Ibid., 1 June and 15 June 1789.
88. Colebrooke, *Remarks*, p. 84.
89. IOR, BRP, P/71/10, 26 May 1789.
90. Ibid., P/71/16, 22 June 1789; Colebrooke, *Remarks*, p. 32; Marshall, *Bridgehead*, p. 22.
91. Ibid., P/71/10, 26 May and 15 June 1789.
92. Ibid., 31 May 1789.
93. Ibid., 15 June 1789.
94. Ibid.
95. Ibid.
96. IOR, BRC, P/51/15, 28 January 1787.
97. IOL, Ms. Eur. D. 75, vol. 2, book 4, fols. 41-2.
98. IOR, BRP, P/71/10, 1 June 1789.
99. Ibid., P/71/16, 3 July 1789.

100. Ibid.
101. Ibid., P/71/10, 15 June 1789; ibid., P/71/17, 21 December 1789.
102. See for instance, Colebrooke, *Remarks*, pp. 22-5; Tennant, *Indian Recreations*, 2, pp. 11ff.
103. F. Buchanan, 'An Account of a Journey', fol. 2.
104. D.H. Grist, *Rice*, London (5th edn.), 1975, p. 38; F. Bray, 'Patterns of Evolution', p. 9.
105. See IOL, Ms. Eur. F. 95, fols. 30-5 and IOL, Ms. Eur. D. 75, vol. 2, book 4, fols. 82-3.
106. IOL, Ms. Eur. D. 75, fol. 82; see J.D. Chambers and G.E. Mingay, *The Agricultural Revolution*, London, 1966, pp. 62-3 for marling in eighteenth century English agriculture.
107. Marling as a device to improve the quality of lighter soils was not an extensive feature of English agriculture before the mid-eighteenth century because of the almost prohibitive costs (see C.G.A. Clay, *Economic Expansion and Social Change, England, 1500-1700, volume 1: People, Land and Towns*, Cambridge, 1984, p. 112). Marling seems to have expanded in the late eighteenth century, and along with manuring, drainage and convertible husbandry, it constituted one of the four major sources of agricultural improvement in that period (Chambers and Mingay, *Agricultural Revolution*, p. 62).
108. *Indian Recreations*, 2, p. 16.
109. Chambers and Mingay, *Agricultural Revolution*, p. 49.
110. Ibid., p. 49; also W. Abel, *Agricultural Fluctuations in Europe: From Thirteenth to Twentieth Centuries*, trans. Olivia Ordish, London, 1980, pp. 109, 208; C.G.A. Clay, *Economic Expansion and Social Change*, p. 113.
111. IOR, HM, vol. 47, 4 September 1762, p. 43.
112. Significantly these were precisely the considerations which were increasingly reducing the importance of the fallow in English agriculture from the late eighteenth century (see Chambers and Mingay, *Agricultural Revolution*, *passim*).
113. IOL, Ms. Eur. F. 95, fols. 36-7.
114. IOL, Ms. Eur. D. 75, vol. 2, book 4, fol. 84; emphasis added.
115. Chambers and Mingay, *Agricultural Revolution*, p. 49.
116. W. Abel, *Agricultural Fluctuations*, pp. 207-8.
117. IOR, BRC, P/51/15, 28 January 1787; also IOL, Ms. Eur. F. 95, fols. 30-5.
118. IOL, Ms. Eur. D. 75, vol. 2, book 4, fol. 83.
119. Chambers and Mingay, *Agricultural Revolution*, p. 61.
120. William Tennant, *c.* 1797, gave a much smaller proportion of the cultivated lands occupied by these cash crops: 'most valuable crops' like 'tobacco, sugar, indigo, cotton, mulberry and poppy' did not occupy more than 'twentieth part of the land' under cultivation (*Indian Recreations 2*, p. 11).
121. Rajat Datta, 'Merchants and Peasants', pp. 39-92.
122. IOR, BRC, P/49/44, 21 December 1773.

123. IOR, BRP, P/70/40, 18 April 1788.
124. Martin, *Eastern India*, vol. 3: Appendix C; also IOL, Ms. Eur. G. 11, Table 40.
125. IOR, BRP, P/71/10, 25 June 1789.
126. IOR, BRC, P/51/27, 28 November 1788.
127. Ibid., IOR, P/50/66, 7 April 1786.
128. See IOR, HM, vol. 401, pp.138-41 for figures.
129. See for instance, Colebrooke, *Remarks*, pp. 27-30; Tennant, *Indian Recreations 2*, pp. 8ff.; Francis Gladwin to Council, IOR, BRC, P/53/55, 31 January 1798; W. Roxburgh, 'An Account of the Hindoo Method of Cultivating the Sugar Cane and Manufacturing the Sugar and Jagery in the Rajahmundry Sircar', IOR, HM, vol. 210, p. 150.
130. IOR, BRC, P/51/7, 8 June 1787; emphasis added.
131. C. Blume, 'Short Sketch of the Measures Adopted for the Introduction of Indigo and Promotion of Agriculture in Bengal Between 1779 and 1790', IOR, HM, vol. 443, p. 600.
132. Ibid. The cultivation of poppy for opium was also largely unattractive to the average peasant in our period because poppy was an extremely precarious crop: 'the poppy is of so delicate a texture before it arrives at maturity that an accident of the weather such as too dry a season, or a tempest of wind and hail coming on before the plant has gathered sufficient strength to resist its force may utterly destroy the produce of a whole pergunnah' (IOR, BRC, P/49/55, 11 August 1775).
133. Thus according to William Roxburgh, 'amongst the natives of India, the transitions from one stage of improvement to another are so exceedingly slow, as scarce to deserve the name . . . however when they see a certain prospect of gain, with little additional trouble, they have been frequently known to adopt our practices' (IOR, HM, vol. 210, p. 150).
134. The discussion of potatoes in eighteenth century is based on the evidence in IOR, BRC, P/53/55, 29 January 1798, 12 February 1798, and 23 March 1798; ibid., P/53/58, 28 September 1798. In Midnapur the cultivation of *shakkarkand* (sweet potatoes) was an important part of agriculture. The seeds for this tuber were sown in the month of *Kartic* (October-November), and reaped in *Pous* (December-January) (ibid., P/51/15, 28 January 1788).
135. *Remarks*, p. 67.
136. M. Atkinson, Resident, Jangipur Factory, 21 November 1792, IOR, HM, vol. 375, p. 318.
137. IOR, HM, vol. 385, 15 June 1789, p. 325.
138. IOR, BRP, P/71/10, 15 June 1789; emphasis added.
139. Ibid., P/72/7, 19 September 1792; emphasis added.
140. IOR, HM, vol. 375, 9 June 1794, p. 118.
141. Ibid., p. 288.
142. F. Buchanan, 'Survey of Ronggoppur', IOL, Ms. Eur. D. 75, vol. 2, book 2, fol. 42.

143. BM, Add. Ms., 19286, fol. 4.
144. WBSA, CCRM, vol. 1, 11 November 1771.
145. IOR, CCR, P/68/7, 4 May 1781.
146. J. Paterson to Board, IOR, BRC, P/51/40, 15 July 1789.
147. Martin, *Eastern India*, vol. 3, pp. 831-2, 870-1, 864-5.
148. *Remarks*, pp. 62, 65, 75, 92.
149. IOR, BRP, P/71/16, 3 June 1789.
150. Ibid., P/71/10, 25 June 1789.
151. Ibid., 25 June 1789.
152. Ibid., 15 June 1789.
153. Ibid., 26 May 1789.
154. Ibid., 15 June 1789.
155. BRP, P/72/10, 26 December 1792.
156. *Bengal Sugar*, p. 53.
157. WBSA, PCR, Burdwan, vol. 1, 30 May 1774.
158. *WBDR, ns, Birbhum*, 16 August 1787.
159. IOR, BRC, P/50/66, 7 April 1786.
160. IOR, BRP, P/71/10, 15 June 1789; ibid., P/71/22, 22 March 1790.
161. Ibid., P/72/7, 19 September 1792.
162. *Arzi* of 'Sundry Ryotts of 24 Pergunnahs', in IOR, CCR, P/68/3, 6 March 1780.
163. IOR, BRP, P/71/10, 25 June 1789; emphasis added.
164. Ibid., 25 June 1789.
165. Ibid., P/71/11, 25 June 1789.
166. IOR, BRP, P/71/16, 26 November 1789; emphasis added.
167. In an earlier attempt to interpret the data, I had incorporated all data points, including famine and dearth prices, in the series (Rajat Datta, 'Peasant Production and Agrarian Commercialism'). The result was a substantial flattening of the real trend. By excluding famine prices while calculating these indexes in this exercise, the graphs can be said to be closer approximations to real price trends. This and the subsequent graphs are, therefore, significantly re-modified from those given in the earlier essay.
168. It should also be stressed that the prices of commodities like opium and, especially, indigo are not being considered in this discussion as, despite being cash crops, these were cultivated by the peasants largely under compulsion, and the prices offered for these crops by the Company were mostly dictated and arbitrary.
169. *Studies in Indian Agriculture* (K.N. Raj et al., eds.), Delhi, 1988, Chapter 1.
170. IOR, BRP, P/72, 8, 29 October 1792, and ibid., P/72/10, 26 November 1792.
171. J.W. Mellor, 'The Subsistence Farmer in Traditional Economies', in C.R. Wharton (ed.), *Subsistence Agriculture and Economic Development*, p. 219.
172. Eaton, *Bengal Frontier*, 194-207; also, Marshall, *Bridgehead*, pp. 3-4.
173. IOR, FR, Murshidabad, G/27/4, 20 December 1771.

174. IOL, Ms. Eur. F. 95, fol. 44 a.
175. IOR, BRC, P/52/40, 13 January 1792.
176. Verelst Manuscripts, IOL, Ms. Eur. F. 218/15, fols. 30, 31; for the famine see Chapter 5.
177. IOR, BRC, P/49/41, 1 October 1773; also ibid., P/49/50, 27 January 1775.
178. Ibid., P/49/54, 18 July 1775.
179. IOR, BRC, P/49/52, 5 February 1775.
180. Discussed in Chapter 4.
181. Discussed in Eaton, *Bengal Frontier*.
182. See IOR, BRP, P/71/26, 4 June 1790; Beveridge, *The District of Bakarganj: Its History and Statistics*, Calcutta, 1876.
183. Eaton, *Bengal Frontier*.
184. Marshall, *Bridgehead*, p. 22.
185. *The Bengal Frontier*, pp. 219-20.
186. Coast & Bay, IOR, E/4/5, 4 January 1754.
187. IOR, BPC, P/1/44, 14 November 1769.
188. Rajat Datta, 'Merchants and Peasants', pp. 385, 397.
189. Eaton, *The Bengal Frontier*, pp. 221-2.
190. The second and third social developments are described in Beveridge, *Bakarganj*, pp. 194-203.
191. See Westland, *The District of Jessore: Its Antiquities, its History and Statistics*, Calcutta, 1871, p. 226 for the connection between supply of rice to Calcutta and extension of cultivation in the Jessore Sundarbans.
192. IOR, BRC, P/50/63, 6 August 1784.
193. Ibid., P/51/9, 9 August 1787.
194. Ibid., P/50/61, 30 September 1785.
195. Ibid., P/50/63, 17 February 1786.
196. J.W. Westland, *Jessore*, p. 228.
197. IOR, BRC, P/51/20, 7 May 1788.
198. Westland, *Jessore*, p. 135.
199. IOR, Letters Received, E/4/24, 31 December 1758.
200. Ibid., fol. 7.
201. F. Buchanan, 'An Account of a Journey', fol. 19.
202. IOR, BR Misc., Committee of New Lands, From Harry Verelst and T. Rumbold to Henry Vansittart, 5 June 1761.
203. *BDR, Chittagong*, vol. 1, 31 August 1770, p. 125.
204. IOR, BRC, P/49/41, 17 August 1773.
205. *The Bengal Frontier*, p. 249.
206. F. Buchanan, 'An Account of a Journey', fol. 28.
207. IOR, BRC, P/49/41, 17 August 1773.
208. F. Buchanan, 'An Account of a Journey', fols. 19, 28, 79.
209. IOR, CCR, P/68/5, 21 September 1780.
210. Ibid., P/68/4, 6 April 1780.

211. IOR, Letters Received, E/4/24, 31 December 1758.
212. Compare, IOR, CCR, P/72/2, 30 September 1779 and IOR, BRP, P/71/22, 1 March 1790.
213. See Chapter 5.
214. W.W. Hunter, *Statistical Account of Bengal, volume 1: 24-Parganas and the Sundarbans*, London, 1875 (Indian rpt. Delhi, 1973), p. 77.
215. See Chapter 4.
216. *BDR, Midnapur,* vol. 1, no. 261, 10 November 1767, p. 191.
217. J.C. Price, *Notes on the History of Midnapur,* Calcutta, 1867, p. 104.
218. IOR, BRC, P/51/25, 1 October 1788.
219. L.S.S. O'Malley, *Bengal District Gazetteers, Midnapur,* Calcutta, 1911, p. 88.
220. BM, Add. Ms. 29088, fols. 110, 199.
221. From Henry Vansittart, Comptroller of the Manufacture of Salt in IOR, BRC, P/50/35, 30 September 1781.
222. IOR, BRC, P/51/25, 3 December 1788.
223. Ibid., P/51/25, 1 October 1788.
224. Ibid., P/51/25, 3 December 1788.
225. IOR, BRP, P/71/26, 4 June 1790; also see Chapter 3 for a detailed discussion.

CHAPTER 2

The Peasantry and the Question of Peasant Stratification

THE REVENUE records of the East India Company are full of terms like 'ryott', 'reiatt' and 'ruatt' which, in continuation of the Mughal tradition, were generically used to denote the status of an individual as a cultivator and a revenue-payer (the *raiyat*). These terms were essentially corrupt usages of the Persian term *raiyat* which, in Bengal, meant 'the immediate occupant of the soil, whether . . . a proprietor or tenant' who tilled the land and paid a 'rent'. These *raiyats* were 'a numerous & inferior class of people, who [held] and cultivate[d] small spots of land on their own account'.[1]

Agricultural production was overwhelmingly organized on small peasant farms which were operated by individual peasant families depending on their domestic labour. It was generally accepted that a plough being worked by a pair of draught animals was capable of cultivating between 15 *bighas* (5 acres)[2] to 25 *bighas* (8.3 acres).[3] The extent cultivable by a plough obviously depended on the nature of the soil. Where the soil was hard the land cultivated by a plough and a pair of bullocks was less. Thus, in pargana Swaruppur 'the usual quantity of land cultivated by one plough requiring two oxen, a man and a boy may be fairly computed on an average of 8 *bighas* (2.6 acres). . .'.[4]

Scanty area statistics for the eighteenth century make it difficult to present statistical profiles of peasant holdings for the entire province. In Dinajpur, there were peasants who held up to 165 *bigha* (55 acres) of land, but such landholdings were generally not replicated elsewhere.[5] Peasants having 7 to 10 acres of land were categorized as the 'influential ryott' in eastern Bengal[6] whereas in western Bengal a peasant could be classified as such with lands ranging from 20 to 10 *bighas* (6.6 to 3.3 acres), depending on the geographical location of the individual's holding and its commercial importance.⁷ Thus in Midnapur, 'influential ryotts' were those who held 20 *bighas* or more. In contrast were their compatriots in villages around the city of Murshidabad (a centre of urban consumption and the capital of the province before 1757), who could still be influential and relatively affluent with 10 *bighas* (3.3 acres) of land at their disposal. It must, however, be stressed that peasants having 10 *bighas* or less were the poorer peasants (the 'inferior ryott') in the rest of the province. The situation in Dinajpur was therefore unique and does not reflect the state of peasant holdings at the provincial level. The reasons why this was so will be analysed later in this chapter.

Positioned below these 'influential ryotts' were those peasants who are identified in our sources as the 'poorer' or the 'inferior class' of *raiyats*. They held less than 5 acres of land, possessed only a pair of bullocks and a single plough, cultivated with the use of their family labour alone and had no reserve stocks to tide over even one season of scarcity.⁸ These comprised 80 per cent of the rural population in the villages of Swaruppur surveyed in 1790 by the Company's revenue officials.⁹ They formed nearly 70 per cent of the cultivating population of Dinajpur in the beginning of the nineteenth century.¹⁰ J.H. Paterson, commissioner at Commilla, was convinced that 'this class of ryotts are by far the most numerous',¹¹ and in Rangpur even those peasants who had not incurred major debts had practically no reserves 'even of a stock sufficient for their subsistence . . .'.¹²

There is another social group whose existence in the eighteenth century needs to be explained. These were the sharecroppers (the *bargadar* or the *adhiar*) whose existence is relatively well documented in our period. The existence of a group of people providing their labour to cultivate another person's land for a share in the produce may at first glance appear paradoxical in a society where land was so easily available. Yet examples of sharecroppers being viciously exploited by zamindars, merchants and revenue-farmers (*ijaradars* and *qutquinadar*) on zamindari lands are numerous, and this needs an explanation. It is perhaps extremely significant that in the extensive zamindari of Burdwan, the village headman (*mandal*) was sufficiently happy to limit his activities to giving periodic agricultural loans to the peasants in need 'to purchase implements of husbandry and for their immediate subsistence',¹³ but it was the zamindar who used sharecroppers (*sanjhadars*) to cultivate his personal (*khamar*) lands on terms which were highly oppressive.¹⁴ Rural inequalities in land distribution and the need felt by the richer peasantry to reduce supervision costs¹⁵ may not be adequate in explaining the prevalence of a sharecropping peasantry in a situation of land abundance (as in eighteenth century Bengal). Their existence in our period perhaps becomes explicable only where persons wanting to use them had more land than could be cultivated by family labour alone, or where merchants required lands to be cultivated for purposes of trade. In Bengal, the former occurred in the case of the zamindars and the *jotedars*, but the latter (i.e. merchants getting lands cultivated for trade) does not seem to have developed to an extensive degree. Merchants created separate instruments for exploiting the small-peasants.

The other peasant group which requires detailed analysis is the *jotedar*. Discussions of the *jotedari* in Bengal centre around the extent to which it was a specific peasant relation of production and rural power. These are concerned also with analysing the connection between the *jotedars* and the evolution of a type of tenurial arrangement which subsequently became most prevalent all over eastern India, viz., the sharecropping (*adhiari* or *bargadari*). The term *jotedar* has been used to express social relations ranging from kulak-landlordism[16] to a 'master-serf relationship' between peasants themselves.[17] The *jotedars* have been treated as a social group 'synonymous with the village landlord'.[18] The specific class nature of the *jotedari* in the nineteenth century (when it apparently became widespread) is beyond the scope of this chapter, but the evidence, discussed below would suggest that it was far from either of the two classifications (i.e. kulak-landlord or village landlord) in the eighteenth. Attempts to locate such a layer for the whole of Bengal on the basis of the descriptions of rural stratification provided by Colebrooke, Sir John Shore (discussed later) and Buchanan-Hamilton's descriptions of the districts of Dinajpur and Rangpur in 1807, seems to equate the existence of a rudimentary form of differentiation within the peasantry with the *jotedars*. With the possible exception of Buchanan-Hamilton,[19] all other contemporary commentators on agrarian matters almost invariably referred to the 'influential ryotts' or to the 'superior ryotts'. The social implications of these strata will be discussed shortly.

The definition of the *jotedar* as a kulak-landlord or even as a village landlord is difficult to sustain before the nineteenth century. The *jotedars* took lands on long-term leases from the zamindars and then had them cultivated either by contracting, on a sharecropping basis, with the 'ryots of the villages most contiguous for their cultivation'[20] or by recalling an 'absconding ryott or settling a deserter from some other place to cultivate for his benefit'.[21] Established peasant rights in the *jotes* were recognized by a set of customary codes, the workings of which will be examined subsequently. The use of the term 'kulak' to describe the *jotedars* is also misleading. In the classical sense a 'kulak' was a semi-capitalist or a small capitalist farmer in the Russian countryside.²² As Bernstein points out, a 'kulak' is an 'all-round agent of the extension of commodity relations'; a 'kulak' combines a number of economic functions: he is a commercial farmer who employs labour power, rents out the necessary means of production, provides local 'merchant's and money-lending capital' and invests in trade and small-scale processing and manufacturing enterprises.²³ This is not substantiated for Bengal by the available historical evidence.

The other concern in discussions of *jotedari* is the question of timing its dominance in rural society. While Chaudhuri sees such *jotedars* as an established fact in rural Bengal only after 1859,²⁴ others tend to push the timing further into the century, even as late as the closing decades of the nineteenth century.²⁵ One point about the emergence of these *jotedars* seems to be clear: their increasing dominance is seen to be closely connected with a growing pressure on land, a noticeable shift towards cultivating lands of marginal productivity and increased sale and transfer of peasant lands, often as a result of distress sales.²⁶ In other words, the systematic consolidation of *jotedari* can be linked, not with the expansion of cultivated land, as in the late eighteenth century, but with a noticeable halt in the process of reclamation and a resultant economic contraction. Regarding the geographical spread of this tenure, Chaudhuri tends to view its prevalence in those regions where 'a few resourceful persons organized large-scale reclamation', thereby making it an all-Bengal phenomenon,²⁷ whereas Bose sees it mainly limited to the northern parts of the province;²⁸ in either case *jotedari* is not seen as a major production/power relation *before* the mid-nineteenth century.

Famine and Bengal's Peasants:
Some Historiographical Considerations

There is a fairly strong consensus in existing historiography that the famine of 1769-70, because the havoc it caused, fundamentally altered the existing structure of peasant society and crystallized the previously nascent position of a rich peasant strata in Bengal. For instance, N.K. Sinha saw in the famine a fundamental rupture of 'the old social life' in rural Bengal. Depopulation meant that lands fell out of cultivation which in turn fostered a competition for labour among the landlords. In this situation, the relative advantages shifted in favour of the non-resident (*pahikashta*) cultivator at the expense of the surviving resident (*khudkashta*) peasant who was ruined by the additional impositions (*najai*) which were levied by the landlords in order to recoup their losses suffered during the famine. The famine additionally 'reduced the value of land'.[29] Sinha's view regarding the post-famine situation closely follows those developed by W.W. Hunter who argued that agricultural recovery was henceforth only possible by attracting the labour of the 'vagrant peasants' (those who had been uprooted from their lands) by offering them underrated leases of land.[30]

B.B. Chaudhuri sees in that famine the greatest single factor in causing long-term decline in agricultural production. Recovery took a long time because of the scale of the devastation and because of several retarding factors such as 'falling agricultural prices for three years after the famine, the enforcement by the government of an increased revenue demand through coercive means while the investment by them towards nursing the ailing economy back to health was negligible, the interruption of the process of wasteland reclamation by migrant peasants which the zamindars encouraged and partly financed, and a series of natural calamities'.[31]

Ratnalekha Ray has provided an interesting interpretation of the post-1770 situation. She disagrees with Hunter and Sinha that the rise of a rich tenantry can be simply looked at in terms of a favourable land-labour ratio after the famine. For Ray, the significance of the famine was situated in the fact that it 'gave splendid opportunities to villagers with capital to increase their holdings by employing their capital to bring waste lands under their cultivation'; this kind of capital could not have come from merchants and bankers, or from the 'territorial magnates' for 'they were in no position to engage in the enterprise of reclaiming waste, for it needed personal and on-the-spot supervision'; agricultural reclamation was therefore the task 'of a class of rich peasants', these being 'an extremely influential and violent class of villagers who had compelled the zamindars to give them under-rated tenancies so as to employ their capital in clearing the waste lands in their own or neighbouring villages'. These were *pahikashta* only in name but represented the combination of the village headman (*mandal*) with some rich and turbulent peasant groups.[32]

The range of issues pertaining to the agrarian consequences of the famine are indeed formidable. They can be effectively reduced to the following:

1. Excess land puts up demand for labour (Sinha, Chaudhuri) and capital (Ray).
2. Value of land is reduced (Sinha).
3. 'Vagrant' and/or non-resident peasants are able to get better terms from landlords in comparison to the resident peasants (Hunter, Sinha, Chaudhuri and Ray). The latter suffer further from the imposition of taxes like the *najai* (Chaudhuri).
4. Agricultural recovery is extremely slow and retarded by a number of institutional constraints (Sinha, Chaudhuri).
5. Traditional social relations are fundamentally shattered (Sinha).

The Pahikashta Peasant after 1770:
A Rich Peasant Class?

The evidence regarding the relationship between the *najai* tax and the further debilitation of the *khudkashta* peasants after 1769-70 is less clear than what Sinha and Chaudhuri have suggested. Undoubtedly the tax was oppressive, being: 'An assessment upon the actual inhabitants of every Inferior Description of the Lands; to make up for the loss sustained in the rents of their neighbours, who are either dead or have fled the country; it is a kind of Security against Desertion, by making the inhabitants thus mutually responsible for each other'.[33]

Hastings also suggests that this tax was not a post-famine phenomenon, since it was 'authorized by the antient [*sic*] and general usage of the country'; he also says that this tax was not oppressive in 'ordinary cases, and while the lands were in a state of cultivation'.[34] The existence of this tax and its apparent regressiveness after the famine would make the supposed advantage of the 'vagrant' or the *pahikashta* peasant very difficult to envisage, if only for the fact that these groups were themselves *resident* peasants in some other area and the very logic of the *najai* tax was intended to restrict the scale of migration, or desertion, of precisely such peasants even under normal circumstances. Under a famine, when the pressure to restrain existing labour was at its greatest, the use of the *najai* operated as the greatest deterrent on the movement of all categories of peasants since they were now made 'mutually responsible for each other'; and the intrinsic bias of this tax was such that it could not possibly have left any particular category out of its clutches.

Moreover, *najai* does not appear to have been levied with the same effectiveness all over the province. Interestingly, this tax does not figure in the post-famine peasant leases (*pattas*) in Rajshahi, Jessore and

Laskarpur,[35] even though they had suffered in varying degrees during the famine. Therefore, the influence of the *najai* on causing greater immiserization of the surviving peasantry cannot be established as an ascertained fact. Additionally, the scale of peasant desertion in the districts worst affected by the famine (discussed in Chapter 6) would certainly indicate that this tax was unable to stop peasants from fleeing when the pressures on them became unbearable.

The central issue in reconstructing the impact of the famine of 1769-70 is the state of the surviving peasantry. Essential to the consensual historiography of that event are three elements: (*a*) that the famine had decimated ten million, or one-third, of Bengal's population, (*b*) the ensuing demand for labour meant that those who survived were in a position to bargain for better terms, and (*c*) recovery from the famine required immense amounts of capital which was available, not from the local landed-proprietors and merchants, but from a class of rich peasants. Our evidence indicates (see Chapter 5) that the figures of famine-mortality are greatly exaggerated. This fact must, therefore, place serious doubts on the extent to which there was a real shortage of labour on a provincial scale.

There is nevertheless some evidence to show that there was a certain movement towards *pahikashta* cultivation after the famine. Such peasants could be seen moving from one talluqa to another in Rangpur obviously in search of a better financial deal.[36] In Rajshahi, the use of *pahikashta* peasants in zamindari lands was being actively encouraged by the state on grounds that such strategies would 'increase industry and population by every possible encouragement and indulgence'.[37] The financial burdens on these peasants in eastern Bengal, is shown in Table 12 which gives comparative assessment of revenue (*jama*) on the *khudkashta* and *pahikashta* peasants in the zamindari of Laskarpur between 1770-1 and 1772-3.

Table 12 does show that after the famine the *pahikashta* cultivators

TABLE 12. *JAMA* OF *KHUDKASHTA* AND *PAHIKASHTA* CULTIVATORS IN LASKARPUR, 1770-1 AND 1772-3 (*Rs. per bigha*)

Jama category	1770-1	1771-2	1772-3
Khudkashta jama	4.69	5.94	7.59
Pahikashta jama	4.51	5.75	7.36

Source: IOR, CCR, P/67/57, 9 July 1773.

were relatively better off, but only marginally; this table, however, also shows that these peasants were not free from subsequent increases made to the *jama* and this raises several questions about the existing consensus among historians regarding such cultivators after the famine. The common understanding that these *pahikashtas* were at a greater advantage after 1770, therefore, no longer appears feasible, and this is so for a number of reasons.

First, it is not at all clear whether the differential rates of assessment which supposedly made the *pahikashta* right attractive to the 'vagrant' peasant actually prevailed all over Bengal. Unfortunately, not much is known about this right in the immediate aftermath of the famine, but subsequent notices in the sources suggest that this advantage may not have been as universal as is believed. The *pahikashta raiyat* in Burdwan and Birbhum appear to have enjoyed such privileges,[38] but in Murshidabad the position was absolutely different. Here the term *khudkashta* was used to designate those peasants who cultivated 'without a patta and paid below official rates [of revenue]', and the zamindars often had to enforce his traditionally sanctioned right '*raj-ul-mulk* [sic]' to coerce them to pay 'the full rates', or to settle *pahikashta* on other lands since they not only paid these 'full rates', but also agreed to fulfil whatever *rasad* (increases in assessment) which the zamindar chose to introduce.[39] In Nadia, lands which had fallen out of cultivation ('*lokshan-jote*') were 'taken possession of indiscriminately . . . by Ryotts of almost every denomination at the commencement of cultivation'.[40] Obviously, the *pahikashta* were given no advantages there. In Dinajpur, the *pahikashta* cultivators were subject to a 'double claim', those from the zamindar where they held *khudkashtkari* rights, and those imposed by the zamindars in their *pahikashta* lands; these peasants, therefore, 'are seldom able, or at least willing to pay both', and in matters of dispute the zamindars in their original villages 'have preference'.[41]

Moreover, the evidence from Burdwan suggests that the position of the *pahikashta* cultivators appears to have undergone a fairly substantial change between the 1770s and the 1780s. In the late 1780s these peasants in Burdwan were considered 'the Ryotts of a Superior Rank that neither pay rent for their dwelling house, nor do they pay so high a revenue for their lands as the khudkasht',[42] but in the mid-1770s they were those who were merely 'exempted from *some* of the abwabs or additional taxes to which the Khudkasht are subject'. Nor could the non-resident cultivator move about at will. Like the *khudkashta*, he was subject to zamindari imposed restrictions on peasant mobility; the only exception was made in the case of those peasants who had a 'joteiccha patta' (cultivating at will) which allowed the peasant 'the liberty to relinquish his ground when he chuses [sic]', and this *patta* was also available to the *khudkashta* cultivator.[43] It was also a practice in the 1770s for small revenue farmers (*qutquinadars*) to sub-divide their farms 'in small portions to ryautts [sic] *who are generally inhabitants of the lands they farm*, and who thus *become in a manner perpetual proprietors*'.[44]

Another reason for doubting the ubiquity of the *pahikashta* right and also its connections with the creation of a 'class' of rich peasants comes from the district of Birbhum where the thrust of reclamation and of settling cultivators on deserted lands was guided by a range of customary codes and regulations which actually favoured the resident cultivator, even when they had deserted. This fact is revealed in two such regulations:

> A deserted ryott on his return shall be entitled to his lands again, & if he declines to reengage for the whole, he shall be entitled to such parts as he will engage to cultivate at an adequate jummah provided that he takes a proportion of the bad land with the good, if any come under the first denomination. If the lands of a deserted Ryot having been divided between established Ryots holding & cultivating other lands and new Ryots, and should a contest subsequently arise for the possession of the whole of the deserted Ryots land, then the *established* Ryots cultivating such lands *shall have preference*.[45]

Moreover, quite often the most productive lands were held by the zamindars as their personal (*nij-jote* or *khamar*) holdings which were extensively cultivated by the use of sharecroppers (discussed later). The distribution of the rest of the village lands was determined as follows: 'In each village are lands of the first, second, third and fourth sorts. At the time of the cultivation, the Ryotts examine the sort of land, agree to a jummah and take out a pottah. The rates of the first sort are the largest, of the second less. This has always been the custom of the Province.'[46]

This evidence regarding the modes of reclamation in Birbhum also suggests that reclaimers, or returning peasants, were expected to take 'bad' (i.e. less productive) lands with the 'good'. It is perhaps more reasonable to suggest that the extension of *pahikashta* rights after the famine was a device framed by the zamindars to improve the state of cultivation by allowing peasants to have access to lands of varying productivity; and those who accepted such rights were either the displaced poor[47] from other districts looking for less burdensome taxes,[48] or they came from the same village moving laterally from their previously held rights as *khudkashta*. Such indeed was the case in Burdwan where we are told of a peasant category called 'nij-gaon paikasht', or those 'who cultivate the land of their own village' as *pahikashta raiyat*.[49] Individual peasants combined *pahikashta* and *khudkashta* rights in the zamindari of Rajshahi.[50]

Nevertheless, these peasants became the source of friction between different zamindars, the best example of which is the tension between the Rajas of Burdwan and Birbhum after the flood-famine of 1788. As the river Ajoy breached embankments and flooded Burdwan, large number of peasants fled to the neighbouring zamindari of Birbhum which suffered less from the floods in 1787. The zamindar of Burdwan subsequently demanded the return of these cultivators on grounds that they were the settled resident-cultivators (*khudkashta*) of his zamindari, and were, therefore, obliged to cultivate land and pay revenue in Burdwan. The counter-claim offered by the zamindar of Birbhum was that these migrants had acquired the legal status of non-resident peasants (*pahikashta*) in his lands, and were therefore entitled to stay and cultivate land in his zamindari.[51] This conflict is significant in so far as it shows the concern of the zamindars to the flight of labour from their lands, and indicates the existence of a favourable land-person ratio in which such flights took place. Additionally, it also reflects the geographical mobility of labour in the late eighteenth century.

Yet, there is very little reason to believe that this mobility served to enhance the economic standards of such peasants in the long-term, or that they represented an improved bargaining power of labour. First, these peasants moved solely because of their scanty resources. Thus the *pahikashta* of Sylhet, 'Have no hereditary habitations, nor are attached to their native home; their dwelling consists of a wretched hovel which together with their families they move at pleasure, and a pergunna covered with small villages today often appears depopulated the next.'[52]

Second, in many cases these peasants were only used seasonally; nor was their use always extensive, even in those districts which had suffered in 1769-70. In Rokunpur, *pahikashta jama* was a mere 11.6 per cent of the *raiyati jama* in 1778.[53] In Dhaka, *pahikashta* cultivators were used in zamindari lands principally for the winter harvest after which 'they return to their own homes'.[54] The evidence from Dhaka is corroborated by that from Commilla where the *pahikashta* cultivators would come and go from the 'surrounding pergunnas' but did not have any land or homesteads in Commilla.[55] Additionally, there was no guarantee that a *pahikashta* would be given the most productive or fertile land to cultivate. In fact the evidence from Birbhum (cited earlier) forcefully indicates that they were given a land of mixed qualities. In Tirhut, these cultivators were invariably made to 'cultivate waste or fallow lands'.[56] Quite often they were given lands of the 'second or third quality' since the most fertile spots were already occupied by the resident and the 'superior' cultivators.[57]

Third, the use of *pahikashta* even on an extensive scale did not necessarily lead to the decline in the position of the resident peasantry. Evidence from western Bengal suggests that these *pahikashtas* cultivated a fraction of the agricultural land. In Burdwan these cultivators accounted for the cultivation of 'near one fourth of the pergunah' in 1793.[58] In Birbhum, there numbers were much less: only 73,000 *bighas* of land were actually tilled by these peasants there in 1791.[59] Compared to this, the scale of *pahikashta* cultivation appears to have been more extensive in the east. In Commilla, for instance, these peasants cultivated 'near half the lands' in 1788,[60] but their utilization was at best seasonal and did not amount to giving them a privileged position among the peasants. The description of these peasants from Sylhet (cited earlier) does strongly suggest that they were essentially the poorer peasants trying to supplement their incomes. In Burdwan, the *pahikashta* were subdivided into various categories. There were those, called *thika pahikashta* who 'undertake for a certain quantity of land at a fixed rate'; in contrast were those, called 'comar' *pahikashta*, who were essentially sharecroppers being used on a zamindar's personal holdings called *khamar*.[61] Existing agricultural practices and norms also ensured that the rights of the settled cultivators were protected, and in many cases *pahikashta* and *khudkashta* rights were jointly held by individual peasants. Finally, *pahikashtas* were also used by the *khudkashta* peasants on terms closely resembling those offered to the sharecropper. Thus in Birbhum, 'the Khoodkasht Ryotts do not plough the land themselves, but cultivate them by means of Kersaans [peasants] from elsewhere [who are] daily labouring husbandmen who receive one third share of the crops as recompense for their labour'.[62]

Therefore, *pahikashta* is best understood as a term reflecting a holder's status in regard to a particular holding among other possible kinds of holdings and does not stand as a description of the social standing of a 'class' of people. Despite the evidence of the increasing use of such peasants for cultivation in the late eighteenth century there is no reason to see a shift in the positions occupied by them in relation to the *khudkashta*. On the contrary, the *khudkashta* continued to occupy a relatively privileged position in village society by virtue of their paying higher rates of revenue.

The Amini Commission (1778) clearly stated that the *khudkashta's* 'right of possession whether it arises from an actual property in the soil or from length of occupancy, *is considered stronger than of the other ryotts* [because] they generally pay the highest Rent for the Lands they hold'.[63] The tendency to have resident peasants cultivating on *muqarrari* (permanent) tenures was a norm closely followed in zamindaris all over the province.[64] The case of Rangpur provides a revealing example of this tendency. Here the *muqarrari* tenure seems to have achieved greater stability between the 1770s and 1780s. In 1778, these tenures were made to 'pay the rent of no more land than the Ryott cultivates [but] a measurement is made by the zemindar once a year, and the land found to be in cultivation is re-assessed'; by 1787, the same tenure was paying 'a fixed rent pr. annum, subject to no taxes whatsoever . . . nor will [the peasant] suffer the zemindar to measure his land'.[65] By 1807, a *muqarrari raiyat* had become vested not only with perpetual possession, but also with the right to sell and transfer the lease, though the previous (1778) zamindari power to measure and re-assess such lands had once again been restored.[66]

Peasant Society and the Village Mandal *(Headman)*

Historians who argue the existence of a 'village oligarchy' in this period point to the increasing influence of the *mandal* in rural affairs as constituting the core of that oligarchy. These headmen are seen as the vehicles through which the 'rich' peasants controlled the politics and external affairs of the village communities, often with violence, in order to increase their economic power inside the village. This 'rich' peasant-*mandal* combination, it is argued, was thus able to get major financial concessions both from the local zamindar and from the state.[67] There are essentially two objections to this view. First, it assumes that these headmen had become powerful all over the province. Second, this view does not consider the potent role of zamindari officials in the countryside. As will be discussed, shotly these officials were crucial in the relations between the zamindars' *sadr* (headquarter) and the *muffassal*, and used their positions to their own advantage. The nature of the village *mandal* therefore needs to be analysed.

According to the Amini Commission, the *mandal* in Mughal and Nawabi Bengal was also known as 'muqaddam'. Though that report described the *mandal* as the 'chief ryott of the village', it is now well known that the *muqaddams* in north India were more like petty zamindars than peasants in the strict sense of persons who actually tilled the land by their own, and their domestic, labour.[68] In Bengal, the *mandal's* duties were: 'To act as mediator between the Ryots and the petty collectors of revenue, to assist them [the cultivators] in selling their crops, and in raising money to pay their Rents, and to settle or accommodate the little disputes which arise in the neighbourhood.'

These duties give a distinct impression that *mandals* of Bengal were village officials set up by the peasants themselves to manage their day to day affairs with the wider world. In fact they were '*chosen* from amongst the oldest or most intelligent inhabitants' and their 'influence and services' depended '*solely* on the *good opinion of the Ryotts*'. They could be removed by the zamindar in case they lost the 'confidence' of the villagers.[69] The report of the Amini Commission (formed to survey the agrarian conditions in the aftermath of the famine of 1769-70, and submitted in 1778), must surely indicate that the principles initially established in the seventeenth century, still guided the late eighteenth century *mandal*, at least in the 20 districts of the province actually visited by the members of that Commission.[70]

The notion that these *mandals* were essentially the richer peasants rests on the assumption that their holdings were larger than those held by the other cultivators. Unfortunately, there is very little evidence to support that. The following figures pertain to the lands held by 16 *mandals* in pargana Anwarpur, *zilla* Nadia in 1794. Here a total of 16 *mandals* possessed 289.45 *bighas* of land in the following fashion:[71]

Amount of land (*bighas*)	Number of *mandals*
Below 10 *bighas*	1
From 10 to 20 *bighas*	9
From 20 to 30 *bighas*	4
Above 30 *bighas*	2

These figures are revealing in so far as they show a striking conformity to the general pattern of peasant landholdings in the province. The fact that 10 out of 16 *mandals* held lands similar to those possessed by the 'poorer ryotts' does not support the view which sees these people as the 'rich' peasant in action.

Nevertheless, there is evidence to show that in certain districts these *mandals* had apparently become powerful. In 1788 the *mandals* of Rajshahi, 'By making themselves the pretended guardians of the Ryotts have obtained an irregular and very dangerous power, destructive of subordination, the remedy of the evils of which is very difficult. They can incite the Ryotts to commotion in an instant at pleasure, without appearing themselves, and by that means compel even the zemindaree power to bend to them.'[72]

In Dinajpur, these people, along with the 'principal inhabitants of the villages, under plea of real or fictitious desertions of the Riauts, have got into their possession considerable tracts of ground at an under-rated assessment . . . and the established rates of Nirk [assessment] has [sic] become obsolete'.[73] In Birbhum, 'the mundulls derive from their number and mutual support such influence, that a success of a settlement in general measure . . . depends entirely on their pleasure'.[74] These descriptions form the crux for the historians who argue the formation of a 'village oligarchy' and a 'powerful tenantry' in this period. The question is: do these descriptions actually support such formulations? It is difficult to say that they do for the following reasons.

First, the timing of these descriptions is critical. These descriptions are without exception for the year 1788 which was a year of a major famine.[75] Massive floods had ruined agricultural production in southern, northern and eastern Bengal. Peasants had been uprooted from land leading to a pressure on the landed proprietors to make concessions. It is likely that a part of the pressure was created by the peasants bargaining for better terms, at least in the short-term. There is very little reason to believe that such pressure was due to the concerted actions of the 'village oligarchy', or that they were necessarily a persistent feature of peasant society. The evidence we have of the *pahikashta* cultivators certainly shows a picture contrary to the creation of a 'rich' peasantry in the long-term. Second, these descriptions are spatially selective and are not confirmed by similar statements for other districts either before or after the famine of 1788. This is significant because it shows that the so-called 'power' of the *mandals* was not a feature common in all areas. Third, the district where the *mandals* had apparently made good display have certain similarities. As discussed earlier (Chapter 1) Birbhum, Rajshahi and Dinajpur had clearly moved downwards in economic terms during this period. They had declined from stable (and surplus) economic centres to areas of chronic instability and declining production during the course of the century. This may have caused some realignments in rural society in favour of the village headman, though all indications show that these shifts were unlikely to effect major changes in social structure.

• In general *mandals* in the late eighteenth century continued to remain as 'a kind of civil superior *elected* by the tribe over whom they respectively preside':⁷⁶ a description which confirms the one provided by the Amini Commission in 1788. It is also significant that the power to dismiss them rested with the villagers and with the zamindar. The latter also had the right to confirm (or reject) the villagers' choice of a *mandal* who had to pay a 'fixed tribute to the zemindars' to acquire the necessary confirmation. Apart from assisting the villagers in their day to day affairs, the *mandal* had to oversee 'the conduct of the inferior members, chiefly to prevent any offence against their caste', and to be 'present at all other marriages and other civil ceremonies', for which he received a stipulated fee.⁷⁷ The exact amounts of such fees are not known, but revenue-free (*baz-i-zamin*) lands were sometimes given to meet the *mandals*' services. However, these lands were not extensive enough to conjure up visions of a powerful village oligarchy. In Birbhum, for instance, such lands amounted to 38.4 *bighas*, which constituted a mere 0.04 per cent of the total *baz-i-zamin* lands (96,032.57 *bighas*) in 1789.⁷⁸

Jotedari: Some Emerging Trends

Much of the discussions of *jotedari* in the eighteenth century centred around three districts of eastern Bengal (Jessore, Rangpur and Dinajpur). Interestingly this term is specifically used almost exclusively in connection with Jessore where there appears to have been a systematic attempt to extend cultivation in the late eighteenth century, and the other term used almost synonymously with *jotedari* is *ganthidar*.⁷⁹ In either case the terms were used specifically to denote a lease granted by zamindars to individuals to bring stipulated portions of land back into cultivation at concessional terms of revenue. A *jote* 'properly so called is granted to a ryott for purposes of bringing deserted land back into cultivation at rates usually fixed at half the assessment on such lands',⁸⁰ whereas *ganthis* were *pattas* [leases] 'granted by zemindars, and sometimes by farmers (*ijaradars*) but confirmed by the zemindar to . . . perpetual leaseholders (*ganthidars*) for entire villages. . . .'⁸¹ In either case the grantor was the zamindar who did so with the dual purpose of raising money and expanding the long-term productive capacity of the zamindari, and these tenures often co-existed with other sub-tenurial arrangements made at the zamindar's discretion in his personal (*khamar* or *nij-jot*) lands.⁸²

The active role of the zamindars in the formation of such tenures can be seen in the creation of the *gutchdari* tenure in Purnea. These *gutchdars* were leaseholders similar to the *jotedars* of Jessore and undertook to cultivate stipulated tracts of uncultivated land partly by the use of their own and their families labour, and the rest by sharecroppers (*adhiars*). The crucial factor in determining the *gutchdar's* tenure was the zamindar, always bargaining for more remunerative terms by procuring rival offers for *gutchdari*. Quite frequently 'the zemindar not being satisfied . . . requires a higher rent than the gutchdar is willing to undertake, and the gutchdarry is relinquished in consequence; the zemindar if unable to provide another tenant, then engages with his under ryotts on the gutch. . . .'[83]

A zamindar's financial consideration was the prime mover in the evolution of the *jotedari patta*; the other consideration was the desire of an individual zamindar to concentrate agricultural production in relatively larger holdings as a device to ensure better monetary returns in the face of a rigid revenue demand and falling real incomes engendered by high prices and periodic harvest failures.[84] The creation of *jotes* and *ganthis* was deemed to represent 'clear profits to the proprietor' since the lands where such tenures were created produced nothing before.[85] *Ganthidars* 'hold considerable farms in the villages and pergunnas . . . cultivating the major part of these lands by their own gauntee [*ganthi*] ryotts, and by that means the zemindar *collects almost as much revenue upon these lands thus united as he would were* [these] *lands held by separate persons. . . .*'[86]

For the East India Company the creation of *jotedari* on concessional terms of revenue was not an entirely desirable prospect as 'the more jotes there are in a cultivated state the less rent it [*sic*] affords'.[87] This viewpoint was that of a state bent on extracting the maximum possible from the agrarian economy; it was not necessarily shared by the zamindars who, regardless of the state's fears, persisted in creating *jotes*, *ganthis* and *gutchdaris* within their territories. By the late 1780s, Jessore was said to have 'not a single pergunnah free from lands of this description'.[88]

The other factors which influenced the shape and size of such tenures in the eighteenth century were the existing land-labour ratio, the extent of reclaimable land available within a zamindari and the nature of the zamindari in which such tenures were patterned. They also influenced the extent to which such holders could improve their positions *vis-à-vis* the zamindar at one level and the inferior cultivators at the other. The suggestion that the *jotedars* were powerful in 'regions characterized by substantial landholders owing their landed position largely to their role in the reclamation process'[89] needs re-examination. In 24-Parganas,[90] which had no substantial landholders,[91] we find evidence of holders of *patitabad patta* (reclaiming leases) undertaking to cultivate 32,650 acres

(97,950 *bighas*) of waste land in 1779.⁹² Chaudhuri's statement would also not apply to areas like Rangpur and Purnea which were generally recognized as being held by a number of small zamindars⁹³ and where the formation of tenures similar to that of the *jotedari* can be traced in the eighteenth century. In Purnea, the *gutchdari* expanded in the wake of the famine of 1770 which is said to have wiped out nearly one-third of its agricultural population and converted nearly a similar quantity of agricultural land into cultivable waste.⁹⁴ In some parganas of Rangpur (like Bodah and Patcoom), small zamindars 'permit a reclaimer to cultivate as much land as he chuses [*sic*] upon a large plain, for which he pays a fixed rent per annum. . . .'⁹⁵

In a large zamindari, where the estate belonged to a single and traditionally powerful zamindar, the extent to which such reclaimers could go to was determined not only by the financial necessities of the landholder, but also by the customary sanctions regulating the cultivation of waste or deserted lands. For instance, in the zamindari of Birbhum, the nature of reclamation and therefore the hold of the reclaimers appear to have been influenced by two customary practices, the details of which have been provided earlier in this discussion. The first was the stipulation that reclaimers paying concessional rates had to forego their lands on the original holder's return, who was then to have the lands restored at the 'former jummah' (original rate of assessment) which he paid prior to his desertion. The second was the enduring rule that the *established* peasants were to have preference in cases of disputes between them and 'new ryotts' coming from elsewhere.

Such customary codes were not typical of the larger zamindaris alone, nor where they limited of Birbhum. It was natural for all categories of landholders to attempt to protect the interests of peasants who had settled to cultivate on long-term leases.⁹⁶ The protection offered to *khudkashta* and *muqarrari* tenures by such zamindars (discussed earlier) is a clear indication of such tendencies. The obvious rationale behind the establishment of such permanent peasant tenures was the zamindars' need for steady incomes which was possible only by ensuring a definite permanency of agricultural production. This need became more pressing because of the two famines and a number of years of scarcity which Bengal suffered after 1765. In such situations, reclaimers could only hold on to the lands of their choice, or increase their holdings, not by their intrinsic economic or social power, but by conniving with zamindari officials at the village level,⁹⁷ or by appealing to the zamindar to apply his discretionary powers on their behalf.⁹⁸

Turning to the economic side of these reclaimed lands we find the following picture. Since reclamation involved relatively large amounts of capital and a substantial deployment of labour, a *jotedari* had to ensure fair returns on both to be a viable enterprise. A part of the capital costs involved were reimbursed by the nature of such tenures. They were initially given at nominal rates of assessment (usually fixed at half the normal *jama* on lands of comparable qualities) and the reclaimers always attempted to get these originally favourable terms frozen for as long as possible.[99] The organization of production was predominantly sharecropping, with the reclaimer providing the necessary financial investment and the sharecropper contributing the necessary labour. The division of the crop was the sphere where the *jotedar* used his superior bargaining position in order to take more than the agreed proportion of the harvest, usually half, as his share. This was achieved by the coercive strategy of counting the advances of seed or money to the sharecropper, not as an advance of capital but as a loan on which a high rate of interest was charged.[100] When paid in cash, the sharecropper faced the following situation: 'The usual money advanced is 6 rupees; and this is called a year's wages . . .; but in fact the term for which he labours is always extended beyond a year, and from 1 to 10 months are added for interest, according to the proportion between the number of employers and those who consent to take service.'[101]

Control over labour was also maintained by another device, mainly administrative in nature, which was delegated by the zamindar to the reclaimer as another incentive for extending cultivation. This was the right to grant *pattas* to the subordinate *raiyat*. In Jessore, for instance, the *pattas* granted by the *ganthidars* to the cultivators were adhered to by the zamindar and the revenue-farmer (*ijaradar*) *even* when the *ganthidar* had vacated his lease.[102]

But the *jotedars* were not alone in shaping the first forms of *bargadari* in Bengal. The zamindars used sharecroppers with equally exploitative zeal on their own lands. Nor did the modes of coercion used by the *jotedars* lead to their ubiquitous domination over all cultivators within the lands they reclaimed. In Jessore, peasants described as *khas raiyat* were not subject to the *jotedar* or the *ganthidar*; they dealt directly with the zamindar, a privilege they enjoyed because of their being the 'ancient Ryotts' of the zamindari who paid 'unchanging rents for many generations'.[103] In Rajshahi, a peasant category called *khush bash* (residing at pleasure) were allowed easier terms and permitted to pay a mutually negotiated amount both by the zamindar and by the reclaiming leaseholder.[104] In Bakarganj, the original reclaimers (*abadkar*), the first phase of reclamation being over, seldom interfered with the way in which the subordinate holders managed their cultivation. This resulted in a proliferation of sub-tenurial rights engineered by the cultivators themselves.[105] Finally, in Birbhum where the *pahikashta* had to compete with the resident cultivators for better terms of revenue, reclaimers were often hard hit in their efforts to maintain a close control over labour in their enterprises.[106]

The *jotedar* of the eighteenth century was not a 'village landlord', nor was he a 'kulak-landlord'. It was a recognized fact that even those holding less than 10 *bighas* (3.3 acres) of land could get *jotedari pattas* from the zamindar.[107] This would make the *jotedars* of the eighteenth century at best 'middling' peasants who had been able to collect sufficient resources to enable them to invest in reclamation using the labour of those who were less favoured. The resources they had may have arisen from their being able to save during favourable agricultural seasons or from profitable sales of their produce. These could also emerge from colluding with the zamindari official in order to juggle their own revenue accounts. The overwhelming portion of these resources, however, came from the 'profits' they made during the distribution of the product between them and the sharecroppers. But 'accumulation' of such resources should, however, not be taken to imply that an individual *jotedar* could manage to extend his *jotedari* unchecked, or that he could create a new power configuration in society.

It was the zamindar's interest which determined the boundaries of the *jotedari* in the last instance. Moreover, none of the *jotes* were extensive enough to conjure visions of 'large farms' being managed by a 'kulak'-type farmer. Very rarely did individual *jotedars* take leases of farms beyond the supervisory capacity of one person.[108] This limitation would tend to make the eighteenth century *jotedari* an agglomeration of small or medium sized leases, designed by the zamindars to bring about an extension of cultivation, and to concentrate production in the hands of those people who had relatively larger funds at their disposal when compared to the poor peasants and the artisans. This is not to say that the *jotedar-bargadar* relationship was not viciously exploitative; but it was not a mode of exploitation which differed substantially from those prevailing in zamindari areas or the ones shaped by the merchants in their dealings with the poor peasant. Nor had the *jotedars* been able to create the necessary conditions for the displacement of the zamindar's power from the matrix of rural society. Such displacements, and the techniques of replacement had to wait till the middle of the nineteenth century before being realized.

Sharecropping in the Eighteenth Century

There are essentially two views of sharecropping. The first suggests (on the basis of the *kedokan* in Java) that sharecropping in history, and its stubborn persistence in some contemporary societies, are features of a 'pre-capitalist' system of risk-sharing in largely subsistence-oriented economy.[109] The second views sharecropping as a '[semi]-feudal' social relation; and this has been applied in the case of the Philippine *kasama* and *bargadari* relationship in eastern India.[110] Discussions of sharecropping arrangements in pre-colonial (specifically eighteenth century) or early colonial India are few, but they nevertheless incorporate the two dominant views of sharecropping. For instance, C.A. Bayly suggests that in origin at least the institution of sharecropping may have been a system of mutual protection and interdependence in the cash-strapped economies of northern India in the eighteenth century,[111] while another analysis, specific to colonial Bengal, has analysed sharecropping as a semi-feudal social relationship of exploitation of a low-caste (and landless) peasantry under conditions of abject indebtedness.[112] A recent study sees sharecropping as a sub-regional phenomenon (limited to north Bengal) in the context of an agrarian structure sharply polarized between the *jotedars* who controlled the credit and the product markets and the *bargadars* who had 'no recognized long-term rights to the land they tilled'.[113]

These approaches do not consider whether or not sharecropping can be a social typology of a different kind. On the other hand, the Marxian position that sharecropping is essentially a transitional (between feudalism and capitalism) method of surplus appropriation is also problematic because of its insistence that sharecropping will wither away with the growth of capitalist relations in the countryside.[114] Since there is a specifically Eurocentric concept of capitalism ingrained here, there is, therefore, very little effort to come to grips with the specific features of sharecropping in different social-formations. At best, as Husken points out, the nexus between sharecropping and commercialism is seen as a process of incorporation, that of a largely pre-capitalist relation (sharecropping) into a newly developing commercialized village economy.[115] Also, the implicit assumption in most Marxist writings is that wage-labour is economically (and politically) superior than most supposedly non-wage forms of labour, and this has hampered an analysis of sharecropping as a spearhead (or even as a symptom) of agrarian commercialism *sui generis*.[116] It would, therefore, be important to consider the possibility that sharecropping—its proliferation and spread—was the tenurial and social relation arising out of the commercialization of agricultural production in a small-peasant economy. In fact, sharecropping was one of the tenurial arrangements which evolved during the transition from feudalism to capitalism in West Europe, and (as in Spain, Italy and France) it co-existed with the capitalist-farmer and peasant-'proletariat'.

As a relation of production, sharecropping entailed a specific relation between investment and labour, and between investment and redistribution. It required the investment of the necessary productive resources by a social group and the labour power by another.[117] 'The ryotts who cultivate these lands are generally supplied with money by the zamindars to *provide seed, implements of husbandry and for the immediate maintenance of their families*. . . .'[118] The division of the crop was based on a previously negotiated arrangement between the investor and the labourer. In Bengal, the share appears to have ranged from one-third to one-half of the produce which remained *after* the costs of investment had been deducted[119] whereas in Bihar the proportion seems to have been 22.5 shares for the zamindar and 17.5 shares for the *adhiar* of the net produce notionally divided into 40 shares of equal quantities.[120] Other forms of division also prevailed. For instance in Burdwan and Bishnupur the zamindar or the 'farmer [retained] the whole produce and [paid] a certain amount in money [to the sharecropper] calculated upon the price which the same article may bear in the adjacent markets'.[121] In this case, the division of the crop was more than a *physical* parcelling of the produce after the harvest; it also entailed the conversion of the produce into cash which perhaps indicates the commercial dimension of sharecropping.

The documentation of the relations of appropriation in the sharecropping system is significant. In a situation where the share of the crop was not done 'by an estimation of the crop valued at the market price, but by the actual division of the crop' the terms faced by the sharecropper were as follows:

1. No lease or written document was granted to the *adhiar*. He cultivated on a verbal agreement.

2. There were a number of deductions made from the produce *previous* to the division of the crop between the *adhiar* and the zamindar or the *gutchdar*. These were (*a*) 'the quantity of seed sown returned *two-fold* to the person who furnished it', and (*b*) deductions to defray the costs of employing watchmen to 'prevent the ryots from privately reaping the crops', for paying blacksmiths 'for work done on the ploughs', for paying the weighman (*kayal*) for 'measuring off the crop' and other incidental expenses to meet the costs of bringing in priests for making ritual sacrifices 'for the mutual benefit of the tenant and the cultivator'.
3. The costs of meeting the expenses in the category 2(*b*) ranged from 7.6 *seers* to 5.75 *seers* for every 2 maunds of grain produced (or between 9.5 and 7.18 per cent of every 2 maunds).
4. 'When the foregoing deductions have been made, the remainder of the produce is equally divided between the Gutchdar and the Addhea Ryott. . . .'[122]

The act of treating the advances of working capital as a loan on which a two-fold rate of return was charged[123] made sharecropping an extension of the general prevalence of usury in the countryside. In situations where the *bargadar* was paid after the produce had been converted into cash, the farmer made sure of a profit from the seasonal variation in prices. The produce was sold immediately after the harvest when prices were at their lowest, while the 'contracts' were negotiated *before* the land was sown at prices which were at their peak in the seasonal swing.[124] The sharecropper invariably got a lot less in real terms owing to the seasonal shift in agricultural prices: 'at the reaping of the harvest a very small part of it falls to the addhea ryott; but by the division the zemindars become possessors of the greater part of the produce of these lands'.[125] In Purnea, the zamindar could 'by these means draw without risque [*sic*] and expense on labour, four rupees from a bega [of land] for which he paid only a few annas'.[126]

This type of appropriation can partly be explained by the economics of interest-bearing capital. But the element of non-economic coercion was critical; it even dominated where sharecroppers were used by the zamindars to cultivate their *nij-talluqas*. In these lands, the zamindar used his prescriptive social status to arbitrarily 'tax the advances with an heavy interest' and/or 'fraudulently' devalued 'the market price of goods, or products of the land in which' the sharecropper was paid.[127] The following description of sharecropping in a zamindar's *khamar* is perhaps a revealing example of the modes of non-economic coercion which went into the making of sharecropping. The description comes from pargana Mandalghat, one of the revenue divisions in the extensive zamindari of Burdwan. The zamindar had farmed the *khamar* lands in this pargana to a small revenue-farmer (*qutquinadar*, also called *kutkenadar*) and the situation was as follows:

In the kummar [*khamar*] lands it is the custom of the pergunnah for the Ryot to receive his share in the proportion of 9 in 20. With this equitable [*sic*] mode of distribution the Ryot would well be contented if it was faithfully adhered to, but the kutkennadar usually over-rates the produce by which means the Ryot is left with a very inadequate recompense for his labour. In addition to this imposition the *Ryot is obliged to pay in ready money such proportion of the crops as belong to the zemindar and kutkennadar*. He suffers another violent oppression by an arbitrary valuation. For example, if a Ryot has to pay for 100 maunds of rice & if agreeable to the price at *which he sells* it at the Bazaar it produces 100 r[upee]s, the kutkennadar enhances the value of Rice by making the Ryot pay him back at the rate of 1 [rupee] for 32 seers by which the Ryot sustains a *loss* of 8 seers in every maund & is compelled to pay 125 [rupees] for rice which he could only sell for 100 Rupees.[128]

Additionally, physical force (confinement and flogging) was used by the zamindars to tilt the balance in their favour in case of disputed shares.[129]

The existence of large numbers of sharecroppers in a situation of land abundance, and in a context where absolute landlessness did not exist appears paradoxical. Yet the evidence from Rangpur and Dinajpur where 52 per cent (in Rangpur) and 77 per cent (in Dinajpur) of the agricultural population were sharecroppers in 1807,[130] and the information cited earlier indicate that sharecropping was situated as a major relation of agricultural production in our period.

The systematic use of sharecropping in our period cannot be explained either in terms of landlessness or in terms of the caste. The decision of a person to engage as a sharecropper was dictated entirely by the constraint of resources. The general insufficiency of land in individual peasant-households was the central resource constraint. The other pressure was the acute shortage of working capital. A good description of a peasant family in eastern Bengal in such circumstances was provided by J.H Paterson from Commilla in 1789:

A poor riatt during the time he can spare from his field hires himself out as a labourer while his wife employs herself at home spinning cotton. A Riatt of this description can not afford to cultivate above 2 Cannees [5.5 acres][131] of land at the farthest and to do that he must be allowed tuccavy [as] the produce of his lands without the above occasional resources could not maintain him especially if he has a family. He resides generally upon the lands of some other riatt as a part of his family by which he secures an exemption from the payment of any Bhitee jumma [tax on house]; and by these aids he makes a shift from hand to mouth and pay the rent of his two Connies. *This class of riatts are by far the most numerous.*[132]

This statement, in association with the state of the *adhiars* of Purnea and the *sanjhadars* of Burdwan (discussed earlier) must surely indicate large numbers of resource-constrained cultivators engaged as sharecroppers as a device to delay the prospects of protracted immiserization. Sharecropping, in other words, was a strategy of survival and of coping with continuous adversity by a majority of small-peasants. The other important reason necessitating such strategies was the cluster of famine and dearth years in the late eighteenth century, the full implications of which will be discussed later. Briefly, a famine or a dearth put immense strains on the already meagre resources at the disposal of these peasants. Additionally, each cycle of adverse weather, or harvest failure, threw up a fresh crop of uprooted peasants. Land was abundantly available but the main problem was that of productive absorption. The urban sector was unable to fruitfully absorb even a fraction of the displaced, thereby giving them no alternative but to stick to land as sharecroppers or agricultural labourers. Even though absolute landlessness was not a major agrarian problem of the late eighteenth century, the objective constraints on agricultural production, nevertheless, led to the creation of a sharecropping peasantry on a scale which can only be described as extensive.

Sharecropping in the eighteenth century emerged out of the poor peasant's need to ensure the reproduction of their household-based economies in the face of scanty resources, crises of subsistence and the financial demands of the state. It is no accident that the 'inferior ryotts' in Purnea, holding *insufficient* land, were *also* the *adhiars* there.[133] It is also significant that zamindars in Rangpur enticed these peasants to work as sharecroppers by providing the necessary advances of seed and implements in addition to small pieces of land 'on which to reside and cultivate pulse' and vegetables 'at a rent paid in money'.[134] In Chittagong, Buchanan noted that there was 'hardly such a thing as a farmer in this part of the country'. Every 'Artificer, Boatman, Labourer or Servant rents a small piece of ground; he pays his rent, cloathing and religious or festival expences by the wages he gets, and by his wife's spinning; and he has the produce of his ground for food and for raw materials.[135]

Even though, sharecropping emerges as a conscious strategy adopted by the small peasant either to circumvent the possibilities of protracted immiserization or to acquire a supplementary income which could then be used for a variety of consumption or production oriented purposes, there were three dimensions which reinforced the relationship between sharecropping and agrarian commercialism.

First, as a relation of production, it entailed a specific relation between investment and labour, and between investment and redistribution of the agricultural product. Investment was the job of one social group (zamindar or merchant) while the labour power utilized belonged to another (the *bargadar* or *adhiar*). Economically, this separation was significant. Since the capital invested was often a loan with heavy rates of interest, it introduced the potential of expropriation (realized in the nineteenth century) within the *bargadari* relationship. This separation between the two productive functions was crucial in the distribution of the product, for the investor invariably managed to get the greatest share of the produce through the debt-servicing mechanism.

Second, inasmuch as sharecropping involved remuneration for labour provided, it entailed the utilization of a notional wage, located in the distribution of the agricultural product, the terms of which were naturally regressive and biased against the sharecropper. This emerges clearly in cases where sharecroppers were remunerated in cash (itself a significant commercial fact) in place of the product in kind. In Burdwan and Bishnupur, not all shares contracts were produce-based. Some were paid '*a certain amount of money calculated upon the price which the same article [i.e. the crop sown] may bear in the adjacent markets*'.[136] Francis Buchanan's description of similar arrangements in Rangpur is revealing. Here, 'the usual money advanced is 6 rupees; and this is called a year's wages. . .; but in fact the term for which he labours always extended beyond a year and from 1 to 10 months are added for interest, according to the proportion between the number of employers and those who consent to take service.'[137]

Third, the sharing of produce between the two agents did not necessarily mean a physical parcelling of the rice heap. The evidence from Burdwan (cited earlier) shows that the product had to be sold, the responsibility of which was the sharecropper's; and the value of the share (established at the beginning of the share contract) had to be paid in cash immediately after the harvest. Therefore, the close connection between the market (where the produce was sold), the sharecropper and the conversion of the product into cash is clearly an instance of the triadic integration (peasant-household, agriculture and trade) with the market which has been highlighted as one of the central features of agrarian commercialism in our period.

Therefore, sharecropping cannot be seen solely as the legacy of the post-Permanent Settlement land-tenure carved out by the British in Bengal as has indeed been suggested by some historians.[138] By strengthening the zamindari, that Act certainly intensified the exploitative relationships embedded in sharecropping; but it was in no way responsible for its genesis.

Nevertheless, its spread during our period needs explanation. Cooper notices a rise in the incidence of share contracts during the high food price and recurring food shortage years of the 1930s and 1940s in Bengal.[139] The case of eighteenth century China is also significant in this context. Here sharecropping became the dominant tenurial relation in the midst of a secular upward trend in the price of rice and a steady aggregate expansion in the output of food.[140] These instances are important for they indicate a connection between sharecropping and the state of agricultural prices, which were on the rise in eighteenth century Bengal, though this is not to imply that sharecroppers emerge because small peasants are able to take advantage of an upward swing in prices by switching to sharecropping.

The Question of Peasant Stratification in Bengal and the Elusive 'Rich' Peasant

Peasant stratification in eighteenth century Bengal cannot, therefore, be explained in terms of massive differences in landholding alone. The situation in Dinajpur, where Buchanan-Hamilton noted the existence of peasants holding 165 *bighas* of land was clearly an aberration and should not be taken as the crucial indicator of peasant stratification for the entire province, as has been done by Ray and more recently by Taniguchi.[141] In the adjacent district of Rangpur peasants with 30 *bighas* of land were identified as the 'superior ryotts' of that district in 1790;[142] in Dinajpur 30 *bighas* would classify a peasant as a 'poor farmer'.[143] The peculiarity of Dinajpur can be explained by the silting up of its major riverain systems and the apparent economic contraction which this engendered.[144] Large 'farms' may have been a device of people with resources to extend cultivation either at the behest of the zamindars, or by small peasants leasing out to their larger counterparts and then working their lands on a sharecropping basis.

In a recent essay, S. Taniguchi has attempted to show the existence of a sharply stratified peasantry on the basis of some evidence from north Bengal.[145] He sees Bengal's peasantry divided into four categories on the basis of unequal resources ('agricultural stock'), and amidst this the 'wealthy farmers, or the rich peasants or jotedars' were those 'who had large agricultural stock and rented extensive lands from the zamindar'.[146] Yet in the same breath, Taniguchi declares that 'the upper line of demarcation, between self-sufficient and rich *raiyats* cannot be so clearly drawn as it varied according to family size'.[147] If control of resources was the key variable in sustaining the pace and the form of stratification, then the demographic constraint would automatically act as a scissor on any further accumulation of resource. In such a situation further stratification would be impossible. Taniguchi derives his evidence from the surveys of Buchanan-Hamilton (the veracity of whose data on Dinajpur he himself questions)[148] and from one conducted in the zamindari of Swaruppur, in 1790, by J.H. Harrington.[149]

The major findings of Harrington regarding peasant lands in three *mauzas* of Swaruppur are as follows:

Size of Holdings (in Calcutta *bighas*)	In *mauza* Radhanagar (peasants)	In *mauza* Maheskal (peasants)	In *mauza* Raghunatpur (peasants)
30 and above	3	1	2
15 to 30	4	2	1
10 to 15	2	1	0
5 to 10	1	1	1
2.5 to 5	10	0	3
1 to 2.5	14	2	0
0 to 1	13	3	1

If the dividing line between a self-sufficient peasant farm and one which had to hire-in labour (the latter being an index of resource control) was 30 Calcutta *bighas*,[150] then the evidence from Swaruppur does not support his claim of having established 'beyond doubt'[151] the existence of a class of rich peasants even for his own subregion. One may also add that Harrington had already written about the dismal state of the peasantry of Swaruppur in 1789 in which he had emphasized their inability in getting 'no specific engagements being executed between the zemaindar and the Ryotts individually' as being the cause of their 'rents being indeterminate'. They were burdened with additional *abwab* and *mathot*. They were given no receipts for their payment, so 'they are much in the power of the Mofussul officer'. Additionally, 'Their Rents being liable to annual variations, the zemindar may avail himself of every favourable circumstance attending to produce or price to demand an increase, in consequence of which they are discouraged from their lands by manure, *as well as from extending their cultivation by additional labourers and implements. . . .*' Finally, not having the benefit of having their 'monthly instalments determined by the time of reaping & selling the produce, the Ryots are obliged to borrow. The rate of interest at which Ryots . . . borrow are excessive. The Zemindar for his security attaches the principal crop, by which means the Ryots are disabled from selling it & are also subjected to the loss of a part of their grain.'[152] Harrington also gives us a detailed account of land use, output of selected crops per *bigha* and their money value which may be used to shed further light on the nature of Swaruppur's peasantry. The total land under *cultivation* in this pargana was 45,202.25 *bighas*. Of this 2,375.58 *bighas* were situated in *mauza* Radhanagar. The patterns of production in that village were as follows:[153]

Crop	Land under cultivation (*bighas*)	Output per *bigha* (average/maunds)	Value of output (Rs. per *bigha*)
Winter rice	1216.6	10.72	3.5
Spring rice	103.15	7.3	1.5
Mustard	68.01	1.87	1
Arhar dal	64.8	1.5	1.25
Masoor dal	0.6	1.98	0.75

Aside from these, the major cash crops had the following profile in the land they were grown in the village of Radhanagar:

PERCENTAGE OF TOTAL LAND UNDER CULTIVATION

Crop	Area	Percentage
Cotton	3.15 *bighas*	0.132
Mulberry	16.3 *bighas*	0.694
Tobacco	1.9 *bighas*	0.079

Such evidences surely militate against Taniguchi's notion of a resource rich peasantry in this region.

The situation was perhaps no different elsewhere in northern Bengal. In Rangpur, Buchanan's testimony does not provide a firm impression of *peasant* stratification, at least not the kind Taniguchi has supposedly found in this area. Here, under the *dewanaya* system, small farmers would organize themselves in work-gangs under a 'chief man, who settles the whole of their transactions with the agents of the landlord; and they are entirely guided by his opinion'. A *dewanaya* was a person with a stock of 3 to 5 ploughs. Certainly this would make him a peasant in comfortable circumstances, but the end purpose of assembling these work-gangs was to cultivate the lands of 'persons who have professions *totally distinct from that of a farmer*'. Buchanan goes on to add that 'many Brahmons, Scribes [Kayasthas], Physicians, Kazis, and other persons . . . lease large farms, and in fact a *great part of the large farms belong to these descriptions of men. . .*'.[154]

Buchanan's description indicates *inter alia* that a distinction must be made between the notion of a stratified peasant society and the actual construction of a sharply polarized *rural* society; and in the case of Bengal the latter is a more meaningful historical problematic. This is borne out with greater clarity from the following evidence of landholdings in the village of Dakninchar, pargana Calcutta in 1776:[155]

Name	Occu-pation	Total ground (*bighas*)	Paddy (*bighas*)	*Bastu* land (*bighas*)	Straw houses (numbers)	Garden (*bighas*)	Fruit trees (numbers)
Dulal Mohun	*raiyat*	2.65	2.65				
Mansur Mullah	*raiyat*	3.95	1.95	2.00	2		
Balaram & Ramsundar Bandopadhay	*La-kharaj holder*	7.35	4.10	2.00		0.85	13
Krisna & Attaram Datta	Officials	18.6	7.6	1.00		10.00	257

Quite clearly, the social concentration of resources corresponded to a social hierarchy in which non-peasant groups occupied the apex slot.

Therefore, given the favourable land-labour ratio and the *general* smallness of peasant holdings, it becomes difficult to find a connection between peasant stratification on the basis of unequal landholdings alone in the eighteenth century. This is adequately testified to by the following data pertaining to the landholdings of five cultivators of *mauza* Gopalpur, pargana Saidpur in 1776:

Name of *raiyat*	*Raiyati* Holding (*bighas*)
Hari Dass	10.1
Gopal Dass	9
Kishen Dass	6.05
Jagannath	2.1
Narhari	2.05

Obviously, Jagannath and Narhari were the 'inferior' *raiyat* in village Gopalpur, compared to Hari Dass whose *patta* stated that he had two *bighas* of land under 'cupass' (cotton), though in 1776 he lessened the land under cotton to one *bigha* and added an extra *bigha* for the cultivation of rice on grounds of the 'dearness' of cultivating 'cupass'.[156] Therefore, the example of Gadadhar Mandal a 'rich *raiyat*' possessing 60 *bighas* of land in Burdwan, devoting 5 *bighas* to the cultivation of sugarcane, cited by Taniguchi to generalize for the province,[157] remains a very singular exception. *Raiyat* Sultan Ghazi of *mauza* Muradpur, zamindari Nadia, possessing 50.95 *bighas* of land in 1794 did not grow sugarcane or cotton on even one *bigha* of his land, nor did Sultan Ghazi's brother with 44.15 *bighas*; whereas their neighbour Jafar Ghazi Mandal possessing 20.5 *bighas* devoted a mere 0.45 *bigha* to 'cupass'.[158]

Peasant resources were traditionally measured in terms of draught animals and ploughs (and not by the amount of land held), and it is perhaps significant that persons having one or two ploughs were considered 'poor' peasants, whereas those possessing between three and five ploughs employed sharecroppers.[159] In Jessore, people with a mere 10 *bighas* (3.3 acres) could take *jote* from zamindars and get them cultivated by *adhiars*.[160] 'The circumstances of a Ruatt [*raiyat*] are known by the number of ploughs he employs . . . but there are those who have no oxen, and who are obliged to hire a plough, either by money or work. Two ruatts who have no oxen are enabled to cultivate a *doon* (6.6 acres) *by joining their cattle and using them alternately* . . .' is a revealing description of small peasants working under a shortage of resources.[161] What these descriptions also tell is that even the slightest variation in resources would make a major difference between relative affluence and relative poverty in rural society.

Identifying the rural poor with the 'untouchable landless groups in the villages'[162] is also problematic. Similarly, the explanation of the upper strata in Bengal as a 'high-caste rural gentry' fails to resolve the issue of non-Hindu zamindars or that of the relatively affluent Muslim agriculturists of Bakarganj who sometimes held three or four separate rights on different lands in the village.[163] Caste cannot explain the activities of the Muslim peasant of Commilla who (in 1789) were said to be dictating the appointment of Qazis in the village mosques and also looking after their maintenance.[164] Untouchability and landlessness are inadequate in explaining social stratification even among the Hindu peasants. Sanyal's study of 'lower' caste peasant groups like Sadgops in Burdwan and Hali-Kaivartas of Midnapur indicates the manner in which these castes were stratified through a process of internal differentiation leading on the one hand to the formation of Kaivarta and Sodgop zamindaris, and of poor peasants of the same caste on the other.[165]

Moreover, landlessness in absolute terms does not appear to have been a major agrarian problem in the eighteenth century. The picture of an uprooted peasantry roaming through the countryside in the wake of the famine of 1769-70 'in search of employment'[166] is largely overdrawn and certainly would not apply to eastern Bengal[167] which did not suffer its ravages to the extent endured by the districts in western Bengal. Therefore, the picture of the eighteenth century peasantry as one polarized between a 'rich peasant class' and 'untouchable landless groups' tends to provide a false image of rural stratification. Given the *general* smallness of peasant holdings, the virtual absence of absolute landlessness and a seemingly widespread shortage of resources we can perhaps only speak of two broad strata in the peasantry: the poor and the middling.[168] The poor were those peasants unable to reproduce themselves through household production because of insufficient land and/or other means of production. The middling peasant had lands which were considered adequate for their social and economic reproduction as well as for the generation of a small surplus which they presumably invested for some improvements in their lands or for the cultivation of better grade crops,[169] or both, depending upon the nature of the current agricultural season. They may also have used their surpluses to give loans to their poorer counterparts in times of need.[170] Thus, when Taniguchi says that a '*raiyat* holding 16 local *bighas* could get along with ease, and accumulate a certain amount of capital if he was cautious in his expenditure [and] could take up extra hands',[171] he is describing not a process of capital accumulation by the rich peasant, but a slight accretion of resources in the hands of a middling peasant of eighteenth century Bengal.

Yet, our sources do give the impression that a section of the peasantry were causing concern both to the local zamindar and to the state. John Shore described these peasants as those who 'cultivate land, of which there is no account, and hold them in greater quantities than they engage for; . . . they hold lands at reduced rates by collusion, obtain grant of land fit for immediate cultivation, on the reduced terms of waste land; . . . their power and influence over the inferior ryotts is great and extensive [and] if any attempt is made to check their abuses, they urge the ryots to complain and resist. . . .'[172]

This description has influenced historians into reconstructing Bengal's peasant society as elaborately stratified into a 'very powerful tenantry' and poor peasants. The former are also seen as a 'village oligrachy' on grounds that they combined superior economic power with political clout. These formulations are questionable for a number of reasons. First, one has to consider the *geographical* location of the so-called powerful tenantry. John Shore based his arguments on the situation prevailing in Dinajpur and Rajshahi in eastern Bengal and on Birbhum and Purnea in the west.[173] I have already drawn attention to the peculiarity of Dinajpur's rural society. The three remaining districts were those which had suffered extensive depopulation during the famine of 1769-70 and had subsequently become areas of unstable production.[174] That famine took its greatest toll from the poorer peasants and the artisans thereby reducing large tracts of land into wastes (*pateet*), a situation which seems to have persisted even when the worst had passed and other districts had managed to make substantial post-famine recoveries.[175] Moreover, the modes of reclamation were governed by a number of customary rights of the established peasants even when they had been forced to desert. These facts leave nagging doubts about the validity of the argument of a large-scale takeover of land by the 'superior ryotts' or by a 'privileged tenantry'.

Second, the reasons for characterizing these so-called 'village oligarchs' as peasants are unclear. Reclaimers in Chittagong were quite often the zamindari *diwan* or *faujdar*.[176] From the point of view of the zamindar the problem does not appear to have been one of peasants colluding among themselves as much as it was the collaboration *between* the zamindari officials and the so-called 'superior ryott'. It is no accident that in almost all types of zamindaris, the universal complaint was the collusion of the zamindari *amil* with the 'influential ryotts' in frauds pertaining to land or revenue. Such cases are numerous and need not be listed in all their details. There is however one complaint which is most representative of the entire genre of such complaints and therefore needs to be cited. This is the grievance expressed by the zamindar of Rajshahi (one of the largest and most famous of the eighteenth century zamindaris) to the Board of Revenue in May 1790 and needs to be quoted in full:

> Many ryotts in this district, thro' collusion with the moffussul karramcharee [official in the village] have caused a considerable part of their jummah [assessed] lands to be set down in the accounts as pulotoka or forsaken [lands] & fresh leases at the lowest rate of assessment have been granted for lands to their own dependents, which is fact to themselves. When they are called on for the payment of their original jummah, they produce these fictitious leases & thereby evade payment of a just demand. The karramcharee ... in my zemmindarry have annually obtained considerable remissions from the fixed jummah of their lands [and] these lands [when] brought under cultivation yield a vast profit to them. Many persons holding leases of joat [cultivation] have possessed themselves of more lands than specified in their pottahs [leases]. Several Individuals of substance have by collusive means obtained remissions of revenue, *which the lower class of ryotts are obliged to make up for by contributions.* The consequence of which oppression is the desertion of numbers....[177]

This indicates the anxiety of one of the most powerful of the eighteenth century zamindaris and shows the overt influence of the zamindari officials in matters pertaining to the distribution of lands in the village and the grant of leases of cultivation (*pattas*) to *all* categories of peasants. That these people were essentially non-peasants is suggested by the absence of any evidence to show that they *actually* tilled the land with their own, or with their families labour, and they, along with the landed proprietors, constituted the gentry of rural Bengal.

Moreover, the description provided by Colebrooke of the use of hired labour in agricultural production[178] does not necessarily demonstrate the existence of a coterie of improving peasants 'cultivating lands by using sharecroppers and agricultural labourers'.[179] As Colebrooke himself noticed, such practices were resorted to by those who 'are restrained by [caste] prejudice from personal labour', these being the Brahmin[180] and presumably the other upper castes. Nor does (what could only have been a very restricted) use of hired labour automatically indicate the 'richness' of the peasant-employer.'[181] These would, for instance, include the indigent Brahmins of Rajshahi and Burdwan who had to eke out supplementary incomes by allowing the women in their households to spin cotton thread[182] in a society where spinning of thread and weaving of cloth were generally recognized to be the occupations of 'the lower class of people'.[183]

The participation of the indigent Brahmins in agricultural production short of actually holding the plough, has been noted by Ray[184] but the case for the use of hired agricultural labour for production purposes has been overstated. The case of Kayasthas may perhaps have been different than that of the Brahmins. They had the necessary social and political authority by virtue of their place in the zamindari administration and it is likely that their lands were cultivated with a greater degree of economic efficiency, but there is practically no evidence to press this point further. Additionally, these were the paid employees of the zamindars who used their official positions to put pressures on the poorer peasantry in order to elicit bribes, and collaborated with those peasants with better resources by juggling their revenue accounts in order to acquire certain financial advantages. The link between the officials and the economy was located more in the realms of economic and political manipulation of the levers of appropriation than in the sphere of agricultural production. The nineteenth century developments seem to have perpetuated this feature of rural society. In a recent study, Sugata Bose has suggested that 'outside the north Bengal bastion of large jotedars', zamindars dealt, not so much with 'land-controlling rich peasants', but with 'village leaders, often known as mathbars' who 'should more appropriately be seen as constituting a seigneurial sergeant class, ready and able to manipulate the rent-collecting mechanisms in a period of enhancements in the nineteenth century'.[185] Bose also suggests that by the middle of the nineteenth century, the peasantry of Bengal was divided into the following agrarian typologies: 'The village-landlord/ rich farmer-sharecropper system . . . in north Bengal; the peasant smallholding system . . . in east Bengal . . . , and the peasant-smallholding-demesne labour complex . . . in west and central Bengal.'[186] Finally, there was the 'tribal society on the fringes of west Bengal [in which] mandals negotiated the terms of village settlement on behalf of the whole community'.[187]

Such variations were absent in the eighteenth century. The situation then was more fluid. The Amini Report of 1778 describes three categories of peasants: *hari, fasli* and *khamar*.[188] The *Risala-i-Zira't* (1785) broadens the net by indicating four categories: *muqarrari* or *patti, fasli, pahikashta* and *kalijana*.[189] *Hari, fasli, muqarrari* held permanent rights on the land they cultivated and paid revenue according to a given schedule of rates in the deeds (*pattas*) given to them. *Pahikashta* were peripatetic, and their firm location in the *Risala* compared to the silence in the Amini Report indicates the rapid increase in the numbers of such peasants in the late eighteenth century. *Khamar* and *kalijana* were generally sharecroppers, or they could also be those who, as the *Risala* explains, 'till land as a subordinate of another cultivator'.

With the possible exception of Dinajpur, the *jotedar-bargadar* nexus had still to take firm roots even in north Bengal. Extensive reclamation in Jessore, however, had resulted in the rise of the *ganthidar*, a *jotedari*-type tenurial arrangement in a part of eastern Bengal which had otherwise an overwhelmingly 'peasant-smallholding' profile. In western and central Bengal, the cultivation of *nij* lands of individual zamindars had shaped a distinctive form of sharecropping: the *khamar* cultivator. This was a direct offshoot of the prevalence of larger zamindaris in the western parts of the province compared to the smaller size of zamindaris in the east. This however did not interfere with the full development of independent peasant proprietorship in the form of *muqarrari, fasli* and *pahikashta* cultivators. Therefore the impression that one gets of rural Bengal in the late eighteenth century is of a society composed overwhelmingly of peasant-smallholders.

The search of a 'rich' peasant 'class' in eighteenth century Bengal is therefore fraught with a number of difficulties, some empirical and some conceptual. The available evidence does not allow a division of peasant society into rich, middle and poor peasants. What we do get is a picture of a rural society where peasants were roughly divided into two strata: those who could reproduce their domestic economies and those who could not, and for the sake of analytical convenience I have divided them into 'poor' and 'middling' categories. 'Rich' peasants, defined as the people with enough accumulated surpluses to invest in production 'through the purchase of superior means of production and/or labour power' and who could 'initiate and maintain a cycle of extended reproduction'[190] were largely absent in society.

A paradox remains. How does an argument for the rapidity of agrarian commercialism in rural Bengal square with the picture of a peasant society so rudimentarily differentiated? The richness of a particular section of the peasantry could perhaps be explained if one is able to show that they cultivated 'cash' crops (like mulberry or cotton) almost exclusively on their own, and that the surplus which accrued was accumulated independently by them and subsequently used for expanded reproduction. But this was not so. The cultivation of mulberry trees (and the production of silk), closely associated with the economic interests of the state, was financed by the East India Company. Cotton was grown in small quantities in tiny plots of land *subsidiary* to the cultivation of rice and other food-crops and the difference between 'rich' peasant and the 'poor' in terms of the portion of an individual's land each could devote to its cultivation was minimal: 'the richer class of ryotts may possess a begah [0.3 acre] each of cotton & the poorer sort from 8 to 10 cuttahs [0.16 to 0.20 acre], the *produce of which is generally converted to family consumption*' states a report on the state of cotton cultivation of western Bengal in 1789.[191] In the northern Bengal zamindari of Swaruppur where Taniguchi finds an entrenched coterie of rich peasants, out of a total cultivated area of 48,430 *bighas*, cotton was grown on 2.70 *bighas* (0.005 per cent), tobacco on 196.95 *bighas* (0.41 per cent), mulberry on 124.75 *bighas* (0.26 per cent) and sugarcane on 481 *bighas* (0.99 per cent).[192]

The evidence also shows that the major cleavage in peasant society was not between the landowner and the landless (nor was it between the high-caste and the untouchable), but between the owners of adequate and inadequate land, or resources. The existence of *khudkashta*, *fasli* and *muqarrari* cultivators shows quite conclusively that a certain customary privileges were attached to particular categories of rights which could become advantageous levers in their hands at certain junctures. This is apparent in the way in which some of them colluded with the zamindari officials to wrest favourable terms of revenue for themselves. The modalities of land reclamation (Chapter 1) also suggest the centrality of such 'resident' peasants in the process. Of course, an inter-regional market for food, an aggressive mercantile presence and the continuing participation of the landed gentry, was the larger context in which the peasants were induced to reclaim land.

Therefore the extent to which peasants were directly involved in the processes of commercial growth or the manner in which they benefited from it depended on a number of extraneous factors, particularly those of exchange and patron-client relations (as in the case of land reclamation induced by the rural gentry) and the creation of intermediate interests between them and the provisioners of large-scale resources. For instance, sugarcane, tobacco and even rice were seldom grown by peasants with their own resources. They were cultivated by a wide variety of advances taken by the cultivator from the merchant or the landowner. This firmly indicates a wider social participation in the processes of agricultural production. It also suggests a plenitude of resources and the existence of substantial accumulated surpluses in the hands of rich social strata situated outside peasant society.

NOTES AND REFERENCES

1. Final Report of the Amini Commission, 1778, IOR HM, vol. 206, pp. 345-6; BM, Add. Ms. 29086.
2. Colebrooke, *Remarks*, p. 64.
3. James Grant in *FR 2*, p. 276.
4. J.H. Harrington to Council of Revenue, IOR, BRC, P/52/10, 12 May 1790.
5. A peasant category called 'principal farmers' by Buchanan-Hamilton held between 60 and 165 *bighas* of land in Dinajpur in the early nineteenth century and they constituted about 6 per cent of the cultivating population (Martin, *Eastern India*, vol. 3, p. 906).
6. For Rangpur (IOR, BRC, P/52/10, 20 March 1790) and for Jessore (ibid., P/51/22, 23 July 1788, p. 131; also W.W. Hunter, *Bengal Ms. Records*, vol. 1, London, 1894, p. 52). Interestingly, a person holding 10 acres in Dinajpur was categorized as a 'poor farmer' in Dinajpur.
7. WBSA, PCR, Burdwan, vol. 1, 6 June 1774; IOR, BRC, P/51/32, 4 February 1789.
8. Rajat Datta, 'Merchants and Peasants', pp. 391-4.
9. IOR, BRC, P/57/50, from J. Harrington to Council, 20 March 1790.
10. Buchanan-Hamilton, *A Geographical Description of Dinajpur*, Calcutta, 1833.
11. IOR, BRC, P/51/40, 3 April 1789.
12. F. Buchanan, 'Survey of Ronggoppur', IOL, Ms. Eur. D. 75, vol. 3, book 4.
13. WBSA, PCR, Burdwan, vol. 18, 17 January 1777.
14. Ibid., also IOR, BRP, P/72/4, 6 July 1792.
15. Rashid Pertev explains sharecropping as a device adopted by the richer peasantry to cut costs of supervision in a situation dominated by small landholdings and rural inequalities in land distribution ('A New Model for Sharecropping and Peasant Holdings', *JPS*, vol. 14, no. 1, October 1986).
16. A. Ghosh and K. Dutt, *Development of Capitalist Relations in Agriculture: The Case Study of West Bengal, 1793-1971*, Calcutta, 1977.
17. Rajat Ray and Ratnalekha Ray, 'Zamindars and Jotedars: A Study of Rural Politics in Bengal', *Modern Asian Studies*, vol. 9, no. 1, 1975.
18. Ratnalekha Ray, *Change in Bengal Agrarian Society, c. 1760-1860*, Calcutta, 1979, p. 54.
19. Even Buchanan-Hamilton's descriptions of tenurial arrangements in the districts of Dinajpur and Rangpur in 1808 do not refer to *jotedars* as a specific term apparently in widespread prevalence. For Dinajpur, the term used is 'principal farmers' to describe agriculturists who held more than 55 acres (165 *bighas*) of land, possessed capital assets ranging from Rs. 5,000 to Rs. 20,000 and cultivated with the use of sharecroppers. In the case of Rangpur, he describes a variety of such tenures without once

using the term *jotedar* (see 'Survey of Ronggoppur', IOL, Ms. Eur. D. 75, vol. 2, book 4, fols. 102-4, 111-13).
20. This prevailed in Burdwan (IOR, BRC, P/51/21, 16 July 1788).
21. This was the situation in Purnea, see IOR, BRP, P/71/28, 7 July 1790.
22. Terry Cox, 'Class Analysis of the Russian Peasantry: the Research of Kristman and his School', *JPS*, vol. 11, no. 2, January 1984, p. 48.
23. Henry Bernstein, 'Concepts for the Analysis of Contemporary Peasantries', *JPS*, vol. 6, no. 4, July 1979, p. 43; emphasis added.
24. B.B. Chaudhuri, 'Rural Power Structure and Agricultural Productivity in Eastern India, 1757-1947', in M. Desai, et al. (eds.), *Agrarian Power and Agricultural Productivity in South Asia*, California, 1984, p. 119.
25. See for instance, W. van Schendel and A.H. Faraizi, *Rural Labourers in Bengal, 1880 to 1980*, Rotterdam, 1984.
26. Chaudhuri, 'Rural Power', p. 121; Schendel and Faraizi, *Rural Labourers*, pp. 29-31; Sugata Bose, *Peasant Labour*, pp. 24-9, 93-5.
27. Chaudhury, 'Rural Power', p. 119.
28. Sugata Bose, *Agrarian Bengal: Economy, Social Structure and Politics, 1914-1947*, Cambridge, 1986, pp. 11-18; also Sugata Bose, *Peasant Labour*, pp. 93-5.
29. *Economic History 2*, pp. 62, 64.
30. W.W. Hunter, *The Annals of Rural Bengal*, London, 1897 (rpt. New York, 1970), pp. 62-3.
31. *CEHI 2*, pp. 299-300.
32. *Agrarian Change*, pp. 56-7.
33. Warren Hastings, 'The State of Bengal in 1784', in G.W. Forrest (ed.), *Selections from State Papers*, p. 265.
34. Ibid.
35. For the absence of *najai* in the listed taxes of peasant *pattas* after the famine in Rajshahi, see IOR, HM, vol. 122, pp. 767-8; in Jessore, see ibid., pp. 765-6; and in Laskarpur, IOR, CCR, P/67/57, 23 June 1773.
36. *BDR, Rangpur*, vol. 1, 23 June 1770.
37. WBSA, CCRM, vol. 5, from Boughton-Rouse to Committee, 4 June 1771.
38. See IOR, BRC, P/51/21, 20 May 1788 for Burdwan, and IOR, BRP, P/71/24, 15 April 1790 for Birbhum.
39. IOR, BRP, P/70/22, 21 December 1786.
40. BM, Add. Ms. 29076, fol. 3.
41. WBSA, PCR, Dinajpur, vol. 8, 17 June 1777.
42. IOR, BRC, P/51/21, 16 July 1788.
43. WBSA, PCR, Burdwan, vol. 1, 20 June 1774; emphasis added.
44. This was the distinctive feature of *khudkashta* cultivation. WBSA, PCR, Burdwan, vol. 15, 22 June 1776; emphasis added.
45. IOR, BRP, P/70/41, 18 January 1788; emphasis added.
46. WBSA, PCR, Burdwan, vol. 15, 22 June 1776.
47. It is perhaps significant that *pahikashta* cultivators in Dinajpur were used

primarily on lands producing *boro* rice which was an inferior and intermediate crop produced between the two of the major (the winter and spring) harvests [WBSA, PCR, Dinajpur, vol. 8, 17 August 1777].
48. See IOR, BRC, P/49/39, 26 March 1773 for the situation in Rangpur where 'the poorer riotts' constantly moved about from 'one tallook to another' in search of slightly better terms.
49. IOR, BRC, P/51/21, 16 July 1788.
50. Ibid., P/49/51, 1 March 1775.
51. Ibid., P/51/21, 4 June 1788; and ibid., P/51/34, 8 April 1789.
52. Ibid., P/52/5, part 4, 24 November 1787.
53. *Raiyati Jama:* Rs. 3,65,090, *pahikashta jama*: Rs. 42,359 (BM, Add. Ms. 29087, fol. 136).
54. IOR, BRC, P/53/58, 23 October 1798.
55. IOR, BRP, P/70/45, 1 August 1788.
56. IOR, BRC, P/50/65, 10 March 1786.
57. IOR, BRP, P/71/48, 18 January 1792.
58. IOR, BRC, P/53/58, 15 November 1793.
59. IOR, BRP, P/71/41, estimate of Asad-ul-Zaman Khan, zamindar of Birbhum provided to the Collector, 25 July 1791.
60. Ibid., P/70/45, 1 August 1788.
61. IOR, BRC, P/49/63, 2 August 1776.
62. IOR, BRP, P/71/41, 25 July 1791; emphasis added.
63. IOR, HM, vol. 206, p. 350; emphasis added.
64. For *muqarrari raiyat* in Bengal see, Amini Report, IOR, HM, vol. 206, p. 350; IOR, BRC, P/50/10, 3 July 1778; IOR, BRP, P/70/26, 13 April 1787; ibid., P/70/30, 22 June 1787; ibid., P/70/41, 6 May 1788; ibid., P/72/17, 19 June 1793.
65. Compare, IOR, BRC, P/50/10, 3 July 1778 and IOR, BRP, P/70/27, 8 May 1787.
66. Buchanan, 'Survey of Ronggoppur', IOL, Ms. Eur. D. 75, vol. 2, book 4, fol. 100. The reason for zamindari laxity in 1787 may lie in the massive flood in that year which perhaps forced them to make temporary concessions.
67. These are in essence the principal arguments of Ratnalekha Ray, *Agrarian Change*, Chapter 2.
68. IOR, HM, vol. 206, p. 35; N. Hasan, 'The Position of the *Zamindars* in Mughal India', *IESHR*, vol. 1, no. 4, 1964; B.R. Grover, 'Nature of the *Dehat-i-Taaluqa* (*Zamindari* Villages) and the Evolution of the *Taaluqdari* System during the Mughal Age', *IESHR*, vol. 2, nos. 2 and 3, 1965; Irfan Habib, *The Agrarian System of Mughal India*, Bombay, 1963.
69. IOR, HM, vol. 206, p. 35; emphasis added.
70. BM, Add. Mss. 29086 to 29088; also R.B Ramsbotham, *Studies in the Land Revenue History of Bengal*, Oxford, 1926, pp. 31-3; B.B.Chaudhuri, in *CEHI*, p. 300.
71. IOR, BRC, P/53/19, 14 May 1794.

72. Ibid., P/51/29, 28 November 1788.
73. IOR, BRP, P/70/40, 11 April 1788.
74. IOR, BRC, P/51/20, 14 May 1788.
75. See Chapter 5.
76. IOR, P/52/16, 28 July 1790; emphasis added.
77. BR, Misc., P/89/37, 1 April 1791.
78. IOR, BRP, P/71/13, 12 February 1788.
79. IOR, BRC, P/51/22, 25 June 1788.
80. Ibid., P/53/55, 16 January 1798.
81. IOR, BRP, P/72/17, 12 January 1793.
82. IOR, BRC, P/51/22, 25 June 1788.
83. IOR, BRP, P/71/26, from H.T. Colebrooke to Board, 26 April 1790.
84. These considerations seem to have motivated the zamindars of Jessore (IOR, BRP, P/51/22, 25 June 1788) and Birbhum (ibid., IOR, P/70/41, 18 January 1788).
85. IOR, BRC, P/53/55, 16 January 1798.
86. Ibid., P/51/22, 23 July 1788, p. 121; emphasis added.
87. Ibid., P/53/55, 16 January 1798.
88. Ibid., P/51/22, 25 June 1788.
89. Chaudhuri, 'Rural Power', p. 118.
90. This area was situated southward of Calcutta and was therefore in close proximity of its urban pull. After having undergone what appears to have been a phase of agricultural contraction during the early eighteenth century (Letters Received from Bengal, IOR, E/4/24, 31 December 1758) it was experiencing a fairly rapid reclamation during the latter part of the century (see Chapter 1).
91. IOR, BRC, P/49/70, Minute of Warren Hastings dt.18 April 1777.
92. IOR, CCR, P/72/2, 30 September 1779. The land taken up for reclamation in 1779 amounted to nearly 21.53 per cent of the total cultivated [*hasil*] land, 1,51,601.3 acres [4,54,804 *bighas*] in 1758 (Letters Received, IOR, E/4/24, p. 96).
93. The districts of Rangpur and Purnea, lying in the frontiers of the province, were *faujdari* areas under the Mughals and the Nawabs of Bengal. These were held by a number of small zamindars whose affairs were supervised by a military-bureaucratic official (*faujdar*) who was also responsible for pacifying the frontiers of the province. In striking contrast were the districts like Birbhum, Burdwan, Nadia (in western Bengal) Dinajpur and Rajshahi (in the eastern parts) which were recognized as substantial zamindari areas where the state had to pattern different administrative arrangements (cf. James Grant 'Historical Analysis of the Finances of Bengal, 1789', in *FR 2*).
94. WBSA, CCRM, 5 December 1771, p. 109; IOR, BRP, P/71/26, 18 June 1790.
95. IOR, BRP, P/70/27, 8 May 1787.
96. See for instance, IOR, BRP, P/71/25, 28 May 1790; Robert Kyd, 'Some

Remarks on the Soil and Cultivation on the Western Side of the River Hooghly', MS. Eur. F. 95, fol. 58; F. Buchanan 'Survey of Ronggopur', IOL, MS. Eur. D. 75, vol. 4, book 2, fols. 98-100; 'State of Bengal in 1786: A Report Compiled by Mr. Beaufoy', IOL, HM, vol. 382, p. 78.

97. Thus in Rajshahi, reclaimers operated by paying 'bribes to the Putwaree, Halsanah & Cutwals of the villages [all zamindari revenue officials], the connivance of all of whom is differently necessary for them' (IOR, BRP, P/70/26, 31 March 1787).
98. This was a marked feature of the *gutchdari* tenure in Purnea, see ibid., P/71/26, 18 June 1790.
99. For Rajshahi, IOR, BRP, P/70/26, 31 March 1787; for Jessore, IOR, BRC, P/70/26, 16 January 1798; for Bakarganj, H. Beveridge, *The District of Bakarganj: Its History and Statistics*, London, 1876, p. 194.
100. IOR, BRP, P/71/26, 26 April 1790.
101. F. Buchanan, 'Survey of Ronggoppur', fol. 112.
102. IOR, BRP, P/72/17, 12 January 1793.
103. Ibid.
104. Ibid., P/70/26, 31 March 1787.
105. Beveridge, *Bakarganj*, pp. 194-6.
106. IOR, BRP, P/70/41, 18 January 1788.
107. This tendency was noticed in Jessore (IOR, BRC, P/51/22, 23 July 1788, p. 133). Ray also noted that 'even a poor peasant holding five acres on a direct lease from the zamindar could describe himself as a jotedar' (1979: 54) but failed to note the contradiction between this and the crux of her thesis that *jotedars* were 'village landlords'.
108. Buchanan, 'Survey of Ronggoppur', IOL, Ms. Eur. D. 75, vol. 2, book 4, fol. 103.
109. B. Hart, W. Turton and B. White (eds.), *Agrarian Transformation: Local Processes and the State in South East Asia*, Berkeley, 1989, p. 256.
110. For a discussion see F.V. Aguilar Jr., 'The Philippine Peasant as Capitalist: Beyond the Categories of Ideal-Typical Capitalism', *JPS*, vol. 17, no. 1, October 1989. For the Indian context see the influential essay of Amit Bhaduri, 'A Study of Agricultural Backwardness under Semi-Feudalism', *Economic Journal*, March 1977.
111. 'Indigenous Social Formation and the "World System": North India since *ca*. 1700', in S. Bose (ed.), *South Asia and World Capitalism*, p. 121.
112. Rajat Ray and Ratnalekha Ray, 'Zamindars and Jotedars'.
113. Sugata Bose, *Peasant Labour and Colonial Capital*, pp. 93-5.
114. For a good but reiterative discussion in this perspective, see R. Pearce, 'Sharecropping: Towards a Marxist View', *JPS*, vol. 10, nos. 2 and 3, January/April 1983.
115. F. Husken, 'Landlords, Sharecroppers and Agricultural Labourers: Changing Labour Relations in Rural Java', *Journal of Contemporary Asia*, vol. 9, no. 2, 1979, p. 147.

116. Caballero has in fact convincingly argued that sharecropping may in fact be a more effective way of utilizing labour than pure wage-labour because the latter has a greater propensity of lowering profits through acts of shirking and wastage by the workers. Under sharecropping, shirking lowers the output which defrauds the sharecropper to the same extent as it does to the landowner. Under conditions of wage labour, where the worker has no share in the output, shirking does not necessarily lead to a contradiction in wages but raises the costs of supervision thereby raising labour and product costs. This may lead to a reduction in labour hiring. It is therefore entirely feasible that the 'wage system may finally result in less-effective labour actually being put in than under sharecropping' ('Sharecropping as an Effective System', *JPS*, vol. 10, no. 2, January/April 1983, p. 108).
117. BRC, P/52/10, 12 May 1790; IOR, BRP, P/71/22, 20 March 1790.
118. IOR, BRC, P/52/40, 15 December 1792; emphasis added.
119. IOR, SCC, P/A/9, 16 August 1769; F. Buchanan, 'Account of Ronggopur' IOL, Ms. Eur. D. 75, vol. 2, book 4, fol. 109.
120. IOR, BRC, P/52/40, 13 January 1792.
121. Ibid., P/51/21, 20 May 1788.
122. IOR, BRP, P/71/26, from Henry Colebrooke in Purnea to Board of Revenue, 26 April 1790.
123. In Rangpur the 'head ryot give as much grain as may be necessary for the *perja* [sharecropper] for seed, and receive from him double the quantity when the crop is cut' (IOR, BRC, P/52/10, 20 March 1790).
124. IOR, BRP, P/71/26, 18 June 1790.
125. IOR, BRC, P/52/40, 13 January 1792.
126. Ibid., P/51/6, 19 February 1790.
127. IOR, SCC, P/A/9, 16 August 1769.
128. IOR, BRP, P/72/4, 6 July 1792; emphasis added.
129. Ibid., P/70/33, 13 September 1787.
130. F. Buchanan, 'Statistical Tables of Ronggoppur', IOL, Ms. Eur. G.11, Table 30; Martin, *Eastern India*, vol. 3, p. 906.
131. 1 *kani* of land = 8 *bighas*, vide., WBSA, PCR, Dacca, vol. 15, p. 360.
132. IOR, BRC, P/51/40, 3 April 1789; emphasis added.
133. IOR, BRP, P/72/22, 22 March 1789.
134. IOR, BRC, P/52/10, 12 May 1790.
135. 'An Account of a Journey Through the Provinces of Chittagong and Tiperrah, 1798', BM, Add. Ms. 19286, fol. 4.
136. IOR, BRC, P/51/21, 20 May 1788; emphasis added.
137. F. Buchanan, 'Account of Ronggoppur', IOL, Mss. Eur. D. 75, fol. 112.
138. For a recent reiteration of this old view, see W. van Schendel, *Three Deltas: Accumulation and Poverty in Rural Burma, Bengal and South India*, New Delhi, 1991.
139. A. Cooper, 'Sharecroppers and Landlords in Bengal, 1930-1950: The

Dependency Web and its Implications', *JPS*, vol. 10, nos. 2 and 3, January/April 1983, p. 231.
140. Compare T. J. Byers, 'Historical Perspectives on Sharecropping', *JPS*, vol. 10, nos. 2 and 3, January/April 1983; Yeh-Chien Wang, 'The Secular Trend of Prices during the Ch'ing Period (1644-1911), *Journal of the Institute of Chinese Studies (Hong Kong)*, vol. 5, no. 2, 1972, p. 231; James Lee, 'Food Supply and Population Growth in Southwest China, 1250-1850', *Journal of Asian Studies*, vol. 41, no. 4, August 1982, pp. 739ff.
141. Ray, *Agrarian Change*, p. 64; S. Taniguchi, 'The Peasantry of Northern Bengal in the Late Eighteenth Century', in Peter Robb, Kaoru Sugihara and Haruka Yanagisawa (eds.), *Local Agrarian Society in Colonial India: Japanese Perspectives*, Delhi, 1997.
142. IOR, BRC, P/52/10, 20 March 1790.
143. M. Martin, *Eastern India*, vol. 3, p. 906.
144. See IOR, BRP, P/70/42, 30 May 1788; WBSA, Grain, vol.1, 15 October 1794; BDR, Dinajpur, vol. 2, no. 367, p. 231.
145. S. Taniguchi, 'The Peasantry of Northern Bengal in Late Eighteenth Century',
146. Ibid., p. 152.
147. Ibid., p. 154.
148. Ibid., pp. 172, 194.
149. I base myself on Harrington's report as appended to IOR, BRC, P/42/10, 20 March 1790.
150. Taniguchi. 'The Peasantry of Northern Bengal', p. 154.
151. Ibid., p. 150.
152. J.H. Harrington, 'Remarks on the Several Collectorships in Bengal in the Year 1788 and 1789', IOR, HM, vol. 385, pp. 291-2.
153. IOR, BRC, P/51/50, 4 September 1789, from D.H. MacDowall, Collector; ibid., P/52/10, from J.H. Harrington, Commissioner, 12 May 1790.
154. F. Buchanan, 'Survey of Ronggoppur', IOL, Ms. Eur. D. 75, vol. 2, book 4, fols. 97, 102, 111; emphasis added.
155. Vide, IOR, BRC, P/49/60, 28 February 1776.
156. IOR, BRC, P/49/65, 12 November 1776.
157. 'The Peasantry of Northern Bengal', p. 150.
158. IOR, BRC, P/53/19, 15 August 1794.
159. F. Buchanan, 'Survey of Ronggoppur', IOL, Ms. Eur. D. 75, vol. 2, book 4, fols. 101-3.
160. IOR, BRC, P/51/22, 23 July 1788.
161. Ibid., P/51/40, 15 July 1789; emphasis added.
162. Ray and Ray, 'Zamindars and Jotedars', p. 84.
163. Beveridge, *Bakarganj*, p. 197.

164. IOR, BRP, P/71/14, 24 July 1789.
165. Sanyal, *Social Mobility in Bengal*, Calcutta, 1981; also L.S.S. O'Malley, *Bengal District Gazetteers: Midnapur*, Calcutta, 1911, pp. 58, 60; also see Lal Behari Day, *Bengal Peasant Life* (ed. M.P. Saha, Calcutta, 1969) for an account of intra-caste stratification among the Sadgop peasants in Burdwan in the late nineteenth century.
166. Ray, *Agrarian Change*, p. 56; also, A. Nagchaudhury-Zilly, *The Vagrant Peasant: Agrarian Distress and Desertion in Bengal, 1770 to 1830*, Weisbaden, 1982.
167. See the descriptions of that famine in IOR, SCC, P/A/10; also Chapter 5. The famines and food shortages which struck the province after 1770 did not have demographic consequences as severe as that of the former year.
168. I have extended the distinction between the poor and middle peasants as provided by Bernstein ('Concepts for Analysis', p. 431) in order to classify these peasants in eighteenth century Bengal.
169. These were peasants who could afford to cultivate crops like mulberry, hemp and tobacco along with high quality rice 'so that in case of failure of one, from casualties of weather, the whole together may yield a profit' (Correspondence & ca. Relative to the Cultivation of Hemp and Flax in Bengal from 1791 to 1799, HM, vol. 375, 21 December 1792, p. 318).
170. Rajat Datta, 'Merchants and Peasants', p. 393.
171. 'The Peasantry of Northern Bengal', p. 37.
172. *FR 2*, pp. 56-7.
173. Incidentally even Henry Colebrooke's assessment was heavily influenced by the situation in Purnea having served there as the collector in the 1780s.
174. By the mid-1770s Purnea was said to be an 'impoverished country' with a 'lowered value of lands and their produce'. The corresponding decline of Rajshahi was 'too evident a melancholy truth' (cf. Chapter 1).
175. According to the findings of the Amini Commission (1788), *pateet* lands constituted 35.96 per cent of cultivable land in Birbhum and 25.99 per cent in Purnea. These were the two districts with the biggest areas of cultivable wastes when compared with the rest of the province (for details, see Table 56).
176. BM, Add. Ms. 19286, fol. 19. The biggest reclaimer in Chittagong was Gocul Ghosal, a prominent salt merchant-cum-contractor, who enjoyed the status of a *no-abadi* zamindar of Joynagar (IOR, BRC, P/49/41, 17 August 1773).
177. IOR, BRP, P/71/25, 28 May 1790; emphasis added.
178. Henry Colebrooke describes three types of peasants in Bengal: (i) those 'applying the labour which they give to husbandry, solely to ground used on their account', (ii) those who were 'monopolizing land to re-let it to the actual cultivator at an advanced rent, or for half the produce', and

(iii) those who 'call in the assistance of hired labour to assist their own' (*Remarks*, p. 60).
179. Ray, *Agrarian Change*, p. 53.
180. *Remarks*, p. 60.
181. Even those peasants who had some surplus resources did not employ hired agricultural labour. For instance, cotton was grown by 'ryotts who employ no labour but manage the cultivation themselves' (IOR, BRP, P/71/10, 15 June 1789).
182. IOR, BRP, P/71/17, 21 December 1789; also ibid., P/71/10, 15 June 1789.
183. See for instance the proceedings in IOR, BRP, P/71/10 and P/71/17 for descriptions of spinning.
184. Ray, *Agrarian Change*, Chapter 2.
185. Sugata Bose, *Peasant Labour and Colonial Capital*, p. 74.
186. Ibid., pp. 79-97.
187. Ibid., p. 79.
188. Add. Ms. 29086.
189. Harbans Mukhia (tr.) in *Perspectives on Medieval History*, Delhi, 1993, p. 269.
190. Bernstein, 'Concepts for Analysis', p. 431.
191. IOR, BRP, P/71/10, 15 June 1789; emphasis added.
192. IOR, BRC, P/51/50, 4 September 1789.

CHAPTER 3

The Rural Elite: Landed Property and the Landed Gentry

THE RURAL ELITE were those who enjoyed an intrinsically superior status in rural society by virtue of the lands personally held by them and by the power vested in them by the state to collect revenue, a part of which they themselves retained. As will be discussed subsequently, these financial perquisites comprised the major proportion of their income which were supplemented by the sale of produce from their personal land in markets, which were often established by them. Therefore, what these proprietors owned were assets, both moveable or fixed, which could be used to provide them with an income, or, if necessary, could be sold for a lump sum. The latter occurred when these proprietors sold their rights of taxation and their personal land, and the Company's testimony indicates that such sales had increased substantially after 1765.[1] They also possessed the right to claim shares in the economic output, that is, over the surplus produced by the peasant or the artisan.[2]

Nevertheless, not all of them enjoyed the right to levy or collect taxes on behalf of the state. Our sources indicate the widespread existence of a social category who held fairly extensive lands called the *la-kharaji* or *baz-i-zamin*. These lands were donated by other landed proprietors or the state on a permanent and hereditary basis. Since these lands were free of all revenue assessments, they provided the incumbents with a secure, and often lucrative sinecures. Additionally, such lands were often sold, thereby making them a distinct form of landed property. These were the people who, Eaton says, constituted a 'religious gentry combining piety with land tenure' in many parts of eastern Bengal, especially in Bakarganj and in Chittagong.[3]

A discussion of the landed elite necessarily entails sifting through the often confusing welter of rights and perquisites over land and its produce which constituted the essence of landed property on which such elites were based. The fact that revenue-collecting structures and property rights were combined almost inseparably adds to the problems, as it did for the early Company administrators, who were also under immense pressures to formulate a workable revenue regime after the initially disastrous revenue experiments and the famine of 1769-70. Naturally opinions differed, and the question of landed property became the source of one of the most enduring, and acrimonious, debates of that time.

To a great extent, a study of such intersecting and overlapping rights is bound to get involved in making sense of various revenue terminologies. The attempt necessarily, therefore, is to cut through the plethora of such terms in order to see the correspondence, if any, between the literal meaning of a term and its social practice. In other words, being an elite in rural society was not merely the function of a label—zamindar, talluqdar, etc.—assigned to a person in revenue literature. An excessive preoccupation with such literal assignations is a major limitation of what Peter Robb has called the 'old revenue histories [whose] worst feature was an uncritical listing of regulations without regard to enforcement and of categories without reference to actuality'.[4] Older revenue histories also insisted on a major separation between management of revenue and management of production. One aim of this chapter will be to demonstrate the major fallacies of such a separation.

An alternative perspective is therefore to see the zamindar, talluqdar and the *la-kharaj* holder collectively constituting a landed gentry in Bengal. Prior to the Permanent Settlement, these people operated under extremely fluid circumstances given the shifting nodes of political patronage under the Nizamat[5] and the speed of the political transition after 1757, the rapidity of the expanding agrarian frontier and the dynamism of a growing cash-nexus which agrarian expansion and European commerce had activated.[6] In such a situation, landed elites had a closer interface with the economic processes, not merely as appropriators but as agrarian patrons, as managers of agricultural production and as actively in the establishment of rural markets where the products of their personal farms were sold, sometimes by precluding competition, to visiting merchants. These market places often became points of conflict bewteen neighbouring magnates.

The Zamindari in Bengal: A Brief Overview

A zamindar, stated the Amini Report in 1778: 'Whatsoever right his tenure or office may convey is the Superior of a District of which, unless his authority is suspended, he collects the Rent, and for which he pays a Revenue to Government. He is the First in point amongst the several Landholders.'[7]

The superiority of the zamindari was partly because of the factor of caste. Most zamindars were Kayasthas, Brahmins and Kshatriyas; but caste alone is inadequate in explaining the social prestige of the zamindars. So-called 'lower caste' Kaivarta, Bagdi, Tili and Sadgop zamindaris co-existed, and expanded, during the course of the century.[8] A trading caste (Khatri) from northern India was able to establish the Burdwan Raj, one of the largest of the eighteenth century zamindaris,[9] while that of Birbhum belonged to a Muslim family who founded the zamindari by dispossessing the previous Rajput family some time in the late seventeenth century.[10] In general, the larger zamindaris belonged to the upper-caste Hindus, while the smaller ones were in the hands of the lower castes. With the exception of the Pathan family of Birbhum, most Muslim zamindaris were of a much smaller scale.[11]

The suggestion that the term zamindar in pre-British Bengal denoted a revenue collector and incorporated the semi-independent potentates (including the *bara bhuinyas*), who had usurped large territories during the period of transition from Afghan to Mughal rule,[12] provides only a partial explanation of the formation and expansion of such rights in the seventeenth and eighteenth centuries. Usurpation of territories was only one of the many ways in which these landed-properties evolved. The histories of some of the larger zamindaris compiled by the Company in the 1780s show that collaboration with the state in matters pertaining to revenue settlements or the subjugation of local rebellions provided the initial opening into the world of territorial acquisitions.[13] It is perhaps significant that lands usurped by individuals and later added to their zamindaris by payments of *nazrana* (gifts) to the state were in theory only to be held as long as the state willed; they could be resumed if a rightful heir could make a just claim.[14] Therefore, neither usurpation nor collaboration are adequate enough to explain the establishment of an elaborate system of landed property encapsulated under the generic term zamindari, a form of property which came to prevail all over the province in a variety of sizes.

The assertion that the zamindars were essentially revenue-administrators who had no control over land or its operation, and who had virtually no role in agricultural production[15] is difficult to sustain, at least for the eighteenth century. There is evidence (discussed later) to show that their participation in agricultural production was closer than what has been proposed by historians. Moreover, the manner in which such properties were formed does show a fairly close connection with the agricultural economy, particularly with the extension of cultivated land: a point which needs elucidation. According to Rai Rayan Rajvallabh: 'Zemmindarries are of various kinds, some are obtained by inheritance, some by clearing the country of wood, some by the ejection of the possessor for ill-behaviour, some by purchase and some in trust.'[16]

This description, though rather cryptic, does indicate that the formation of zamindaris in Mughal and Nawabi Bengal was by a combination of state action and individual initiative. One of the earliest English writers to focus on this combination was Boughton-Rouse who, in 1791, saw the origins of *milkiyat* (right of property) in Bengal in three sources: (*a*) *jangal-bari* (clearance of waste-land for cultivation), (*b*) *inteqali* (transfer of right) and (*c*) *a-hukami* (by order of authority). *A-hukami* operated when the state removed a zamindar and settled another in his place; but the other two sources clearly indicate the connection between agricultural reclamation and the creation of zamindari. *Jangal-bari* referred to those zamindaris created by reclaiming land for the first time, whereas *inteqali* occurred where: 'The land is in a good state of cultivation, and productive to the amount of revenue; yet, on account of the neglect of the incumbent, or for want of heirs to the land, another person has with the permission of the Emperor, or the Government delegated by him, obtained a sunnud [order] for his office in his own name.'[17]

Muhammad Reza Khan also made a clear distinction between zamindaris created by *jangal-bari* or purchase (*kharida*), and those who had been established by the state whom he called the *sanadi* zamindars. Included in these *sanadi* zamindars were also people who had been empowered by the ruler to bring spots of waste land into cultivation. These were not *jangal-bari* zamindars in the strict sense, but functioned on a similar footing. He further added that the relationship between the state and the zamindar varied according to the specific nature of each one's origin. *Jangal-bari* and *kharida* zamindars enjoyed permanent and inviolable rights of property, and the state had no further right 'other than receiving rents'. *Sanadi* zamindars were subject to closer state control; some of them were more in the nature of temporary offices to be bestowed, and revoked, by the rulers at their discretion.[18] These descriptions by experienced revenue officials and John Shore's opinion (in 1788) that 'most of the considerable zemindars in Bengal may be traced within the last century and a half' and that 'the extent of their jurisdictions has been considerably augmented during the time of Jaffier [Murshid Quli] Khan and since through a mixed process of 'purchases, acquisitions and confiscations'[19] indicate that a zamindari in pre-British Bengal could emerge from a number of directions, not all of them were related to the will of the state. What appears crucial is the creation of zamindaris from below when individuals became zamindars either by reclaiming or by purchasing land: in the former case what obviously occurred was the extension of agricultural land and in the latter situation one can perhaps speak of the creation of an extensive land-market in the province.

These zamindars from below constituted, what James Grant called, 'small, or single pergunnah zemmindarries [in] districts and petty mahls, dispersed throughout Bengal' whose numbers had 'so prodigiously increased' since the administration of Murshid Quli Khan, and whose composition was constantly changing 'in denomination, extent or possessory rights [so] as to become now [in 1789] a work of considerable labour to trace their revolutional [*sic*] progress'.[20] A reason for this growth appears to lie in the concessions given by the state to such proprietors to manage their internal affairs with very little interference.

One of the most important devices of political intervention adopted by the Mughals was that the zamindars were obliged to acquire the state's sanction in cases of inheritance and alienation of the zamindari by division or sale. These sanctions were apparently strictly enforced in the case of the 'superior zemindars' (such as those of Burdwan, Nadia, Rajshahi and Dinajpur), who, in case of inheritance or succession, were 'accustomed to receive on payment of nuzerranah, paishkash & ca. a Dewanny Sunnud from Government', whereas 'the zemindars of a middle and inferior rank . . . and the talookdars . . . at large have held their lands to this day by virtue of inheritance alone'.[21] Such non-interference allowed the evolution of a whole range of such zamindaris. Some, like the chaudhuris of Jalasor, sub-divided the pargana into a number of portions by a process of internal fragmentation. Each chaudhuri enjoyed a share in the *rasum* (perquisites) allowed to the landed proprietors and managed the affairs, including the payment of revenue, in their separate portions.[22] Others, like the ones in Jalalpur (in eastern Bengal) held zamindaris which were 'not contiguous to each other but intersected and divided by other districts'; though the state realized that such 'separation makes it extremely difficult for one man to superintend the collection of the whole [revenue]',[23] there was very little done to prevent such spatially dispersed rights from developing in pre-British Bengal.

The zamindari was the dominant form of landed property, but as a social organization it was remarkably stratified. At the apex were those who were, in the parlance of the East India Company, the 'principal' or 'capital' zamindars. These were the zamindars who had been able to carve out near-principalities and ruled over extensive territories. Positioned below them in the social hierarchy were the smaller zamindars with territorial jurisdictions over a few parganas and even over a few villages. Districts such as Midnapur and Chittagong had 1,500 to 3,000 zamindars in the 1780s.[24] This stratification notwithstanding, the zamindars were the most powerful social class in the countryside, and also the most affluent. This affluence arose from their rights to tax and, as I will discuss, from their participation in agricultural production, which made these people much more than mere state functionaries. They also had a privileged social standing based on a combination of caste, prescriptive status and local power. In fact the social prestige of a zamindar in Bengal was measured by an inseparable mix of territorial jurisdiction, amount of the assessed revenue (*jama*) and by the extent of patronage which an individual zamindar could provide to the local temples, mosques and other religious orders who sought his help.[25] This latter function not only bestowed upon these zamindars a great degree of local prestige and power, it also allowed the formation of another kind of landed property, the *la-kharaj*, whose nature and social implications will be discussed subsequently.

The Talluqdari: *Evolution and Nature in the Eighteenth Century*

Positioned below the zamindars in the hierarchy of landed property were the talluqdars. In the revenue terminology of the eighteenth century, a talluqdar, in the words of John Shore, 'literally means a holder or possessor of a dependency'[26] of two major types: the *huzuri* (or *khas*) and *mazkuri*. The *huzuri talluqdars* held their rights under, and paid revenue, directly to the state, whereas the *mazkuri* 'hold their tenures under a zemindar or chowdhuri to whom they pay their rents'.[27] In the complex networks of landed-rights the talluqdar was recognized as one specific form, but ascribed an inferior social position relative to the zamindari. Its recognition as a specific landed property is in evidence from Muhammad Reza Khan's statement that both in zamindari and talluqdari 'laws of inheritance, property, sale, purchase and donation have been observed in this country from the dawn of civilisation', but on the other hand, its relatively inferior status can also be seen from his note that 'a big zemindar is called a Raja, while petty zemindars are called zemindars and chaudhuris [and] in a zemindarri there are several talookdars'.[28] Reza Khan's description is also corroborated by the one provided by Rai Rayan Rajvallabh for whom talluqdars: 'In regard to the rights of property and inheritance are the same [as the zamindars], but there is a difference [between the two] in point of revenues, dignities & privileges arising from a difference in extent of territory.'[29]

In general, a talluqdar was recognized as 'inferior in rank and title' to the zamindar. Nevertheless it was a distinct form of property and the proprietors possessed 'their little territories on a tenure full as secure [as the zamindari], and at a revenue generally more fixed'.[30]

Talluqdari differed from the zamindari on four specific aspects. First, as has been noted, there was the basic difference between social standing of these rights: the zamindari being considered the superior right and the talluqdari the subordinate one. Second, the act of zamindari lay in the payment of revenue (*malguzari*) to the state which included the payment of talluqdari revenue within their jurisdictions. In other words, a zamindar was at the *khidmat* of the state, whereas the talluqdar had no such service obligation.[31] Even *huzuri* talluqas, which in theory were entitled to pay directly to the state, were, nevertheless, placed by it under a larger zamindar who collected and paid revenue on their behalf. This was considered an administratively convenient arrangement and was in vogue, for instance, in the *chakla* (revenue circle) of Murshidabad.[32] Third, as the *milkiyat* of a zamindar in Bengal could grow out of three major sources: *jangal-bari*, *inteqali* and *a-hukami*, but the talluqdari right was primarily limited to *jangal-bari*. Finally, a zamindari was an inherited right which was, at least in theory, dependent upon the sanction of the state, whereas talluqdari arose by purchases or grants. One would consider this to be a crucial difference between these two rights for reasons which will be enumerated shortly. When purchased, a talluqa was known as *kharida* and those formed by grants, made either by the state or the zamindar, were known as *pattai*.[33]

The crucial fact about talluqdari was that it was essentially purchased, either from the state or from the zamindar,[34] or granted by either of these two agencies on some specific considerations.[35] Regarding the formation of *huzuri* talluqas John Shore expressed the opinion that these 'appear to have been originally portions of Zemindarries, sold or given by the Zemindars; and to have been separated from their jurisdiction, either with their consent, or by the interest of the Talookdar with the Governing power'.[36] What John Shore describes seem to be talluqas which were initially *mazkuri* and then converted into *huzuri* by the authorities; and this may have been one way in which such talluqas were formed during Mughal Bengal. But this could not have been the most prevalent form for two reasons. In the first place, it was widely recognized that the zamindars resisted any act of converting *mazkuri* in to *huzuri* (or *khas*) talluqas, and sometimes this resistance could turn violent as, for example in Dhaka, where a *mazkuri* talluqa could not be 'released' (i.e. made *huzuri*) without 'proving the violence of the Zemindars'.[37] Zamindari opposition was natural as these talluqas were important in the former's financial structure, first, because they were generally purchased for ready cash, and, second, as noted by the Amini Commission: 'A talook comprehends only a few villages, or a small tract of ground, and the possessor is able to attend to the cultivation of every part of it. It improves by his care; the rents of it increases & it becomes populous and valuable', which meant that a zamindar could look upon it as a steady source of income, even if it sometimes amounted to plundering the resources of these talluqdars.[38] The fear that 'the separation of the talookdars would reduce many of the [zamindars] from affluent circumstances, to a state of indigence, and the titles of Rajah & Zemindar, which they enjoy, will become mere empty names'[39] provided the crux of their opposition to all attempts to convert *mazkuri* into *huzuri* talluqas: a fact which must indicate the importance of these talluqas in zamindari finances.

The second reason why the pattern described by Shore cannot be ubiquitously applied to the formation of *huzuri* talluqas in Bengal stems from the fact that the kind of talluqdars, Shore was describing, formed a tiny portion of the social category called *huzuri* talluqdars. Shore's description pertains only to those people who were given such rights (also called *khas*) by virtue of being 'some favourite or underling of Government'.[40] These grants were limited to the main administrative centres, they 'comprised small mehauls' [*mahls*], and were 'generally given to the dependents of the court'.[41] According to James Grant such people

> Were all rich, or favourite individuals in the neighbourhood of the principal Mussullman Capitals, who having obtained small territorial grants... were then to be rated at a fixed annual assessment, subject to no future increase; and as they had probably made some pecuniary compensation by way of purchase of possession, so with the privilege of being exempted from zemmindarry jurisdictions.[42]

The bulk of the evidence pertaining to *huzuri* talluqas indicates that these were formed either by the government with specific administrative and financial considerations in mind, or that they arose from the state's attempt to incorporate a number of 'ancient' talluqdars (i.e. talluqas which were older than the established zamindaris in certain areas),[43] and who were subsequently brought under the direct administration in order to facilitate the collection of revenue.

One of the best examples of the creation of *huzur* talluqa by incorporating such 'ancient proprietors' comes from the *Naibat* of Dhaka which was subdivided, at an early stage of Mughal rule, into a number of small talluqas given by the subahdars (provincial governors) to persons employed in defending the frontiers of the province. The greater number of these were given for the maintenance of the flotilla (*nowarrah*). These lands were then specified in the *jama-i-tumar* (revenue-roll) of Todar Mall in 1588, and the holders were confirmed as proprietors and afterwards made *huzuri* by the Nawabs of Bengal.[44] In Mymensing, *jangal-bari* talluqdars were established much before the establishment of Mughal rule in Bengal; these people acquired lands from 'the Bhuyas [*bara bhuinyas*], or before the (Mughal) King got possession of Bengal' and subsequently became *huzuri*.[45] Elsewhere these *huzuri* talluqdars were placed under the supervision of the local zamindar, but the state made sure of placing a number of restrictions on the latter's power. Thus in Bakarganj the connections between the state, zamindars and talluqdars were shaped in the following fashion:

Previous to the division of the country into Pergunnahs & Tuppas & fixing what is called the Tuxeembundy, many persons undertook to cultivate jungle and waste lands; and when the Tuxeembundy was made, these new lands [called *jangal-bari*] were constituted Talooks & included in the Jummabundy [revenue-roll] of the nearest Zemindar by the Government of that time & if any increase or remission was granted [to] the Zemindar, a proportional part fell to the Talookdar. If the Zemindar withheld from the Talookdar any part of this, he was at liberty to complain to the Government who compelled the Zemindar to allow the Talookdar his proportion of the remission. If the Talookdar died leaving heirs, they got possession of the lands in the same manner as their predecessor & the Zemindar had nothing to do with them but receive his Malguzary [revenue] agreeably to Kistbundy [instalments]; but if there happened to be no heirs, the Zemindar was the manager on behalf of the Government.[46]

Similar strategies were adopted by the state in western Bengal too. A statement from Birbhum (an extensive zamindari) mentions three *huzur* talluqas. First, there were 'talookdars whose talooks were formed before the zemindar, to whom they now pay their rents, or his ancestors succeeded to the zemindarry'; second, 'talookdars the lands comprised in whose talooks were never the property of the zemindar, to whom they now pay their rents, or his ancestors'; and third were 'talookdars who held land under a special grant from government'.[47] In Midnapur, where talluqdari rights had developed extensively in the course of the seventeenth and eighteenth centuries, *huzuri* talluqdars were 'the purchasers of small portions of lands. They are clearly independent and answer the native terms of Khauridge [free] Talookdars. The same independence as they now possess [in 1792], they possessed previous to the Company's accession of the Dewanny'; these talluqdars dealt directly with government officials, called tehsildars,[48] who were 'stationed to collect the Revenue, recorded them as separate renters [*sic*] and fixed separate assessments for their talooks, judging them more responsible for their revenue'.[49]

These instances show that *huzuri* talluqa rights were sanctioned by the state because of a number of considerations, these being: administrative convenience (in Dhaka), easier collection of revenue (in Birbhum) and extension of cultivated land (in Bakarganj). The case of Bakarganj also shows that great care was taken, or sought to be taken, by the government in order to protect the rights of these talluqdars from the zamindars; and this was apparently motivated, additionally, by the official objective of recognizing the talluqdari as a form of landed property completely distinct from the zamindari though occupying a social position inferior to the latter in the landed-hierarchy. The Amini Commission gives the assessed revenue of some *huzuri* talluqas for the year 1778 which is shown in Table 13.

TABLE 13. *HUZURI* TALLUQAS IN BENGAL, 1778

Zamindari	*Jama* (Rs.)
Rajshahi	4983.5
Laskarpur	4495.31
Chandeli	156.37
Cadir Ganj	7.62
Chunakali	629
Curgaon	42.12
Rokunpur	2450
Mahmud Shahi	1285.56
Muscoories	46
Guznainpur	38.19

Source: BM, Add. Ms. 29088, fol. 163.

The other type of talluqa, the *mazkuri*, existed in one form or another in all areas of the province. Muhammad Reza Khan described the formation of *mazkuri* talluqa in the following words:

A zemindar procures a sanad from the sovereign for waste and uncultivated land, which is not included in the jummah and of which there is no owner. He then grants pottas of talookdari to several persons who exert themselves in improving the condition of the land, spend money over it and bring it under cultivation. They pay revenue to the zemindar who is entitled to receive it. Another way is when a zemindar sells . . . a village in his zemindarri. [Then] the purchaser becomes the proprietor of the lands and enjoys the privileges of selling and donating them. The zemindar and the sovereign are to receive the revenue only.[50]

Our sources mention four ways in which *mazkuri* rights could be formed within any particular zamindari: (i) persons could take *jangal-bari pattas*, and after successfully cultivating these lands, they assumed talluqdari rights; (ii) others purchased waste lands from the zamindars and were then recognized as talluqdars; (iii) others acquired cultivated lands directly by purchase as talluqas; and (iv) some procured, or purchased, *la-kharaj* lands and agreed to pay a stipulated revenue to the zamindar.[51] Examples of such talluqas in different areas indicate that these were generally created by the zamindars either to bring about an extension of cultivation, or to acquire ready cash by selling portions of their zamindari to those who could afford, or were willing to buy. Bakarganj had *pattai* and *kharida* talluqdars;[52] similar rights existed in Jessore.[53] In Birbhum, *mazkuri* talluqdars were those who 'purchased their lands by private or publick [*sic*] sale, or who obtained them by grant from the zemindar to whom they pay their rents'.[54] In Midnapur, 'many have obtained talooks, either in former [i.e. Mughal] times by private purchase or in these latter times by public purchase by auction'.[55]

Pattai talluqas were specifically established by the zamindars to bring about an extension of cultivation. Such talluqdars were apparently responsible for 'improving the condition of the land' by spending money and bringing it under cultivation.[56] While doing so, the talluqas were held free of revenue for the first two or three years and afterwards made to pay their share of the 'general rent' of the pargana, along with certain taxes fixed on a customary basis (*nirkh-i-dastur*) payable to the zamindar; these talluqas were inheritable by the family of the holder, but could not be sold, or alienated, without the consent of the zamindar, who also had the first right of rejection.[57] In Bakarganj, for instance, these talluqdars were not meant to pay revenue for the first two or three years, but after that time there was to be a progressive increase (*rasad*) until the stipulated rate of revenue (*pura dastur*) was obtained—the entire process usually took about five years.[58] In Dhaka, a zamindar would take a contract (*ta'ahud*) from a talluqdar for a stipulated sum of money to be paid after five years on a progressive rate of increase.[59]

When *mazkuri* talluqas were created by sale of zamindaris they closely approximated landed property of a private kind since these talluqdars were empowered by the act of purchase to sell, alienate and mortgage their properties without any restrictions.[60] They were also entitled to fragment their rights among members of the family. In this event, and if the talluqa happened to be held by the collateral branches of a family, the descendants of the 'original founder' was to be given an extra share.[61] In Birbhum, talluqdars of this type could be found 'succeeding' to others' talluqas 'by right of purchase, gift or inheritance',[62] and in Bakarganj, these people, called *zer-kharid* talluqdars, were at 'liberty to sell with or without the permission of the zemindar'.[63] Similarly, in Mymensing such 'kharid-farokht' talluqdars, were entitled to 'sell the whole or any part of the talook without the approbation of the zemindar'.[64]

In fact, the sale of these rights had created certain sub-talluqdars in the *Naibat* of Dhaka. In Bakarganj we come across the *ausat*[65] talluqdars, who were described as a 'talookdar within a talookdar'.[66] In pargana Bikrampur, we are told of the existence of *howaladars* (holders of property by transfer) in the following description:

In the pergunnah of Bekrampore a custom prevails that if any talookdar sells any part of his talook to another person, upon receiving the purchase price he calls him a Howalladar of so much land, who pays his rents to the talookdar; but if any dispute arises between the talookdar and the howalladar, he can get his howallah separated from his talookdar & included in some other talookdar's lands. [The howaladar] is subject to increase or decrease of Revenue along with the other renters [*sic*] & the property is hereditary & transferable.[67]

One cannot be certain whether sale of talluqas in other areas of the province created such specifically demarcated sub-talluqdaris as in Bakarganj and Bikrampur in eastern Bengal. In Midnapur, we are told, there existed a sub-category called the *gutch-amini* talluqdar who 'were taxed with the charge of insolvent talookdars and made responsible for the revenue, and the original proprietors remaining unable to answer their Revenue, they continued in other's hands'.[68] This category corresponds closely to the system followed in Dhaka where a number of small talluqdars were placed under a bigger one who collected and paid their revenue; this talluqdar was called *zimmadar*, his own talluqa was called *neez* (personal) and comprised the only real property belonging to him.[69] What we cannot be sure of is whether these intermediary rights were actually purchased.

Nevertheless, the fact that these *mazkuri* talluqas were created by purchase and then were saleable is indisputable. 'All talookdars' noted Shore in 1788, 'unless restricted by the terms of their grants under which they hold, have a right to dispose of their lands by sale, gift or otherwise . . . and indeed this practice prevails in opposition to the condition of their pottahs'.[70] A bill of sale had to be signed by the seller and witnessed by 'his partners, and by the zemindar [along with] the people of the neighbourhood [and then] attested by the seal of the Kazi [Qazi] and the signature of the Canongoes'.[71]

To the extent, however, that *mazkuri* was a purchased right, it implied that the zamindars were under no formal obligation to create such talluqas within their territories. The extent of such sales appears to have depended upon the capacity, or willingness, of the local people to purchase such talluqas, as well as upon the immediate financial requirements of a zamindar. In Midnapur, for instance, almost all talluqas were established by the sale of zamindari lands as the following description shows:

Zemindars were originally proprietors of the whole or some of the Mehals which are now Talookdarry. From extravagance, indolence or neglect of their lands, from badness of season or other causes, being unable to pay their Revenue, or support their extravagance, they were constrained to sell portions of their lands as they were pressed by either one or other of these necessities. In this manner sometimes the whole Mehal was disposed of.[72]

In general, there was a greater concentration of *mazkuri* talluqas in eastern Bengal than in the western part of the province. For instance, 29.47 per cent of the assessed revenue of the zamindari of Muhammadshahi, 12.51 per cent of that in the zamindari of Rajshahi and 8.59 per cent in Yusufpur were *mazkuri*; whereas in western Bengal it was 1.54 per cent of the *jama* in Burdwan and 1.07 per cent in Nadia.[73]

Financial considerations were also accompanied by the desire to bring about an extension in cultivated land. In the latter case, *mazkuri* talluqas were granted by *pattas*. In Muhammad Aminpur, *chakla* Hughli, there were 62 *mazkuri* talluqas in 1778.[74] A survey of the history of 34 such talluqdars in pargana Jalalpur (in 1791) revealed that 14 talluqas were created by purchases, whereas the remaining 20, were established by the zamindars through grants.[75]

Sale and purchase of such talluqas helped in creating a market for landed property in the province. John Shore's opinion in this regard was that should the price of the 'Lands paying revenue [be] less than double of the tuxeem jama, or if the revenue payable to Government as expressed in the deeds of sale, was less than what the lands were let for by the Zemindar in that year, the sales were at undervalue, if otherwise [then] not.'

Table 14 gives some sale prices expressed as a portion of the assessed revenue of some talluqas between 1724 and 1793.

Table 14 seems to go against Shore's assertion that he 'never [knew] of any instance in Bengall in which the price of revenue lands sold, out of Calcutta, exceeded the revenue of two years'.[76] Nor does the evidence make it possible to agree with B.B Chaudhuri's argument that land market in the eighteenth century, especially during the later years, was sluggish, even a depressed one because of (i) their low sale-values, (ii) these (especially the zamindaris) were sold out of distress because of

TABLE 14. SALE VALUES OF TALLUQAS IN BENGAL, 1724-93

Year	Place of sale	1 Assessed revenue (Rs.)	2 Sale price (Rs.)	3 Ratio of 3 to 2
1724	*Tappa* Sunderkool in Selimabad	189	700	1:3.7
1731	*Mauza* Golah Pargana Fatehjangpur	42.56	81	1:1.9
1737	*Mauza* Berole Pargana Mehlind	967	775	1:0.8
1740	*Mauza* Srepur Pargana Alapsing	178.87	22.37	1:0.12
1741	Chaudharai Fatehjungpur	1287	3301	1:2.56
1776	*Mauza* Ghosegaon Pargana Chunakali	1282	800	1:0.62
1778	*Mauza* Mahimpur	898	700	1:0.78
1779	*Mauza* Janmuhammad	1460	2100	1:1.44
1779	*Mauza* Ballaserah	2801	9600	1:3.43
1780	*Mauza* Chandpara	1065	2200	1:2.06
1786	villages in pargana Sundarpaye, chakla Goraghat	8717	11652	1:1.34
1793	5 villages in Jessore	6002.93	39885	1:6.64

Sources: BRC, IOR, P/51/18, 2 April 1788; WBSA, PCR, Murshidabad, vols. xv and xxiii, 12 Feb. 1778 and 17 Feb. 1780; IOR, BRC, P/50/67, 11 July 1786 and ibid., P/53/12, 21 February 1794.

the ruinous impact of the Company's revenue policy after 1765, and (iii) they were mainly mismanaged estates.[77] There must have been distress sales as also mismanagement of individual estates, but on the whole Table 14 suggests a buoyant land market in this period.

The Nature of La-kharaji Property

La-kharaj or charity lands were created by the Mughals, and later the Nawabs by: 'grants [of land] with certain privileges and exemptions annexed to the tenure, whether flowing from the favour of the Prince, or granted as a reward for services, or yielded as the condition upon which the original grantees engaged to reclaim or improve the lands; or finally as pious donations from the respective Princes to the Ministers of the established religion, or for other charitable purposes'.[78]

Once again, like the talluqdari, we find the need to extend the land under cultivation emerging as one of the major considerations behind the creation of *la-kharaj* rights in pre-British Bengal. The major difference between such rights and those of the zamindari or talluqdari was that lands held *la-kharaj* did not have to pay any revenue since they were initially given as charity or in lieu of service. Thus these grants embraced two social groups. The first were service-grants (called *chakeran*) which comprised territorial assignments made for meeting the expenses of servants and employees. The second category, called *baz-i-zamin*, signified all kinds of charitable support to 'religious edifices and colledges [*sic*]' since the support of such institutions was 'deemed sacred under the feudal Mogul Government'.[79] Such grants were also known as *kharij-jama* (revenue free), and in the revenue records this term is often used interchangeably with *la-kharaj*.

The pressing need to extend cultivation and its connection with the creation of various types of landed-properties is nicely borne out by the spread of *la-kharaj* lands in the frontier regions of the province in the seventeenth and eighteenth centuries. The reason why the best spots of land in Chittagong were held *la-kharaj* in 1788 was ascribed to the need to pacify that area after Aurangzeb defeated the Maghs in 1666. The upheaval caused by the Mughal invasion had caused 'persons having the means of subsistence to live elsewhere'; the district 'thus became the resource of needy persons who on applying to men in office under the Government found no difficulty in procuring charity grant for such lands as they were disposed to reduce to a state of cultivation'.[80]

Sylhet provides another example. Here the state's need was to prevent the 'incursions of the hill people' into Bengal, for which a *faujdar* was appointed from 'the Nabob's nearest relations, or confidential friends, to whom it was in fact a jaghier, and little more was expected by government than a few elephants, some chunam, oranges and birds of handsome plumage'. The task of settling such tracts of land was entrusted to the *faujdar* who, 'Being followed by a multitude of needy dependents was liberal in his gifts without expense to himself, and never failed to bestow upon them large tracts of land upon a charity tenure under his own seal'.[81]

As in the case of talluqdari, the zamindars were quick to follow the example set by the state. Large chunks of revenue-paying (*hasil*) and waste (*pateet*) land were given out by the zamindars and, to a lesser extent, by the talluqdars as *chakeran* and *baz-i-zamin* 'in the names of their friends, relations and dependents'.[82] Such grants were apparently made with the approval of the state before 1765.[83] For instance, the zamindars of Bishnupur were apparently assessed at a very small *peshkash* (tribute) under the Mughals on grounds of the zamindari 'being a frontier of the Province' which 'enabled [them] to lay by considerable sums of money from the revenues of the Country [Bishnupur], which for the major part appear to have been expended for religious purposes'.[84] The Company was deeply suspicious of all such grants on the grounds that these were detrimental to the state's financial interests since they did not pay revenue[85] and that zamindars: 'Have carried this practice [i.e. granting *la-kharaj*] so far as to have the malguzarry [revenue paying] land totally incapable of yielding the jummah and suffered the nominal zemindarry to be sold for the balance, retaining the bazee zemin, a profitable estate, for their support.'[86]

Table 15 shows the distribution of *baz-i-zamin* in some zamindars before and after 1765, for which comparable estimates are available from the assessment made on behalf of the Company.

The *baz-i-zamin* lands were meant for the upkeep of a whole range of dependants ranging from, as in Rajshahi, 'Bramins, Mohunts, Sannasies, Gungabassies [widows], Fakirs, Pirs & ca'[87] to (in Coch Behar) 'servants and muttassudies [revenue officials] . . . poor Bramins, Rajahs & the Kitmutgars [*khidmatgar*, or servant] & Bearers'.[88] Unfortunately, there is very little evidence to compare these lands to *raiyati* (land held independently by the peasants). Nevertheless, the evidence collected by J. Sherburne, the collector of Muhammadshahi in 1787 is extremely significant in this respect. A list of the registered *baz-i-zamin* is given in Table 16.[89]

TABLE 15. GRANTS OF *BAZ-I-ZAMIN* BY SOME ZAMINDARS, BEFORE AND AFTER 1765 (*in bighas*)

Zamindari	Before 1765	After 1765
Rangpur	2,45,760.35	83,880.80
24-Parganas	3,18,113.50	70,763.15
Nadia	7,43,144.25	62,283.35
Burdwan	12,629.00	7,184.90
Bishnupur	93,130.86	11,506.25
Muhammadshahi	2,04,653.75	12,516.30
Birbhum	91,927.80	17,933.60*

Note: *The figures for Birbhum are for the year 1770.
Sources: Rangpur: IOR, BRC, P/52/17, 6 August 1790; 24-Parganas: ibid., P/52/21, 1 December 1790; Nadia: IOR, BRP, P/71/12, 13 August 1789; Burdwan: IOR, BRC, P/52/21, 1 December 1789; Bishnupur: ibid., 1 December 1790; Muhammadshahi: IOR, BRP, P/71/13, 27 August 1789; Birbhum: FR, Murshidabad, IOR, G/27/1, 5 October 1770.

TABLE 16. *BAZ-I-ZAMIN* HOLDERS IN MUHAMMADSHAHI, 1787

Land (*bighas*)	Number of holders	Total land (*bighas*)
1 to 50	3,421	66,739.70
51 to 100	58	39,708.35
101 to 150	216	26,926.20
151 to 200	68	11,910.70
201 and upward	275	62,803.20

Source: IOR, BRP, P/70/24, 6 February 1787.

This list also provides a very rare comparison of *baz-i-zamin* with the raiyati or *malguzari* lands in four villages of Muhammadshahi in the fashion shown in Table 17.

Another estimate of such lands made in two parganas (Amberabad and Bhuluah) in eastern Bengal showed the distribution of land (Table 18).

Pargana Apole in the zamindari of Dinajpur had its lands divided in the manner shown in Table 19.

A statement regarding the distribution of lands in seven villages in Jalasore in 1792 contains some important informations which have been shown in Table 20.[90]

Land given as *chakeran* constituted the other component of these revenue-free properties. Table 21 gives a comparative picture of *chakeran* lands in some zamindaris for which data are available. It must be borne

TABLE 17. *BAZ-I-ZAMIN* AND *RAIYATI* LANDS IN MUHAMMADSHAHI, 1787 (in *bighas*)

Mauza	Total land	Raiyati	Baz-i-zamin
Jagkarna	318.26	73.50	244.76
Gopinatpur	378.86	38.90	339.96
Byraddey	515.06	56.65	458.41
Jaikispur	614.25	58.65	555.60

Source: IOR, BRP, P/70/24, 6 February 1787.

TABLE 18. DISTRIBUTION OF LAND IN AMERABAD AND BHULUAH, 1775 (in *bighas*)

Total land	4,08,504
Forests	2,408
Daria Shikast or carried away by rivers	23,808
For salt production	8,456
La-kharaj	55,352

Source: IOR, BRC, P/49/46, 3 November 1775.

TABLE 19. DISTRIBUTION OF LAND IN PARGANA APOLE, 1777 (in *bighas*)

Total land	1,12,375
Waste since 1770	20,276
Forests	1,401
La-kharaj	34,377
Cultivated	56,321

Source: Testimony of Sibprasad Tagore, chaudhuri of the pargana in IOR, BRC, P/50/1, 1 August 1777.

in mind that the data in this Table pertain to zamindaris after they had been quite extensively de-militarized by the Company's regime, which means that the extent of land alienated as *chakeran*, or the number of officials who held such lands, must have been far greater under the Nizamat.

It was a recognized fact that most zamindars had large establishments of officials and other servants, including a sizeable militia, who were maintained by grants of extensive tax-free lands in lieu of pay. These persons were:

The Zemindar's own particular servants, paid by him in land and not in specie.... Only a small part of them are at fixed stations, such as are appointed to guard and protect the Cutcherries and Gunges and those placed over the Zemindars Kummar [*khamar*] and his fruit trees. The far greater number are employed under the mandals [village headmen] and other officials in the mofussil [country] to collect Rents from the Ryotts, and in measuring lands and a variety of occasional duties. It is the business of some to carry orders from the Head Cutchery [zamindari courts] and circulate them throughout the country [i.e. the zamindari], whilst others escort treasure going from and coming from the Sudder [zamindari headquarters]. In fine they are intimately connected with every branch of the Revenues.[91]

The arrival of the Company meant a substantial reduction in the number of zamindari troopers,[92] but what vexed the Company greatly was that this de-militarization was not accompanied by a corresponding reduction in the quantity of land alienated as *chakeran* in various zamindaris. In fact one of the major suspicions harboured by Company

TABLE 20. DISTRIBUTION OF LAND IN JALASORE, 1792 (*in bighas*)

Total land	Baz-i-zamin	Pateet (waste) land	Raiyati land
6,150.75	859.90	1,109.95	4,180.90
1,433.50	165.25	76.00	1,192.25
7,977.70	1,178.90	2,170.70	4,628.10
2,007.40	305.40	22.70	1,679.30
2,320.45	334.90		1,985.50
1,123.50	211.65	40.05	871.80
1,628.35	185.01		1,443.34

Source: IOR, BRC, P/52/43, 20 April 1792.

TABLE 21. *CHAKERAN-ZAMIN* IN SOME ZAMINDARIS (*in bighas*)

Year	Zamindari	Total land	Militia	Officials
1766	Burdwan	3,58,516.00	2,39,238.00	1,19,278.00
1770	Muhammadshahi	1,54,629.35	29,873.60	1,24,755.75
1770	Bishnupur	1,55,681.96	96,987.90	58,694.06
1770	Birbhum	1,50,237.35	86,413.65	63,823.70
1770	Dinajpur	2,73,316.15	1,52,117.70	1,21,198.45

Sources: For Burdwan, WBSA, Proceedings of the Select Committee, vol. 2, 28 October 1766; for Muhammadshahi, Bishnupur and Dinajpur, WBSA, CCRM, Appendix to Proceedings, vols. 1 and 7; for Birbhum, FR, Murshidabad, IOR, G/27/1, 10 October 1770.

officials was that lands 'secreted' under this denomination had actually increased after 1765. Thus in the seven years between 1777 and 1783, 44,975 *bighas* 'of all most good land' were alienated as *kharij-jama* in 24-Parganas, whereas between 1769 and 1776 such grants had totalled 9,972 *bighas*.[93] In Rajshahi, 39,280 *bighas* had been added to the existing *chakeran* lands between 1778 and 1789; *baz-i-zamin* had increased by 23,205 *bighas* in the same period.[94] In Birbhum the additions to *chakeran* lands totalled 90,790 *bighas* between 1777 and 1787.[95] Between 1773 and 1787 the estimated value of lands alienated as *baz-i-zamin* in Midnapur and Jalasore had gone up substantially, as the figures in Table 22 show.[96]

The supposedly phenomenal increase in such lands after 1765 was ascribed as one of the major difficulties in the collection of revenue from Dinajpur as 'the management of these alienated articles—the collection and expenditure—are placed under separate and distined [*sic*, distinct?] officers, the business transacted privately and secreted from inspection'.[97]

Unfortunately, the proportions of *chakeran* and *baz-i-zamin* in lands alienated as *la-kharaj* cannot be worked out for all zamindaris. Table 23 provides the evidence from some zamindaris which shows the distribution of such lands.

One feature of such grants is certainly quite apparent. These grants definitely allowed small numbers of people, whether officials or militia, to take control of large tracts of land. Thus in Dinajpur, 266 zamindari horsemen jointly held 26,478 *bighas* of land in 1770,[98] which gives us a grant of 99.54 *bighas* per capita. In Burdwan, 710 *huzuri* servants (those attending to the person of the Raja) held 81,240 *bighas* (or 114.42 *bighas* per head) in 1766,[99] and in Birbhum, 204 *thanadars* held 16,385 *bighas* (80.32 *bighas* per person) in 1770.[100] In Burdwan, 438 persons were given 19,000 *bighas* as *baz-i-zamin* after 1765.[101]

TABLE 22. *BAZ-I-ZAMIN* IN MIDNAPUR AND JALASORE
1773 AND 1787

	Year	Value of *baz-i-zamin* (Rs.)
Midnapur	1773	42,821
	1787	1,00,085
Jalasore	1773	10,738
	1787	20,920

Sources: Compare IOR, BRC, P/49/38, 17 March 1773 and ibid., P/51/19, 11 April 1788.

TABLE 23. *CHAKERAN* AND *BAZ-I-ZAMIN*
IN SOME ZAMINDARIS (*in bighas*)

	Year	La-kharaj	Baz-i-zamin	Chakeran
Dinajpur	1770	4,19,436.85	1,66,635.7	2,52,801.15
Dhaka	1778	7,01,572.14	615,468.8	186,104.6
Birbhum	1778	207,043.4	108,780.1	98,264.3
Rajshahi	1789	7,26,324.00	4,52,354.0	2,73,970.00

Sources: For Dinajpur, WBSA, CCRM, Appendix to Proceedings, vol. 2, 13 to 30 December 1770; for Birbhum, BM, Add. Ms. 29088, fols. 7-8; for Rajshahi, IOR, BRP, P /71/12, 30 July 1789.

which gives us a grant of 43.38 *bighas* per person. The fact that the generic term *la-kharaj* amounted to giving extensive territorial control to a privileged few in rural society is revealed in Table 24 which represents the state of *la-kharaj* in 1778 according to the findings of the Amini Commission.

According to the Commission, about 56,00,942 *bighas* of land were held as *la-kharaj* in the areas surveyed by it in 1778.[102] John Shore added another 27,75,000 *bighas* alienated as *kharij-jama* in areas not surveyed by the Amini Commission, and adopting a standardized revenue rate of 1 rupee 8 *annas* per *bigha*, he calculated that the holders of such lands possessed resources worth nearly Rs.1,25,63,913 in 1789.[103] If, as Shore reported, the revenue rate of 1 rupee 8 *annas* per *bigha* represented about 40 per cent of the produce in the 1780s,[104] then the material assets actually controlled by the *la-kharaj* holders could easily be in the range of Rs. 3,14,09,782. This made them effective controllers of at least 9.54 per cent of the province's gross agricultural product in 1790,[105] for which they paid no revenue. It is, therefore, not surprising that it was difficult, both for the state and the zamindar, to dispossess such people from their grants which, over the years, had come to assume distinct features of private property.

First, these lands had become inheritable by the successors of the original grantees as a matter of right.[106] Second, like all things privately owned, these lands were bought and sold. Since they did not pay revenue, *la-kharaj* land-values were not expressed as a ratio of the assessed revenue (*jama*) as was done in the case of zamindari and talluqdari. Thus *la-kharaj* lands were sold as privately owned properties having independent monetary values. Scattered evidence from Midnapur indicates that the value of such lands there had increased from Rs. 2.1 per *bigha* in 1732

TABLE 24. *KHARIJ-JAMA* IN ELEVEN DISTRICTS, 1778

1 District	2 Gross *jama* (Rs.)	3 *Kharij-jama* (Rs.)	3 as % of 2
Nadia	15,85,498	4,78,731	30.19
Hughli	2,76,062	1,16,548	42.21
Jessore	4,83,388	1,19,304	24.47
Bishnupur	5,18,731	2,43,548	46.95
Jahangirpur	3,63,750	38,542	10.06
Rajshahi	29,64,631	6,63,839	22.38
Birbhum	11,44,825	2,35,888	20.06
Purnea	19,09,214	7,34,907	8.49
Pachete	1,54,423	20,481	13.26
Rangpur	16,50,655	1,84,503	11.11
Dhaka	54,63,561	8,01,572	18.59

Source: BM, Add. Mss. 29086, 29087, 29088.

to Rs. 5.01 in 1759;[107] unfortunately the state of prices for such lands, after 1759 are not known. In eastern Bengal such lands were valued at prices ranging from Rs. 13 to Rs. 3 per *bigha*, depending on their location.[108] In 1786 Charles Grant purchased 1,194.45 *bighas* of *madad-i-ma'ash* land in pargana Rokunpur, *chakla* Akbarnagar for Rs. 7,000.[109] In Nadia such lands could be purchased at a 'fair price' of Rs. 8 a *bigha*.[110]

The paucity of evidence does not allow us to make exact estimates regarding the frequency of such sales. It is nevertheless, though tentatively, possible to posit that the sale of *la-kharaj* had become pervasive in the late eighteenth century. This period was certainly marked by a fairly rapid and increasing sales of all kinds of landed property, in which *la-kharaj*, being another such property, in all likelihood formed an important part. It is perhaps extremely significant that *la-kharaj*, when sold, fetched a higher price when compared to the sale values of zamindaris or talluqdaris. In Dhaka, 'an alienated, or Rent free land of 100 r[upee]s produce is valued at 10 years produce when purchased & consequently is equal to 1000 [rupees]',[111] whereas the sale prices of zamindaris seldom exceeded five times their jama.[112] It is therefore entirely feasible that the landed proprietors all over Bengal replicated the example of the 24-Parganas where zamindars, and even *ijaradars*, were happy to put up their revenue-paying (*malguzari*) lands for sale, though they secretly purchased, or acquired, *la-kharaj* which was deemed 'a profitable estate for their support'.[113] The collector of Birbhum was convinced that by the 'annual secret appropriation of malguzzary lands under the head of neez-talooks and bazee zemin' the zamindar had actually cornered nearly 67,035 *bighas* of land between 1770 and 1787 to his 'private advantage'.[114] In fact, one of the most perceptive of the Company's observers, Sir John Shore, was quite definitive that 'proprietors and possessors of Rent free lands' made 'considerable additions to the quantity originally held by them & so as in several instances have doubled the property founded in just title'.[115]

There can be very little doubt that *la-kharaj* holders constituted a set of privileged property owners in rural Bengal. They were both 'charismatic pioneers'[116] as well as the new agrarian entrepreneurs whose existence was vital in the working of Bengal's agricultural economy. As owners of lucrative properties, but having the elite's antipathy to holding the plough, these people were in all likelihood the single largest employers of hired labour in Bengal's countryside. This suggestion is supported by the close correspondence between the proliferation of peripatetic peasant groups and the expansion of such properties in eastern Bengal (discussed in Chapter 2). Their social position also meant the existence of petty zones of high-value consumption for Bengal's agricultural and artisanal produce. This was vital for the vibrancy of the internal market. The Company's intervention disposed a substantial part of the zamindari militia, but was unable to reappropriate the lands which had been given to them in lieu of their service. Therefore, it is unlikely that consumption levels were drastically effected, at least not during the late eighteenth century. The proliferation of rural markets (see Chapter 4) indicates likewise.

Landed Proprietors and the Economy:
the case of Zamindars

That the zamindari was not a unit of production is certainly an acknowledged fact, but to say that the zamindars had practically no role in the production process[117] cannot easily be accepted. Unlike the feudal demesne, the zamindars did not actually organize production in all the territories under their jurisdiction, but a zamindari was an agglomeration of otherwise dispersed production units centred in the villages, and this being so a certain interaction between the landholder and production was bound to take place. The provision of 'soft' agricultural loans (*taqavi*) and the maintenance of irrigational and flood controlling embankments (*pulbandi*) were recognized as the traditional functions of the zamindars. These functions were most pronounced in case of the *jangal-bari* zamindars, who were given *sanads* 'on condition that within the Boundaries the jungles should be cleared, embankments and water courses made, inhabitants brought from other places and land put in to a state of cultivation'.[118] Reclamation in Chittagong was 'often assisted by the Zemeendar . . . with money as a temporary support' to the peasants.[119] The crucial position occupied by the zamindars in the entire network of reclamation of land and the extension of cultivation have already been discussed (Chapters 1 and 2), and their importance can be seen in the fact that in the extensive zamindari of Nadia, *pateet* lands of intrinsically good quality could not be reclaimed because of the financially distressed situation of the zamindar.[120]

These zamindars also had the responsibility of overseeing the maintenance of flood-controlling embankments and the construction of local irrigation networks. There were two forms in which this task was traditionally undertaken: *khas pulbandi*, or embankments made by the government and zamindari *pulbandi*, or the embankments constructed by the zamindar. The costs of the former task were defrayed either by remission of revenue or by levying *mathot* (imposts), while those of the latter were the responsibility of the zamindar 'in whose countries the works are carried on'.[121] The importance of *pulbandi* can be seen from the following contemporary description: 'The Poolbandy, or the management of Pools (which in the Bengal language signifies Dikes, Dams or raised causeways) is a department of vast mangnitude and consequence, as involving in it the fate of the Harvests, and, in consequence, the revenues of the whole province. Some tracts of very low, rich land, equal to large English Counties, have their means of culture upheld by these dams. . . .'[122] The other job, that of irrigation, was another zamindari duty which was of local importance since tanks and wells could become crucial for the survival of high-value crops during dry seasons. In fact the drought of 1769-70 became particularly severe because these devices of subsidiary irrigation dried up owing to a protracted failure of the monsoon in that year.

Most areas had 'public . . . and private tanks . . . dug at the expense of individuals from religious and other purposes, the waters of which are more than sufficient for the preservation of the crops of the proprietors of them'; these tanks were located in *malguzari*, *la-kharaj* and *khamar* lands.[123] Designed essentially as back-up irrigational facilities and as sources of drinking water, these tanks were, nevertheless, important for a number of reasons. First, the very act of digging these tanks meant the clearing of waste lands and certain financial gains for the zamindar: if the tank owner was the zamindar it became his private property, if it belonged to somebody else then the zamindar received a *salami* ranging from Rs. 50 to Rs. 100 per tank. In Birbhum, 'it is the custom, added to the salamy, for the person to pay immediately 9 years rent for the lands occupied by the tanks, after which he holds it rent free'. Second, 'the tanks are a source of great profit to the zemindar, from the fish bred in them; and the produce of Tall [*tal*, or palm] trees planted on the banks; and as they are situated on eminences they supply water for the plantations around them'.[124] Provision of agricultural credit and flood control or irrigational duties could not have been far removed from production, at least in the *khamar* or *nij-jote* lands. Additionally, the clearance of forests and reclamation of cultivable waste (*patitabad*) were rights entrenched with the zamindars, as were those of pasturage (*gaucharee* or *gop mahal*) which provided them with a supplementary income.[125]

The sphere in which the zamindars participated directly in agricultural production was in the cultivation of their personal lands. It is generally recognized that a zamindar's income arose from the financial perquisites sanctioned by the state (*nankar*),[126] but there were other sources of their financial survival and these have not been placed in a proper perspective in the existing historiography of the province. Briefly stated, a zamindar's income arose from a combination of state sanctioned perquisites, a plethora of agricultural and non-agricultural taxes levied from the peasants and artisans, duties collected from market places and local trade and from 'the profits of neez [*nij*] talooks'[127] and their *khamar* lands.[128] It is with the latter, the *nij-talluqa* and *khamar*, that this section is concerned.

Historians acknowledge the existence of these lands in the framework of the zamindari, but they underemphasize their importance on the grounds that these constituted only a small portion of the gross cultivated area.[129] True as this may have been in the overall context of agricultural production, *nij-jot* and *khamar* lands were certainly significant both in extent and in terms of their social importance at the local levels. Estimates of *khamar* lands available suggest that these were substantial. The zamindar of Bishnupur held nearly 21,000 acres as *khamar*, that of Dinajpur occupied 3,885 acres and the lands held by the zamindar of Birbhum were 12,887 acres; smaller zamindars in eastern Bengal had personal lands ranging from 750 to 400 acres,[130] though in Jahangirpur these *khamar* lands were a hefty 23,616 acres.[131] These lands were given over to the cultivation of fruits and other products of a superior quality, which were then sold at the local market places[132] often established by the zamindars themselves (discussed later).

Profits from the sale of such produce contributed to a zamindar's income over and above the customary perquisites allotted to him by the state.[133] The average yearly earnings from such lands of the zamindar of Birbhum constituted 28.97 per cent of his annual income between 1766 and 1771.[134] For other zamindars, income from such lands was equally significant as is shown by the figures in Table 25.

That the income component of these lands could increase in the structure of zamindari finances emerges as a distinct possibility in the light of the evidence from the zamindari of Dinajpur. Here, 'the Rajas Profits, exclusive of his allowances in ready money, arise principally from the Comar [*khamar*] lands and the other farms in his hands'; the income from *khamar* lands was estimated at Rs. 30,000 a year in 1770, which was considered 'much smaller than what most of the [other] Zemindars enjoy';[135] but by 1786 'profits' from these lands were bringing in Rs. 58,889 a year to the zamindar of Dinajpur, thereby constituting nearly 14.78 per cent of the Raja's income of Rs. 3,98,423.75 in that year.[136]

There were other ways in which these lands were of importance in the working of the zamindari in Bengal. It was a universally acknowledged fact that they were held free from any governmental interference in matters of assessment.[137] Therefore, these were the lands where a zamindar could hope to make improvements without inviting the financial

TABLE 25. SOURCE OF ZAMINDARI INCOME, 1786

Zamindari	Total income in 1783 (Rs.)	Income from *nij, khamar*, etc. (Rs.)	% of total income
Mahisadal	48,296.87	12,118	25.09
Fatesing	7,591.6	1,523.7	20.07
Dinajpur	4,10,423.75	145,826	35.53
Satsyka	8,439.12	1,434	16.99
Tahirpore	4,082.15	1,424.6	34.89
Muhammadshahi	32,761.14	12,812	39.11

Sources: IOR, BRC, P/50/66, 7 April 1786; also cf. Sinha, *Economic History*, vol. 2, pp. 328-45.

attentions of the state. The situation seems to have changed somewhat under the closer interest taken by the East India Company in zamindari affairs, but there is not enough evidence to say that the privileges enjoyed by the landholders in their personal lands had been severely eroded in the second half of the century.[138] Moreover, there were no fixed rules about the division of the produce between the zamindar and the cultivator in these lands. This often resulted in an arbitrary division of the crop by which the zamindars raised their share at the expense of the peasants.[139] In terms of appropriation, these were the lands in which the zamindars' claim to the surplus was highest when compared to those arising from the lands cultivated by the independent peasants.[140]

Of greater significance to this discussion is the organization of production on these *khamar* or *nij-jot* lands. Here we find the zamindars not only organizing production but also acting in a fashion which was closely replicated by the *jotedars* on their lands (discussed in Chapter 2). The zamindars worked their lands by sharecroppers, the basis of which was a combined use of economic levers and non-economic coercion: the economic mechanism assumed the form of capital invested, whereas the non-economic dimension was the use of force. The Amini Commission (1778) noted the widespread prevalence of sharecroppers and their use by the zamindars.[141] In Burdwan, a zamindar's *khamar* was 'cultivated by the ryotts under engagement of receiving a share of the crop according to the sort of land they stipulate . . . or by taking a share after gathering in the crop'.[142] In Rangpur, where there were a multitude of petty zamindars, the common practice was to let a sharecropper (*proja*, or *adhiar*) work on the zamindar's lands at a third of the produce plus small plots of land given by the landholders to enable the *adhiar* to construct homesteads and to grow some essential food supplements.[143] Force often entered into this relationship: physical confinement and flogging were the landholders stock in trade in order to tilt the balance in their favour in cases of disputed shares.[144] The use of force was vital in the economics of the zamindari as it was a crucial element in the dynamics of income distribution. Not only did it decide the rent relationship, it also established the distribution of landed property between lord and peasant prior to the actual division of the immediate product of the land between the two.

Significantly, the zamindars of Bengal do not appear to have used unpaid labour (*begar*) for production on their lands. The only recorded case of such labour comes from Birbhum where under the denomination of *sa'ir silpa khana* (tax on artisans), the zamindar:

Could demand the attendance and labour of the Blacksmiths, Carpenters, Goldsmiths and Florists, and each artificer of this description when called was obliged to attend and work for the space of ten or fifteen days in the year, so that no one attended twice in the same year, and assist in the zamindari buildings and other works receiving only a small daily allowance and diet money. Such as did not attend from distance or other cause paid a consideration or fine agreeable to their circumstances, which money was brought to the zemindar's private account.[145]

The above example shows that unpaid utilization of labour was not a salient feature of agricultural production, though the zamindar perhaps was vested with the right to call in the services of artisans for his domestic requirements. Some payments, no matter how small, were made for these services and the artisans had the choice to get their services commuted by paying a 'consideration or fine'. In any case the practice in Birbhum was abolished by the government in May 1790 which, according to the collector, 'diffused the greatest satisfaction' in the district.[146] Customary labour or service claims by the zamindars were practically absent in the rest of the province. For the Company, these zamindars were notorious for taxing 'everything that is carried on a Road or embarked on a River or purchased in a market or brought in a private house',[147] but they were never accused of using corvee labour for production in their lands.

This absence was in striking contrast to the situation in the neighbouring province of Bihar where the Company's officials recorded an apparently widespread use of slave labour in the cultivation of zamindari lands. In Bihar, the terms *mullazadah* and *kahar* were used to differentiate between Muslim and Hindu slaves. In either case, these people were 'considered in the same light as any other property & [were] transferable by the owner or [descended] at his demise to his heirs'; though the practice of selling and buying of slaves had apparently declined in the eighteenth century, slaves, for whom slavery had descended through generations, nevertheless remained, and the *al-tamgha* (permanent and hereditary) lands of the zamindars were actually cultivated 'by the hands of these slaves who also [performed] the menial offices of the [zamindar's] house'.[148] The fact that the estimated annual produce from these *al-tamgha* lands was worth Rs. 8,45,150 in the 1780s[149] must surely indicate the fairly significant use of unfree labour in production by the Bihar zamindars.

The zamindars of Bengal organized agricultural production on their personal lands by striking a specific social relationship with the peasantry.

The use of sharecroppers on their *khamar* lands was a device to commercially exploit the small-holding and resource-constrained cultivators. Through sharecropping zamindars invested capital in agricultural production and attempted to maximize the appropriation of the surplus, the principal mechanisms of which were interest rates and the distribution of the product in their favour.

Landed Proprietors and the Economy: the Talluqdars

Data regarding the actual operation of a talluqdari, its internal management and relations with the peasantry are scarce, thereby making a detailed study of its productive significance very difficult. The evidence of the growth of this form of landed property (discussed earlier) shows that in both of its forms talluqdari was a product of two specific requirements of the agrarian economy. It was, at one level, a direct manifestation of the desire of the zamindars, and of the provincial government, to effect an extension of cultivation. At another level, it was one way in which the zamindars could get hold of ready cash by selling such rights. Both suggest that these talluqdars possessed the finances necessary for the development of agricultural production. Moreover, a talluqa being a small territorial unit, and therefore easily manageable, lessened the administrative workload in a relatively larger zamindari and this was probably another consideration in a zamindar's willingness to establish talluqas in his territory. In any case these talluqdars were financially important for the zamindars. One reason why the zamindars in eastern Bengal were violently opposed to all plans of separating *mazkuri* talluqas from their authority was because such talluqas were used by them as security to get loans from moneylenders, and they feared that 'when the moneylenders will see that the talooks are to be separated from us, they will distress us so much for our loans that life itself will not be worth holding'.[150]

The most important function of these talluqdars, especially those held *mazkuri*, was to organize and oversee an improvement of cultivation. In Jessore *mazkuri* talluqa *pattas* were given to those people 'who appear to have done something in the way of improving their lands'.[151] In Birbhum, when crops failed, or the peasants deserted owing to a succession of bad harvests, these talluqdars reportedly encouraged them to 'come back in their season of cultivation, their lands being granted to them at something lower than the usual rate [of revenue]'.[152] The talluqdars in the zamindari of Buzurgumedpur often borrowed 'considerable sums of money from the merchants and made advances to the ryotts' in order to get their lands cultivated.[153] In Jalalpur 'the expectation of having the lands cultivated and the revenues improved and their own advantages' were three considerations which shaped the nature of *mazkuri* talluqas as 'possessors of hereditary properties'.[154] It is interesting to note that these talluqdars uniformly claimed that they 'spend a great effort in improving their lands & in extending cultivation at a great expense, in advancing money to the Ryotts to clear the lands for cultivation and in devoting their time and attention to superintend & direct the improvement of them'.[155] In the 24-Parganas, the talluqdars were said to be regularly advancing 'tuccavy to their Ryotts, to enable them to commence and continue the Cultivation of the current year, [re]payable around the month of Poos [December-January]',[156] i.e. after the major *aman* (winter) harvest.

In this respect a talluqdari perhaps represented an enterprise with close links with the agrarian economy. Among the officials in the *khas* talluqas of Murshidabad[157] were people who acted as bailiffs so that the cultivators could be supervised closely (see Table 26).

The inadequacy of evidence does not allow us to reconstruct the exact relationship between them and the cultivators. The impression that we get from contemporary accounts is that the peasants in the talluqas were treated more leniently than their counterparts in a zamindari. According to Richard Becher 'The Tenants of Talouks are possessed of so many indulgences & taxed with such evident partiality & tenderness in proportion to the rest, that the Talouks generally swarm with inhabitants, while other parts are deserted. In addition to the natural desire of changing from a worse to a better condition, enticements are frequently employed by the Taloukdars to augment the concourse to their lands.'[158]

TABLE 26. OFFICIALS IN *KHAS* TALLUQAS OF MURSHIDABAD

Officials	Amount of *chakeran zamin* (*bighas*)
Patwari	2.5
Mandals	80
Halsannah: 'Men whose Business it is to make Ryots work'	112
Paiks	113.31
Matchlockmen	2.25
Carriers of Measuring Lines	44.31

Source: BM, Add. Mss. 29088, fol. 166; emphasis added.

Becher's statement is also supported by the Amini Report which says: 'A Talook comprehends only a few villages, or a small tract of ground, and the possessor is able to attend to the cultivation of every part of it. It improves with his care; the rents of it increase & it becomes more populous & valuable than those parts of the District which remain under the management of the Zemindar or his officers.'[159]

Such descriptions suggesting that the cultivators enjoyed greater security in talluqas than elsewhere is also substantiated by an account of *talluqdari-raiyat* relations in Dhaka. This account states that peasants in a talluqa enjoyed greater security, this being seen in the fact that 'they often remain in the same spot, son after father, cultivating the same piece of land' owing to: 'The perfect agreement between them and the talookdar. The Ryott being the immediate cultivator of the land, the talookdar's interest is intimately connected with his, since if he is oppressed he will immediately desert. [Hence] the talookdar . . . is most concerned in the cultivation [*sic*] of them [and] is under a necessity of permitting him to enjoy his small right'.[160]

Yet there is no hard evidence to show that the position of the peasantry actually improved in these talluqas. Not all talluqdars actually were benevolent landed proprietors. Even in the 24-Parganas where some talluqdars provided *taqavi* to fund the winter harvest (discussed earlier), they were amply matched by those who were 'not guided by a similar liberal principle' and from whose territories 'the Ryotts actually fled . . . or were obliged to borrow money at exorbitant interest payable from the produce of the first crop'.[161] *Taqavi* was not always available as 'the talookdars, from their extreme poverty can afford the Ryotts no assistance to the improvement or extension of their cultivation'.[162]

Bankruptcies were common among talluqdars as they were among the zamindars. Talluqas could 'daily go to ruin and [talluqdars] not be able to pay malguzarry and cultivate them',[163] at which point they were sold, often at largely deflated prices, a process in which the zamindar had the first right of rejection. Impoverished talluqdars in pargana Laskarpur, 'immediately abscond when pressed for rents',[164] and in Dhaka they were often forced to 'relinquish their lands by Istuffa [*istifa* or resignation] from a total inability to discharge their rents',[165] in which case these talluqas naturally reverted to the zamindar who usually sold them again.[166] Thus the possibilities of peasants actually improving their positions in these talluqas, or the extent to which these proprietors could provide the cultivators with a way out from the zamindar's control, appears remote. It is perhaps significant that the report from Dhaka (cited earlier) which suggests that talluqdars and peasants operated on a 'perfect agreement' also says that these peasants 'can be scarce considered in any other light than as day labourers, earning little more than is necessary for the maintenance of themselves and [their] families'.[167]

La-kharaji *and the Agrarian Economy*

Proprietors of *la-kharaj* lands were also closely associated with the agrarian economy for, as has been discussed, their origins and future existence (especially for those held as *baz-i-zamin*) depended almost exclusively on their capacity to extend cultivation. *Baz-i-zamin* lands initially comprised both revenue paying cultivated (*hasil* or *malguzari*) and waste (*pateet*) lands.[168] In pargana Muhammadshahi 12,516.30 *bigha* were given as *baz-i-zamin* between 1770 and 1789; of these 2,877.10 *bighas* comprised *hasil* lands while the rest (9,639.20 *bighas*) were situated in *pateet* lands.[169] Of 1,738.04 *bighas* of *baz-i-zamin* granted in pargana Mahisadal in 1790, 930.48 *bighas* were distributed in *hasil*, and the remaining 807.56 *bighas* were located in *pateet* lands.[170] Such data do indicate the close connection between *la-kharaj* and agricultural production. In Sylhet, for instance, 'the property of these waste lands' was fixed on the grantee and the 'quantity [of waste land] contained in each pergunnah [was] divided among the several parties in the same proportion as they enjoy the cultivated land': a practice which was traditionally recognized as one of the best means of inducing an extension of cultivation.[171]

The greatest advantage enjoyed by *la-kharaj* property was that it did not have to pay any revenue or taxes either to the state or to the local zamindar 'however it may be cultivated'.[172] This provided the proprietors with extra funds to invest in extending the land under cultivation. Moreover, these lands were mostly given to individuals 'independent of any partnership' in order to prevent disputes or fragmentation[173] of the type which plagued the zamindars and the talluqdars. A *la-kharaj* was also exempted from defraying 'any charge for the repair of Bunds [embankments] and other publick [*sic*] works necessary for their preservation',[174] which meant that overheads were substantially reduced. The Company did not look upon the holders of such lands as being conducive to an efficient revenue administration, but the continuation of such financial privileges despite the state's disapproval would appear to be in contravention to its attempt to enforce rigorous collections of revenue. This fact is striking and must, therefore, indicate the continuing importance of such properties in the agrarian milieu of the late eighteenth century.

There is unfortunately no evidence to show the patterns of cultivation or the nature of labour use on these lands. Evidence from Muhammadshahi shows that 12,516.30 *bighas* were given as *baz-i-zamin* between 1765 and 1789; of these 9,639.15 *bighas* were *pateet* and 2,877.15 *bighas* were *hasil* lands. The *hasil* lands were cultivated in the fashion as shown in Table 27.

The absence of comparative data from other areas does not allow generalizations regarding the economic content of these *la-kharaj* lands, but it would perhaps be true to say that these lands, and the proprietors, had a significant role in the agrarian economy. Crops like mulberry and lentils were surely cultivated for commerce; so were rice and the other 'inferior crops'. Given the rising demand for food in our period (see Chapter 4) it is inconceivable that the output of rice and paddy from these lands was designed for a purpose other than sale. The fact that *baz-i-zamin* holders were also instrumental in establishing a whole range of markets, from *ganjs* to *haats*, 'solely for their own profit'[175] must surely indicate that these people combined a number of functions which were significant in furthering the pace of commercialization.

As a composite social group possessing substantial resources, these people were crucial to the local economies. The role of the *chakeran zamin* holders was particularly important in generating demand for rural produce and in speeding the processes of commercial exchange. They were given revenue from land in lieu of income. These lands were large, so were the resources. They consumed, which meant that 'many of the poorer classes subsisted by providing them with food, clothing and furniture'.[176]

Their influence on agricultural production was also significant. *Baz-*

TABLE 27. CULTIVATION OF *HASIL* LANDS IN PARGANA MUHAMMADSHAHI

	Land (*bighas*)	% of *hasil*
Mulberry and other 'superior crops'	223.25	7.76
Various types of rice and other 'inferior crops'	2,494.45	86.70
Lentils	159.45	5.54

Source: IOR, BRP, P/71/13, 27 August 1789.

i-zamin holders were entrusted with the task of extending the land under cultivation on extremely lucrative financial terms. After all, no concession could be more lucrative than the freedom from taxation in the late eighteenth century. Given this privilege, it is entirely feasible that their lands were cultivated more efficiently than normal *raiyati* or even zamindari lands. At least John Shore thought that this indeed was the case which: 'Was effected by withdrawing the Ryotts from the Revenue Lands, and inducing them to settle on the Bazee Zemin, which the Proprietors can afford to rent to them on easier terms than a [revenue] Farmer or Zemindar who pay an assessment for the lands held by [them]'.[177] Such indeed was the case in Muhammadshahi where: 'The Bazee Zemin is so extensive that the possessors of it are competitors with Government for the Ryotts; and as they enjoy the Rents of their lands without participation . . . they can afford to let them at a considerable reduced rate [of revenue]; and those [lands] of Government [i.e. *malguzari*] are consequently deserted, while theirs are stocked with numerous inhabitants'.[178] In Birbhum some peasants from *malguzari* lands would move to *baz-i-zamin* as 'They are encouraged to do so from the low terms which the possessors of these places grant to Ryotts . . . which they can afford as the rents arising from the increase in the cultivation of their lands is clear profit to them, and it is certainly a great inducement for a Ryott to desert and cultivate these grounds.'[179]

This report also says that the cultivators of *baz-i-zamin* paid 12 *annas* (Re. 0.75) per *bigha* to the holder in comparison to 'a rupee and a half for malguzarry lands'. In Birbhum such lands totalled 1,09,861.40 *bighas* in 1770 which means that the 'clear profit' from cultivation for these proprietors was at the very least in the range of Rs. 82,396.05 per annum in our period.

The Rural Elite and Local Economies

There are two dimensions of the interaction between these agrarian elites and the local economies. The first was their role in the establishment and maintenance of rural markets, and the second was the implication of their consumption for the working of such economies.

With regard to the first, witness the following description:

Nundcoomar Roy, be it known to you, that in consequence of a proposal made by you for erecting a Bazar on Burmooter [charity] lands appertaining to you, and situated in the pergunnah Jehangerabad, you are hereby authorised to erect a Gunge at that place, upon which Gunge a certain amount of rent will be

LANDED PROPERTY AND LANDED GENTRY   167

fixed. . . . You will build or establish the gunge at your own expense, after which
establishment and when collections are made from it, you are hereby required to
pay annually as Malgoozary [revenue] Sicca Rupees 200, which payment being
made, all the surplus produce [will be] received by you or your posterity [and]
the foregoing assessment will never be increased nor will abwabs [conjectural
imposts] be imposed on you.[180]

This extract (c. 1750) is representative of the manner in which the nucleus of a market as a physical site was created in late medieval Bengal. At the core was the founder who was charged with establishing a settlement (*basati*) over which he had the primary territorial and fiscal claim. This he had by virtue of having built the site at his own expense, albeit with the authorization of a superior authority. This primary claim was therefore hereditary and inviolable. The outer circles were composed of the landed magnate and the state. The intersection between the core and the outer shell was achieved by a downward devolution of authority and an upward transmission of resources.[181]

This was true not merely for the larger markets in towns like Dhaka and Murshidabad, but also for a whole range of small and intermediate markets (*haats* and *bazaars*) which were linked in a pervasive commercial network. Some of these smaller centres of redistribution were created partly by the family members of the Nazims and by their officials. In greater part, however, they were the products of zamindari initiative. For instance, while most of the wholesale forward markets (*ganjs*) in Murshidabad were established by members of the provincial ruling elites over the years (these markets were said to be between 17 and 300 years old in 1793);[182] the *bazaars* and *haats* 'appertain to ancient zemmindarries and the tolls are the rent right of the zemindars'.[183] Elsewhere also, these markets were created and held by 'zemindars and tallokdars from whom it does not appear that any security was ever required to be taken'.[184] In Jessore, the *haats* were subdivided into various shares by the zamindars as 'there are many partners each [having] a share in the haut [*haat*] and paying their revenues in distinct tahoods [contracts] or engagements. . . .'[185] In the extensive tract of Burdwan, markets were established by 'the proprietors of altumgahs [permanent revenue-assignees] and the tenants of jaighurs [jagirs] as well as the holders of every other kind of [revenue] free lands and the amount collected is solely their own profit'.[186] With regard to questions of control and jurisdiction, it appears that the actual operation of these markets was left in the hands of the founding landholder: 'the government of the soubahs [the Nizamat] contented themselves with imposing general regulations for the prevention of undue exactions and occasionally interfered to modify or abolish particular imposts as they occurred or were discovered . . .'.[187] Even holders of charitable land-grants (*lakharaj*) were given governmental sanction to establish 'gunges, bazaars and hats' and were 'authorised to exercise the privilege of collecting duties thereon'.[188] It is therefore not surprising that the zamindars persisted in claiming from the Company 'an equal right in rents arising from gunges and bazaars' as late as 1790.[189]

The jurisdictional boundaries between the magnate and the operative side of the *bazaar* can be seen from the following testimony of Nusrat Jung Sayyad Ali Khan, the Nawab of Dhaka about the markets under his ownership

I have two small gunges situated in the town of Dacca . . . which were erected by my grandfather . . . the revenues of which are about a thousand rupees per annum . . . I also have a small bazar near the door of my house where there are three or four moodies [grocers], shopkeepers, and a few people who sell vegetables, from whom a tax in kind is sent to my kitchen, *and when I want any more I purchase it.*[190]

The state's jurisdiction as well as that of the immediate landed elite centred around the collection of taxes. In fact, the power to levy taxes was a direct expression of political sovereignty, and the downward devolution of this power into the hands of the landed magnates was the parcelling of an essentially royalist prerogative.[191] For the regional state, market taxes were 'numerous [and] of a mixed nature being partly ground rent, partly market tolls, but the greatest part of what is collected on these taxes must be considered taxes on industry'.[192] Specifically they were said to include 'Customs or Duties collected on Merchandize passing through the Country, or sold in the Markets, rents of lakes or ferrys, and fees paid by Brokers or by Weighers'.[193] The state's intervention was also ensured by a set of officials it stationed in a *bazaar*. The market officials of Purnea are listed in Table 28.

While these officials were in the nature of paid watch and ward staff, the person who made a major intervention in market transactions was the official weighman called the *koyal*. Enquiries conducted by the Company in the 1780s revealed that:

Government many years ago found it necessary for the better collecting the duties [*sic*] to keep a number of servants, at a monthly allowance of wages, in order that the quantity might be more easily ascertained. On this account Weighmen or Kyalls were established and placed at different stations, who weighed the goods

LANDED PROPERTY AND LANDED GENTRY

TABLE 28. LIST OF MARKET OFFICIALS IN PURNEA, 1790

Designation of official	Numbers	Pay per year (Rs.)
Daroga	9	1,056
Pasbans or Watchmen	195	3,183
Peons	47	1,692
Dholis or Criers	61	708
Sweepers	59	354

Source: IOR, BR Misc, 13 September 1790.

without distinction of either person or place, and were authorised to collect a duty called Kyalle or Paunee Chitakee.[194] In addition, landed proprietors had their own set of officials. In general they were 'the zemindars own particular servants, paid by him in land and not in specie . . . [of these some] are at fixed stations, such as are appointed to guard and protect the Cutcherries and Gunges and those placed over the Zemindars Kumar [*khamar*] and his fruit trees'.[195]

The consensual view of these landed elites, that of people interested only in maximizing their rental and other tax-based incomes needs to be substantially rejected. No magnate could afford to antagonize merchants in these markets by transgressing limits established by custom: 'the shopkeepers of a Buzar, not being satisfied with their situation, can remove [*sic*] to another Buzar which pays revenue to Government'. Alternatively, they could also 'settle upon a piece of ground where no Buzar has been properly established, and there buy and sell and form a Buzar themselves . . .'. In such cases, they had to take the authorization from the state, failing which 'the magistrate . . . will oblige them to remove and abolish their Buzar'.[196]

The law of the land in such matters was stated as follows: 'If the shopkeepers of a Buzar established by Sunnud, and paying revenues to Governemnt go [to] reside in the Buzar of another Renter from Government, it is a matter of no consequence; but if they go to a new established Buzar where no revenue is paid to Government, *the renter of a deserted Buzar cannot himself seize the shopkeepers who has [sic] left him, but must complain to the Magistrate.* . . .'[197] Such was indeed the case, even as late as 1790, when the proprietors of the markets at Aliganj and Sahibganj complained that in 1786 the zamindar of Rajshahi had established a new market, called Coor Ganj situated 'one coss east' of Aliganj and 'half a coss from Sahibganj'. At Coor Ganj 'the duties at present are very much under the extablished rates', and as a result 'even the old merchants this year have not brought their goods to Allegunge but have carried to the newly established one [and] Allegunge which was a very long established Gunge is now from goods not being brought to the markets neither a Haut nor a Bazar.'[198]

The above episode also reveals another dimension of the problem. It shows the seething competition which existed between magnates over the territory, boundaries and jurisdiction of markets. It was a well known fact that in Bengal, zamindars often clashed over conflicts centring around 'boundaries, lands thrown up by rivers, gunges, rights to fisheries and other privileges'.[199]

In 1775 the zamindar of pargana Chandeli established a *bazaar* at Mohunganj in order to compete with Mahmudiganj 'the previously established market place'. Traders were forced to provide *muchalkas* (undertakings) that they would stop trading with Mahmudiganj, and by these means 'all bioparries [traders] are carried to Mohun Gunge'. Having forced the traders so, the zamindar was then said to have 'prevented the boats that used to come from different places from going to that Gunge'.[200] In a similar fashion, in 1777, the agent (*gumastha*) of the zamindar of Dinajpur 'erected a new Haut at Rajnagar in opposition to Nabobgunj, and the bepparies [traders], who frequented the said Gunge have been seised on the roads and carried to the haut'. Additionally, the cultivators of Rajnagar were asked to give 'mochulkas' that they would desist from taking their goods for sale to Nabobganj.[201]

Finally, consider the following petition (in 1780) of a talluqdar of the 24-Parganas:

In the village of Rajepoor, there was a Haut which was my Talouk descended to me from my Ancestors. Collysunker Zemindar who has a Talouk in the said village established a Buzar there by which the above Haut was destroyed. As this happened before the time of the Company, I never complained against him. In 1184 B.S. [1775 A.D.] I rebuilt the above Haut & from the said year to the 15th of Bhadoun [30 August] a Haut has been held there, and he has also kept a Buzar there daily. On the 16th of the above month, the said Zemindar erected another Buzar near the former one; and has, and is carrying away the Ducans [shops] from the road and my Haut, and has taken a mochulka from all the Ryotts of the Pergunnah and [the aforesaid] village that they would not resort to my Haut. I am a servant of the Government and cannot raise Disturbances in the mofussil [countryside] & such is the oppression and injustice the Zemindar has committed.[202]

Turning to the other aspect, that of the economic significance of elite consumption, we have a detailed account of the expenditures of some zamindars. The details of these cash disbursements have been clubbed under three heads in Table 29.

Table 30 shows a comparative statement of cash disbursements of the zamindar of Mahisadal for two years, 1773 and 1783, for which data are available.

With regard to personal expenses, the profiles of zamindari disbursements are as shown in Table 31.

Table 29 shows that on an average, the landed elites spent 32 per cent of their incomes on personal/domestic consumption, 31 per cent was spent on maintenance of the infrastructure, while the remaining 37 per cent went into religious and charitable expenses. The fact that these disbursements were made in cash make these expenditures a major variable in strengthening the circuits of cash transactions in rural Bengal. Furthermore, the evidence from Mahisadal (Table 30), which shows an

TABLE 29. ANNUAL CASH EXPENDITURES OF NINE ZAMINDARIS (in Rs.)

Year	Zamindari	Expenditure on personal consumption i.e. food, clothing, utensils, etc.	Expenditure on establishment and charitable structure: repairs, wages to servants, workmen, officials, etc.	Expenditure on religious occasions	Total
1773	Birbhum	1,20,600 (50.95%)	38,880 (16.43%)	77,200 (32.62%)	2,36,680
1777	Burdwan[a]	37,170 (25.93%)	53,197 (42.83%)	33,822 (27.24%)	124,189
1783	Mahisadal	25,252 (27.47%)	37,992 (41.33%)	28,669 (31.19%)	91,913
1783	Fatesing	2,630 (38.33%)	1,190 (17.34%)	3,042 (44.33%)	6,862
1783	Dinajpur	85,970 (15.17%)	1,97,971 (34.93%)	2,82,824 (49.9%)	5,66,765
1783	Satsyka	2,787 (16.64%)	10,177 (60.78%%)	3,781 (22.58%)	16,745
1783	Tahirpur	1,835 (39.66%)	1,479 (31.96%)	1,313 (28.38%)	4,627
1783	Muhammadshahi	21,329 (48.92%)	11,401(26.15%)	10,870 (24.93%)	43,600
1791	Rajshahi[b]	8,900 (23.65%)	2,600 (7.49%)	23,200 (66.86%)	34,700

Note: [a]For Burdwan, expenses are for six months: 15 August to 14 February 1777; [b] for Rajshahi, expenses shown are those made privately by Rani Bhawani for the year 1791 as given by her son Raja Ramkrishna to the collector.
Sources: IOR, BRC, P/49/40, 4 June 1773; ibid., P/50/66, 7 April 1786; N.K. Sinha, Economic History, vol. 2, pp. 238ff.; IOR, BRP, P/71/44, 24 September 1791.

TABLE 30. ANNUAL CASH DISBURSEMENTS OF THE ZAMINDAR OF MAHISADAL, 1773 AND 1783 (in Rs.)

Year	For personal consumption	For zamindari infrastructure	For religion and charity
1773	23,697	21,368	30,860
1783	25,252 (+ 6.56%)	37,992 (+ 77.8%)	8,669 (- 7.1%)

Total Disbursements in 1773: 75,925
Total Disbursements in 1783: 91,913 (+ 21.06%)

Source: For 1773, IOR, BRC, P/49/41, 3 August 1773; for 1783, IOR, BRP, P/70/49, 12 December 1788.

overall increase in expenditure of 21.06 per cent between 1773 and 1783 with a 77.8 per cent hike in disbursements for infrastructural maintenance suggests that such spending would have a multiplier effect on cash-flow in the countryside. The data in Table 31 also show the critical impact of elite consumption on the circuits of commerce in the countryside, both at its luxury and the basic-necessity ends. For instance,

TABLE 31. PERSONAL EXPENSES OF BENGAL ZAMINDARS (in Rs.)

Zamindari	'House expenses' including food	'Wearing Apparel'	'House Furniture'
Birbhum (1773)	30,000 (12.67%)	24,000 (10.14%)	15,600 (6.59)
Mahisadal (1773)	10,472 (13.79%)	3,947 (5.19%)	Not available
Mahisadal (1783)	15,164 (16.49%)	2,859 (3.11%)	Not available
Burdwan* (1777)	22,800 (18.36%)	14,370 (11.57%)	Not available
Fatehsing (1783)	1,960 (28.56%)	660 (9.62%)	Not available
Dinajpur (1783)	68,152 (12.02%)	17,818 (3.14%)	Not available
Satsyka (1783)	1,652 (9.86%)	1,135 (6.78%)	Not available
Tahirpur (1783)	Not Available	443 (9.57%)	1,392 (30.08%)
Muhammadshahi (1783)	8,790 (20.16%)	2,965 (6.8%)	2,831 (6.49%)
Rajshahi* (1791)	8,400 (24.21%)	500 (1.44)	Not available

Notes: The figures in brackets are perecntages of total disbursements vide Table 29.
*For Burdwan, disbursements are for six months. For Rajshahi, the amounts spent are the personal expenses of Rani Bhawani alone.
Sources: IOR, BRC, P/49/40, 4 June 1773; ibid., P/50/66, 7 April 1786; N.K. Sinha, *Economic History*, vol. 2, pp. 238ff.; IOR, BRP, P/71/44, 24 September 1791; IOR, BRC, P/49/41, 3 August 1773; IOR, BRP, P/70/49, 12 December 1788; N.K. Sinha, *Economic History, Volume 3, 1793-1848*.

in 1773 the Raja of Birbhum spent Rs. 3,600 on perfumes, exotic fruits and betel-leaf, while in the same year the zamindar of Mahisadal spent a hefty Rs. 4,000 for 'rice at the table'.[203] Similarly, the considerable amounts spent on clothing and furniture certainly contributed significantly to vibrant artisanal activity outside the purview of the Company's commercial control. In fact, the high concentration of artisans in Bengal's countryside (discussed in Chapter 4) would in large part be a result of such patterns of consumption.[204]

An important component of elite spending was the wages paid to its officialdom, entered as 'monthly allowance and servants wages' in the accounts. The data for this are given in Table 32. The fact of these cash salaries introduces another dimension of the economic role of the agrarian elites in rural Bengal. Salaries varied according to the status of the official. In Birbhum, *vakils* and *munshis* (both critical in the zamindari bureaucracy) were paid Rs. 12,000 and Rs. 3,600 per year respectively.[205] 'Menial servants' of the Raja of Burdwan took home Rs. 778 a year, while the 'keepers of elephants, camels, horses etc.' were paid Rs. 1,943.[206] In the zamindari of Mahisadal, 'writers at the sadr' (i.e. *munshis*) received Rs. 7,246.[207] When we consider the fact that a salt manufacturer in Mahisadal was paid Rs. 6.3 for the entire salt producing season,[208] the salaries paid to officials appear extremely high in comparison. The

TABLE 32. CASH SALARIES AND WAGES OF ZAMINDARI OFFICIALS

Zamindari	Salaries and wages of officials	% of Total disbursements
Birbhum (1773)	26,880	11.36
Mahisadal (1773)	10,480	13.80
Mahisadal (1783)	23,668	25.75
Burdwan (1777)	22,042	17.75
Fatehsing (1783)	759	11.06
Dinajpur (1783)	76,055	13.42
Satsyka (1783)	2,927	17.78
Muhammadshahi (1783)	6,536	14.99
Rajshahi (1791)	2,600	7.49

Note: Total disbursements vide Table 29 above. For Burdwan, the statement is for six months. For Rajshahi, the statement pertains to the personal expenses of Rani Bhawani.

Sources: IOR, BRC, P/49/40, 4 June 1773; ibid., P/60/66, 7 April 1786; N.K. Sinha, *Economic History*, vol. 2, pp. 238ff.; IOR, BRP, P/71/44, 24 September 1791. IOR, BRC, P/49/41, 3 August 1773; for 1783, IOR, BRP, P/70/49, 12 December 1788.

overall result was the creation of a sub-elite in the countryside. They constituted another set of affluent consumers as can be seen from Table 33 which pertains to such consumers in Rangpur and Dinajpur.

The Rural Elite: An Overview

The question of the rural elite cannot be understood in the narrow confines of revenue administration alone. The other explanation, that of an almost complete collapse of landed property under the financial pressures of the Company, appears inadequate as an interpretation. It is undeniable that a number of zamindaris disintegrated, but new ones were formed with equal speed. Some talluqdars could become bankrupt, but that did not stop others from bidding for talluqas; nor did it prevent zamindars from attempting to usurp talluqdaris and to oppose governmental plans to separate talluqas from their jurisdictions.

Landed property was not a closed shop. Access to it by birth (or caste) was equally balanced by entry through purchase or office, and when a landed property was sold a whole range of potential buyers (from zamindars to merchants to 'ryotts') would come forward with their bids.[209] Such sales had created a plethora of small landed proprietors. In Dhaka for instance nearly 20,000 talluqdaris had been established by sale of zamindari lands between 1765 and 1790: zamindari officials and

TABLE 33. EXPENDITURE PATTERNS OF OFFICIALS SERVING THE 'GREAT LANDHOLDERS' IN RANGPUR AND DINAJPUR (*per annum*)

Item	Expenditure in Rangpur (Rs.)	% of Total	Expenditure in Dinajpur (Rs.)	% of Total
Lodging	113.25	7.84	56	6.42
Furniture	123.25	8.54	40.81	4.68
Ornaments	302.62	20.96	324.87	37.27
Clothing	96.94	6.71	72	8.26
Food	354.87	24.58	174	19.96
Servants	237.62	16.46	60	6.88
Religion and charity	200	13.87	144	16.53
Sundries	15	1.04	Nil	
Total	1443.55		871.68	

Sources: For Rangpur, F. Buchanan, 'Survey of Ronggoppur', IOL, Ms. Eur. D. 75, Statistical Appendices, Table 39; for Dinajpur, M. Martin, *Eastern India*, vol. 3, book 3, Appendix G to M.

ijaradars bought the larger properties, while others bought lands no larger than average peasant holdings.[210]

Such sales do not necessarily reflect the financially distressed state of the property. Auctioning the property of a bankrupt zamindar or talluqdar was only one form of land sales. Most properties were still sold privately and prices were negotiated by the parties concerned outside the state's purview. In fact talluqdaris were created by sales of this type: each sale deed was properly witnessed, endorsed and registered at the zamindari *sadr*. It is true that land values did not appear high enough to European observers and there were cerainly distress sales, but these have to be seen in relation to the facts that that zamindaris and talluqdaris could fetch five times the assessed revenue and *la-kharaj* could easily be worth ten times its estimated produce, prices which appear strikingly high in a situation where a sale price equal to two years *jama* was considered of good value in the market. This would certainly not indicate a sluggish land market. Moreover, such sales increased during the late eighteenth century, as testified to by Cornwallis' belief that 'purchasers of Zemindarry land are become very numerous since the acquisition of the Dewanny'.[211]

Property in land could also develop by other means. Being a zamindari official was often a passport to landed property. A 'small pecuniary consideration for the zemindars' was allegedly enough to enable zamindari *amils* to acquire *hasil* lands 'recorded in moffussul accounts as bazee zemin' in Jessore.[212] In pargana Bikrampur *qanungoes* and *mohrirs* had managed to obtain 'a greater property in the Pergunnah than the Zemindars themselves now hold . . . and to prevent detection of these and other acts of a like fraudulent nature have either destroyed, or keep concealed, the original papers of the pergunnah by which alone the true jummah can be ascertained'.[213] While corruption was one important factor in allowing officials to establish footholds in landed property (at least this was the official explanation given by the Company while coming to terms with the power of the zamindari officialdom) the situation was certainly more complex. These officals did command substantial resources as paid servants. Equally potent was the role of 'long custom & possession' of offices which in Bakarganj allowed the *tahsildars*, initially appointed by government to supervise collection of revenue from *huzuri* talluqdars to acquire 'a sort of hereditary claim to their management from which they derive pecuniary advantage, as well as influence of consequence'. These *tahsildars*: 'Can for some time to come have no such hereditary claim as the zemindars, tho' there is no doubt that possession may at some future period give them a handle to set up pleas of the same kind.

Thus a new set of men will be set up, who will claim property to which they have no title, at the expense of those who are already admitted to possess, or supposed to have, a right in the soil'.[214]

Finally there were the *la-kharaj* holders who effectively constituted owners of privileged property in late eighteenth century Bengal. Originally encouraged to help pacify the frontiers, largely untouched by the Company,[215] and continued on a reduced scale by the zamindars after 1765, these lands were also used as a device to avoid land revenue, not least (as the Company suspected) by the zamindars themselves.

NOTES AND REFERENCES

1. 'Purchasers of zemindarry land are become very numerous since the acquisition of the Dewanny' wrote Cornwallis about the pace of such sales (IOR, BRP, P/70/23, 18 January 1787).
2. The right to a share of the produce and possibly over labour are the criteria Irfan Habib has used to define landed-property in Mughal India (Irfan Habib, 'Social Distribution of Landed Property in Pre-British India', *Enquiry, New Series*, Winter 1965). My definition differs from Habib's in so far as I see such rights as existing in addition to the ownership of material assets, both in land and money.
3. *The Bengal Frontier*, pp. 219ff, 252ff; quotation from p. 219. Despite the aptness of this characterization, Eaton's work is limited by its exclusive focus on the 'rooting' of Islam in eastern Bengal through the agency of this gentry. In fact the popularity of Vaisnavism and its dissemination in later-medieval Bengal was accompanied by the growth of its own gentry variant who were paid for their piety by a variety of *debottar, brahmottar* and *bishnuwattar* grants of revenue-free lands. The papers of the Amini Commission have enormous information on such grants throughout Bengal (see BM, Add. Mss. 29086, 29087, 29088). If the Muslim gentry's power revolved around the shrine and the mosque, the power of his Hindu counterpart revolved around the *matha* or the various temples—*chala, shikhara* and *ratna*—which were constructed all over Bengal, especially in western Bengal in the same period (see, H. Sanyal, 'Temple Building in Bengal from the Fifteenth to the Nineteenth Centuries', in Barun De (ed.), *Perspectives in Social Sciences: Historical Dimensions*, vol. I, Calcutta, 1977; also see Apala Bhadra, 'Aspects of a Regional Culture: Vaishnavism, Temples and Scroll Paintings in Late Medieval Bengal', unpublished M.Phil. dissertation, Jawaharlal Nehru University, 1996).
4. Peter Robb (ed.), *Agrarian Structure and Economic Development: Landed Property in Bengal and Theories of Capitalism in Japan*, London, 1992, p. 4.

5. See for instance, John McLane, *Land and Local Kingship in Eighteenth Century Bengal* (Cambridge, 1993) for a discussion on the basis of a study of the zamindari of Burdwan.
6. Rajat Datta, 'Markets, Bullion and Bengal's Commercial Economy: The Eighteenth Century Perspective', in Om Prakash and Denys Lombard (eds.), *Commerce and Culture*, but for an excellent long-term perspective of this process, see Richard Eaton, *The Bengal Frontier*, esp. Chapter 8.
7. IOR, HM, vol. 206, p. 344.
8. S. Akhtar, *The Role of zamindars in Bengal, 1707 to 1772*, Dhaka, 1982, p. 23; also H. Sanyal, 'Continuities of Social Mobility in Traditional and Modern Society: Two Case Studies of Caste Mobility in India', *Journal of Asian Studies*, February 1971.
9. IOR, BRC, P/50/66, 7 April 1786.
10. Ibid., P/50/11, 21 July 1778.
11. IOR, BRC, P/50/63, 11 February 1786; ibid., P/51/25, 1 October 1788; ibid., P/52/2, 10 February 1790.
12. Tapan Raychaudhuri, *Bengal under Akbar and Jahangir*, Delhi, 1969, p. 63.
13. John Shore's Minute of 2 April 1788 in IOR, BRC, P/51/28; ibid., P/51/66, 7 April 1786; also FR, Murshidabad, IOR, G/27/1, 16 October 1770.
14. Muhammad Reza Khan to Board, WBSA, Miscellaneous Proceedings of the Board of Revenue, vol. 1, 1772-6, pp. 338-9.
15. As made, for instance, by Ray and Ray in 'Jotedars and Zamindars'.
16. IOR, BRC, P/51/18, 2 April 1788.
17. C.W.B. Boughton-Rouse, *Dissertation Concerning the Landed Property in Bengal*, London, 1791, IOL, *Tracts*, vol. 168, p. 57.
18. WBSA, Miscellaneous Proceedings of the Board of Revenue, vol. 1, 1772-6, p. 350.
19. IOR, BRC, P/51/18, 2 April 1788.
20. *FR 2*, pp. 199-200.
21. IOR, BRC, P/51/18, 2 April 1788, reply of Rai Rayan Rajvallabh to Council, 2 April 1788.
22. IOR, Committee of New Lands, P/98/10, 26 July 1761.
23. IOR, BRC, P/51/51, 9 December 1789.
24. Sinha, *Economic History*, vol. 2, p. 165. Even single parganas could have up to 300 zamindars paying revenue ranging from Rs. 25,000 to Rs. 50 (WBSA, PCR, Murshidabad, vol. 12, 5 May 1777).
25. See J. Harrington to Board, IOR, BRP, P/71/36, 4 February 1791; T. Henckell to Council, IOR, BRC, P/52/20, 27 October 1790; L. Mercer to Council, IOR, BRC, P/52/30; Amini Report, IOR, HM, vol. 206, p. 346.
26. *FR 2*, p. 749.

27. BM, Add. Mss. 29086; IOR, HM, vol. 206, p. 345; also IOR, HM, vol. 68, pp. 732-3.
28. *CPC*, vol. 8, 1788-9, Delhi, 1953, 2 February 1788, p. 420.
29. IOR, BRC, P/51/18, 2 April 1788.
30. IOR, HM, vol. 206, p. 345; also Boughton-Rouse, *Dissertation*, p. 25.
31. Ibid., vol. 68, p. 733; Boughton-Rouse, *Dissertation*, p. 86; also see B.R. Grover, 'Nature of *Dehat-i-Talluqa*', p. 270.
32. WBSA, PCR, Murshidabad, vol. 8, 18 February 1776.
33. IOR, BRP, P/71/25, 24 May 1790; ibid., P/71/25, 4 June 1790; Westland, *Jessore*, p. 84.
34. It is interesting to note that a zamindar could also be a talluqdar by purchase: the zamindar of *tappa* Jaffarnagar was also a talluqdar by purchase of *mauza* (village) Salia in *tappa* Alinagar and part of *mauza* Gonya in *tappa* Haveli, pargana Jahangirnagar in 1774 (WBSA, PCR, Dacca, vol. 3, 20 June 1774).
35. Sometimes, as in Tippera and Mymensing, *jangal-bari* could become *kharida* talluqdars by purchasing the land cleared by them from the zamindar (IOR, BRP, P/71/51, 9 April 1792; IOR, BRC, P/52/13, 11 June 1790).
36. *FR 2*, p. 749.
37. WBSA, PCR, Dacca, vol. 13, 16 January 1776; also *FR 2*, p. 750.
38. BM, Add. Ms. 29086.
39. IOR, BRP, P/71/26, 4 June 1790.
40. IOR, SCC, P/A/10, 16 August 1769.
41. IOR, BRC, P/49/52, 13 June 1775.
42. Ibid., P/51/33, 8 March 1788.
43. According to John Shore, 'the ancestors of many (talluqdars) were in possession of their talooks before the zemindarry jurisdictions in which they are now included were formed, or the families of the zemindars were known' (IOR, BRC, P/52/10, 12 May 1790).
44. IOR, BRC, P/52/4, 10 February 1790; also Taylor, Dacca, pp. 153-4.
45. Ibid., P/52/13, 11 June 1790.
46. IOR, BRP, P/71/26, From W. Douglas, Collector, Dhaka, 4 June 1790; emphasis added.
47. *WBDR, ns, Birbhum*, 24 May 1790, p. 85.
48. *Tahsildars* were also used to collect revenue from some talluqdars in Bakarganj (IOR, BRP, P/71/26, 4 June 1790).
49. Ibid., P/72/2, 25 May 1792.
50. *CPC*, vol. 8, pp. 421-2.
51. WBSA, Khalsa, vol. 1, 4 January 1775; IOR, BRP, P/71/26, 25 May 1790.
52. IOR, BRP, P/71/26, 4 June 1790.
53. Ibid., P/71/25, 24 May 1790; IOR, BRC, P/52/17, 2 August 1790.
54. *WBDR, ns, Birbhum*, p. 85.
55. IOR, BRC, P/51/50, 23 October 1789; also see IOR, BRP, P/72/2, 25 May 1792.

56. *CPC*, vol. 8, p. 421.
57. WBSA, Khalsa, vol. 1, 14 December 1775.
58. IOR, BRP, P/71/26, 4 June 1790; also Beveridge, *Bakarganj*, p. 194.
59. For example a *ta'ahuud* of Rs. 4,294-5 *annas* taken by a talluqdar in Buzurgumedpur (in 1779) shows the following rates of payment to be made to the zamindar:

   Payment in the first year   Rs. 3,001
   Payment in the second year   Rs. 3,326
   Payment in the third year   Rs. 3,651
   Payment in the fourth year   Rs. 3,936
   Payment in the fifth year   Rs. 4,294-5 *annas*

   WBSA, PCR, Dacca, vol. 25, 29 September 1779.
60. Beverigde, *Bakarganj*, p. 416; *CPC*, vol. 8, p. 422; WBSA, PCR Murshidabad, vol. 23, 17 February 1780 and 3 April 1780 for a list of mortgaged talluqas in Murshidabad.
61. WBSA, PCR, Dhaka, vol. 3, 25 July 1774.
62. *WBDR, ns, Birbhum*, p. 85.
63. IOR, BRP, P/71/26, 4 June 1790.
64. IOR, BRC, P/52/13, 26 May 1790.
65. Literally one holding or occupying a middle position.
66. IOR, BRP, P/71/26, 4 June 1790.
67. Ibid.
68. Ibid., P/72/2, 25 May 1792.
69. Taylor, *A Sketch of the Topography and Statistics of Dacca*, p. 154.
70. *FR 2*, p. 750; emphasis added.
71. WBSA, Khalsa, vol. 1, 4 January 1775.
72. IOR, BRP, P/72/2, 25 May 1792.
73. WBSA, CCRM, Appendix to Proceedings, vol. 5; BM, Add. Ms. 29086; also *FR 2*, pp. 333, 415, 480.
74. BM, Add. Ms. 29088, fol. 123.
75. IOR, BRC, P/52/31, 10 June 1791.
76. IOR, BRC P/51/27, 26 November 1788.
77. See Chaudhuri, 'Agrarian Relations: Eastern India', in *CEHI 2*, pp. 93-4.
78. IOR, BRC, P/49/50, 7 February 1775.
79. IOR, BRP, P/70/48, 11 November 1788; also *FR 2*, pp. 268-70.
80. IOR, BRC, P/52/2, 10 February 1790. See E :on, *The Bengal Frontier*, for a detailed account of the connection between *la-kharaji* and the settlememt of Chittagong.
81. Ibid., P/52/5, 24 November 1787.
82. IOR, BRP, P/70/27, 8 May 1787.
83. WBSA, CCRM, vol. 8, 5 December 1771.
84. IOR, BRC, P/52/21, 1 December 1790.
85. *FR 2*, p. 28.

86. IOR, BRP, P/70/27, 8 May 1787.
87. IOR, BRC, P/52/20, 10 October 1790.
88. Ibid., P/49/42, 16 November 1773; emphasis added. The term Raja here obviously refers to the petty zamindars in Coch Behar whose ruler was an autonomous chief under the Mughals.
89. The term registered was used to denote those people who had valid *sanads* and had these registered in the *sadr daftar*. There were others who held lands without any *sanads*, and the Company always thought that they outnumbered the ones who had registered.
90. IOR, BRC, P/52/43, 20 April 1792.
91. Ibid., P/50/25, 9 June 1780; emphasis added.
92. Evidence of the reduction in the numbers of zamindari troopers is available in WBSA, Proceedings of the Select Committee, vol. 2, 28 October 1766 for Burdwan; FR, Murshidabad, IOR, G/27/1, 10 October 1770 for Birbhum and 16 October 1770 for Dinajpur.
93. IOR, BRP, P/70/27, 8 May 1787.
94. BM, Add. Ms. 29087, fol. 133; IOR, BRP, P/71/12, 30 July 1789.
95. Compare *FR 2*, p. 335 and Sinha, *Economic History*, 2, p. 272.
96. Compare IOR, BRC, P/49/38, 17 March 1773 and ibid., P/51/19, 11 April 1788.
97. WBSA, BRFW, vol. 6, 6 November 1786.
98. WBSA,CCRM, Appendix to Proceedings, vol. 2.
99. WBSA, Proceedings of the Select Committee, vol. 2, 28 October 1768.
100. FR, Murshidabad, IOR, G/27/1, 10 October 1770.
101. Sinha, *Economic History*, vol. 2, p. 272.
102. BM, Add. Ms. 29086, fol. 8-9.
103. *FR 2*, p. 28.
104. Ibid., p. 519.
105. According to Colebrooke, the gross agricultural output per year was Rs. 32,91,30,000 in 1790 (*Remarks*, pp. 15-16).
106. IOR, BRC, P/52/20, 27 October 1790.
107. Ibid., P/50/41, 26 July 1782.
108. WBSA, PCR, Dacca, vol. 3, 1 July 1774.
109. IOR, BRC, P/51/1, 9 September 1786.
110. IOR, BRP, P/71/37, 30 March 1791.
111. WBSA, PCR, Dacca, vol. 1, 9 December 1773.
112. G.C. Meyer, Preparer of Reports, Department of Khalsa, in IOR, BRC, P/51/33, 25 March 1789. John Shore believed that, apart from Calcutta, 'the price of revenue land sold [never] exceeded the revenue for two years'; ibid., P/51/27, 26 November 1788.
113. IOR, BRP, P/70/27, 8 May 1787.
114. IOR, BRC, P/51/20, from J. Sherburne, 14 May 1788.
115. Ibid., P/50/40, 3 May 1782; emphasis added.

116. The term is Eaton's (*The Bengal Frontier*).
117. As argued for instance by Ray and Ray in 'Jotedars and Zamindars'.
118. IOR, BRP, P/72/32, 7 July 1794, *arzi* of Debnarain Roy, Sadr Qanungo of Hijli outlining the history of his zamindari since *c*. 1720.
119. BM, Add. Ms. 19286, fol. 19.
120. CCR, IOR P/67/62, 13 June 1776.
121. WBSA, Proceedings of the Controlling Committee of Revenue, vol. 1, 16 March 1771; IOR, BRC, P/49/51, 14 March 1775.
122. Anon. (James Rennell?), 'Poolbundy—A Description of It by a Private Hand', IOR, HM, vol. 47, p. 25.
123. IOR, BRC, P/52/36, 21 October 1791; emphasis added.
124. IOR, BRP, P/71/13, 24 August 1789. The *tal* (palm) trees gave them an additional income because they were usually leased to the toddy-tappers to manufacture liquor (IOR, BR Misc., P/89/36, 29 October 1790).
125. In the zamindari of Coch Behar, *gaucharee* was farmed out to others (IOR, BRP, P/71/35, 25 November 1790). In Birbhum access to pasturage was taxed by the landholder at the rate of Re. 0.063 (1 *anna*) per cow and Re. 0.125 (2 *annas*) per buffalo (IOR, BR Misc., P/89/36, 5 July 1790).
126. *Nankar* was usually rated as 10 per cent of the assessed revenue, which could be paid either as a separate deduction from the revenue in cash or as portions of rent-free lands. I have discussed the importance of *nankar* in zamindari incomes in 'Aspects of the Agrarian System of Bengal during the Late Eighteenth Century', M. Phil. dissertation, Jawaharlal Nehru University, 1980, Chapter 1. *Nankar* lands in Jahangirpur (in 1772) were stated at 42,695.70 *bighas* (see WBSA, Proceedings of the Committee of Circuit, IOR, P/70/15, 8 October 1772).
127. IOR, BRC, P/49/45, 16 March 1774.
128. IOR, FR, G/27/1, 16 October 1770.
129. See, B.B. Chaudhuri, 'Rural Power Structure', p. 104; also R. Ray, *Agrarian Change*.
130. WBSA, CCRM, Appendix to Proceedings, vol. 1, 3 December to 10 December 1770, p. 32; ibid., vol. 2, 13 December to 30 December 1770, p. 170; IOR, BRC, P/49/40, 4 June 1773, p. 1981; WBSA, PCR Dacca, vol. 14, 28 February 1777; ibid., 28 February 1777.
131. Proceedings of the Committee of Circuit, Dacca, IOR, P/70/15, 8 October 1772.
132. WBSA, CCRM, Appendix to Proceedings, vol. 1, p. 32; IOR, BRC, P/49/39, 26 March 1773.
133. IOR, BRC, P/49/45, 16 March 1774.
134. Ibid., P/49/40, 4 June 1773.
135. IOR, FR, G/27/1, 16 October 1770; emphasis added.
136. IOR, BRC, P/50/66, 7 April 1786.

137. Even as late as the 1770s these lands were said to be paying revenue at the original rates established by the Mughals in 1582 (WBSA, CCRM, 30 March to 28 August 1772, Appendix to Proceedings).
138. For the continuation of such privileges in Burdwan see IOR, BRC, P/49/42, 10 December 1773; for Midnapur, see ibid., P/51/20, 5 February 1773; and for Birbhum, see ibid., P/51/20, 14 May 1788.
139. WBSA, PCR, Burdwan, 1 March to 29 December 1775, 29 December 1775.
140. For a similar situation in Mughal north India see Habib, *Agrarian System*, pp. 141-3.
141. See IOR, Home Misc., vol. 201, pp. 335-401.
142. WBSA, PCR, Burdwan, 3 June to 31 July 1776, 22 June 1776.
143. IOR, BRC, P/52/10, 20 March 1790.
144. IOR, BRP, P/71/33, 13 September 1787.
145. IOR, BRC, P/52/14, 7 July 1790, and IOR, BR Misc., P/89/36, 5 July 1790; emphasis added.
146. Ibid.
147. 'State of Bengal in 1786: A Report Compiled by Mr. Beaufoy on 12 April 1786', in IOR, HM, vol. 382, p. 81.
148. IOR, BRC, P/49/46, 16 August 1774.
149. Ibid., P/50/51, 12 March 1784; also ibid., P/50/54, Appendix to Consultations.
150. IOR, BRC, P/52/14, *arzi* of the zamindars of Jalalpur, 23 June 1790.
151. Westland, *The District of Jessore*, p. 80; also IOR, BRP, P/71/25, 24 May 1790.
152. WBSA, PCR, Burdwan, vol. 5, 9 January 1776.
153. Ibid., Murshidabad, vol. 23, 21 January 1779.
154. IOR, BRC, P/52/14, 7 June 1790.
155. WBSA, CCRM, vol. 8, 30 December 1771.
156. IOR, BRP, P/72/6, 1 August 1792.
157. BM, Add. Ms. 29088, fol.166; emphasis added.
158. SCC, IOR, P/A/9, 16 August 1769.
159. IOR, HM, vol. 206, p. 347.
160. WBSA, PCR, Dacca, vol. 16, 21 July 1777.
161. IOR, BRP, P/72/6, 1 August 1792.
162. IOR, BRC, P/51/4, 29 January 1787.
163. WBSA, PCR, Murshidabad, vol. 23, 17 February 1780.
164. Ibid., vol. 6, 24 July 1775.
165. IOR, BRC, P/50/10, 10 July 1770.
166. IOR, BRP, P/71/26, 4 June 1790.
167. WBSA, PCR, Dacca, vol. 16, 21 July 1777.
168. In Bengal wastelands denoted 'land that has not been in cultivation for the last four or five years' (IOR, BRC, P/52/9, 7 May 1790). Since

LANDED PROPERTY AND LANDED GENTRY 183

chakeran zamin was given to officials in lieu of pay, it is quite likely that these lands were situated mostly in cultivated areas.
169. IOR, BRP, P/71/13, 27 August 1789.
170. IOR, BRC, P/52/21, 1 December 1790.
171. Ibid, P/50/24, 28 March 1780.
172. IOR, BRP, P/71/15, 5 October 1789.
173. IOR, BRC, P/50/24, 28 March 1780.
174. IOR, BRP, P/70/24, 6 February 1787.
175. IOR, BRC, P/52/9, 9 April 1790 for Burdwan; ibid., 28 April 1790 and 8 April 1790 for Mymensing and Jessore respectively.
176. Tennant, *Indian Recreations*, 1, p. 264.
177. IOR, BRC, P/50/40, 3 May 1782.
178. IOR, BRP, P/70/24, 9 October 1786.
179. IOR, BRC, P/49/54, 11 July 1775.
180. '*Sanad* of Raja Tilakchand of Burdwan to Nundcoomar Roy as revealed in a *arzi* of Nundcoomar to the Collector of Burdwan' vide, IOR, BR, Misc., 25 February 1794. Raja Tilakchand acquired the zamindari in 1745.
181. The active role of the local agrarian elites in the establishment of market places in western India has recently been emphasized by Sumit Guha ('Potentates, Traders and Peasants, Western India, *c.* 1700-1800', in Burton Stein and Sanjay Subrahmanyam (eds.), *Institutions and Economic Change in South Asia*, Delhi, 1996, p. 81).
182. IOR, BRC, 15 May 1793.
183. Ibid., P/51/23, 3 September 1788.
184. IOR, BRP, 8 August 1787.
185. IOR, BRC, P/52/14, 7 July 1790.
186. Ibid., 9 April 1790.
187. Ibid., 11 June 1790; emphasis added.
188. Ibid., 9 April 1790.
189. Ibid., 9 April 1790.
190. Ibid., 11 June 1790; emphasis added.
191. See P. Calkins, 'The Formation of a Regionally Oriented Ruling Group in Bengal, 1700-1740', *Journal of Asian Studies*, vol. 29, 1970, pp. 799-806; P. J. Marshall, *Bengal: The British Bridgehead*, Cambridge, 1987, pp. 63-4; and J.R. McLane, *Land and Local Kingship in Eighteenth Century Bengal*, Cambridge, pp. 27ff. for the devolution of authority during the Nizamat.
192. IOR, BRC, 28 April 1790.
193. IOR, HM, vol. 206, 25 March 1778, p. 338.
194. This gives a customary entitlement of a sixty-fourth part of a *seer* of a commodity weighed. *Koyals* were however used most extensively in grain markets (IOR, BRC, 11 September 1781).

195. Ibid., 9 June 1780; emphasis added.
196. 'Opinion of Canongoe Kanai Sen on Buzars' is recorded in IOR, BRC, 22 December 1788.
197. Ibid., emphasis added.
198. Ibid., 26 November 1789.
199. WBSA, PCR, Murshidabad, vol. 1, 3 January 1773.
200. Ibid., Murshidabad, 27 November 1775.
201. WBSA, ibid., Dinajpur, 14 January 1777.
202. IOR, CCR, 4 September 1780.
203. IOR, BRC, P/49/40 and P/49/41, 4 June 1773 and 3 August 1773.
204. The extremely high component—37 per cent on an average—of elite spending on charity and religious activities has two major implications for this study. First, it demonstrates the complex nature of agrarian commercialism. Obviously, the institutions or individuals supported by such largesse would in turn become zones of local consumption. This would provide the market and the producer, both agricultural and non-agricultural, with a major feeder of cash. Second, these disbursements demonstrate an intricate connection between a flow of cash and an elaborate system of religious patronage. This would have interesting implications for a discussion of legitimacy and cultural hegemony of the elites in rural society.
205. IOR, BRC, P/49/40, 4 June 1773.
206. Ibid., P/49/68, 14 March 1777.
207. Ibid., P/49/41, 3 August 1773.
208. 'The State of Salt in Moysadul, 1778', BM, Add. Ms. 29088, fol. 197.
209. See IOR, BRC, P/49/40, 16 May 1773; ibid., P/49/55, 5 September 1775; ibid., P/50/61, 17 September 1785; IOR, BRP, P/71/26, 4 June 1790.
210. Talluqdari *jama* ranged from Rs. 76,001 to Rs. 3.5 per annum in Dhaka (IOR, BRC, P/50/61, 30 September 1785 and IOR, BRP, P/71/26, 4 June 1790).
211. IOR, BRP, P/70/23, 18 June 1787.
212. IOR, BRP, P/71/13, 27 August 1787.
213. Ibid., P/70/30, 21 June 1787.
214. Ibid., P/71/26, 4 June 1790; emphasis added.
215. See Sinha (*Economic History,* vol. 2, pp. 274-5) for the failure of the East India Company to exercise control, i.e. get revenue or acquire information about the actual state or extent of these lands before the Permanent Settlement; also P. J. Marshall, *Bridgehead,* p. 126.

CHAPTER 4

The Agrarian Economy and the Dynamics of Commercial Transactions

BEHIND ALL THE developments in rural society were the rhythms of economic life, especially those of the manner, channels and agencies for the consumption and distribution of food. Apart from its importance for the physical survival of people, economic transactions and exchange in agricultural produce had a critical bearing on the structure and organization of internal markets,[1] the modes, methods and formation of mercantile activity, and the relationships between the merchants and the producers. Equally critical were the factors of demand and prices in shaping the dynamics of local trade and in influencing the spatial directions of the flow of commodities. These separate questions when combined together give us a typology of agrarian commercialism, and it is to this construction that we now turn.

Bengal's Economy and the Factor of Demand

The influence of town demand had a clear effect on local trade in agricultural produce. Contemporary observers talked eloquently of the increase in Calcutta's population throughout the century.[2] Dhaka had about 4,50,000 people living within its environs in 1765[3] and continued to be thickly peopled later on.[4] In 1757, Murshidabad was declared as 'one of the richest cities in the world'[5] and, in 1764, Robert Clive described the city of Murshidabad as 'extensive, populous and rich as the city of London, with this difference that there are individuals in the first possessing infinitely greater property than in the last [named] city'.[6]

Apart from these premier cities there were other towns which were positioned at medium levels of consumption and had their combined effect on trade. Bhagwangola, near Murshidabad, handling about 18 million maunds of grain in the 1760s[7] was one such centre. Azimganj in Hughli was another, having grown during the early eighteenth century as a centre of the grain trade between Murshidabad and southern Bengal: 'being one of the first gunges and established under a powerful patronage, it was invested with extensive controul [sic]' was how it was described in 1773.[8] Then there were the towns like Katwa and Kalna in western Bengal which, along with Calcutta, redistributed grain from the eastern parts of Bengal (like Jessore) to other parts of Bengal and Bihar.[9] Chinsura, a prosperous trading settlement, was 'thickly interspersed with houses and small gardens' in 1778,[10] whereas Chittagong was described as a 'very large and extensive town' in 1789.[11]

The demand for food generated by these towns exerted a crucial influence on the direction and movement of local trade. The average monthly consumption of common quality rice in Calcutta was 2,50,000 maunds in the 1780s.[12] Murshidabad needed more than 1,30,000 maunds of rice and 57,000 maunds of paddy for its sustenance in the 1790s.[13] Towns lower down the scale needed proportionate amounts of food. Unfortunately, these needs cannot be worked out even in the most tentative fashion because of the scarcity of relevant statistical information. Quantitative data of the amounts of money involved in the purchase of food and other items of consumption are equally scarce. However, the evidence indicates that the amount of foodgrains available for merchants from different localities for catering to town demand and for exports outside the province (referred to as the 'exportable surplus' in our sources) seldom exceeded 20 per cent of the gross agricultural output under normal circumstances.[14] The remaining portion, excluding 6 to 8 per cent retained as seed stock,[15] was consumed in the countryside.

Unfortunately, how much of this remaining produce entered the rural market, or how much of it was directly consumed within the household, and the proportion of both in the overall structure of rural consumption cannot be worked out due to the limitations of the evidence. A contemporary calculation valued the annual turnover of rice, both traded and untraded, at 43.8 million maunds in 1791.[16] If we see this in relation to the fact that the combined annual consumption of rice in Calcutta and Murshidabad was 4.56 million maunds (based on the data of monthly consumption given earlier), then there is a very strong indication that rural demand was indeed exerting an extremely dynamic influence on the movement of marketed rice in our period. It would perhaps not be incorrect to suggest that this demand probably accounted for 70 to 80 per cent of the local trade in rice.

TABLE 34. ESTIMATES OF VILLAGES IN DIFFERENT DISTRICTS OF BENGAL

Year	District	Number of villages
1760	Burdwan	8,000
1765	Birbhum	6,000
1771	Birbhum	4,500
1772	Nadi	3,499*
1774	Hijli	579
1778	Mahisadal	438
1778	24-Parganas	3,124
1778	Hughli	579
1778	Purnea	5,350
1778	Rajshahi	16,132
1788	Midnapur	4,303
1788	Jalasore	1,891
1788	Baldakhal	1,200

Note: *Indicates the severity of the famine of 1769-70 in this district.
Sources: John Johnstone, *Letter to the Proprietors of the East India Stock*, London, 1776, p. 6; R.K. Gupta, 1977; 47; B.M. Add. Ms. 29076, fol. 11; Add. Ms. 29087, fols. 58, 97, 119; Add. Ms. 29088, fols. 108, 116-22; IOR, BRP, P/70/49, 15 December 1788; ibid., P/70/48, 7 November 1788.

The idea that most of Bengal was an interlinked chain of 'innumerable villages' interspersed with a few towns was originally put forward by James Rennell in 1765.[17] The same theme recurs when W. Hamilton, writing in 1828, says: 'villages from 100 to 500 inhabitants are astonishingly numerous, and in some parts form a continued chain of many miles along the bank of rivers'.[18] The absence of area statistics for this period does not allow us to estimate the number of villages in all areas of the province, or account for their spatial distribution. Table 34 gives the data available.

Obviously clusters of villages would tend to be greater in areas closer to important commercial and administrative centres, but the spatial distribution of these villages does suggest a continuous, and often densely packed, chain constituting most parts of Bengal's countryside. Area figures given by Rennell and the data available in Table 34 show the spatial distribution of villages in our period as shown in Table 35.

Population data for these villages are practically non-existent, though Johnstone's estimate does give us an average of 250 persons per village in Burdwan in the 1760s.[19] Relatively large villages could have more than 250 households around Murshidabad,[20] whereas in a relatively less

TABLE 35. SPATIAL DISTRIBUTION OF VILLAGES

District	Village: square mile ratio
Burdwan	1: 0.64
Birbhum	1: 1.16
Midnapur	1: 1.14
Purnea	1: 0.96
Rajshahi	1: 0.80
Nadia	1: 0.91
Jessore	1: 1.71

Source: James Rennell, *Bengal Atlas: Containing Maps of the Theatre of War and Commerce on that Side of Hindostan*, London, 1781, p. 53 and Table 34.

developed districts like Commilla 'the average is less than 10 persons [per] village'.[21] The crucial fact about these villages is that they jointly accounted for about 70 per cent of the agricultural product available for internal consumption. An interesting example of this phenomenon is provided by the rice of the spring harvest (*aus dhan*). It was generally recognized that this rice was intrinsically of an inferior quality than the one produced in winter (*aman*) and was consumed by the 'lowest and poorer classes'.[22] Rice of this kind would presumably have featured prominently in a system of natural exchange; yet the grain merchants made it a point to purchase the *aus* crop from the cultivators *before* they could dispose of their surpluses independently in local markets.[23] Since the so-called 'coarsest grains' were sown in the *aus* season, the intervention of the merchants here indicates the existence of an exchange economy in the province.

The important bearing this system of exchange had on the working of the provincial economy would come sharpest in focus under conditions of dearth. Witness for instance the following description of the state of the food-market during the dearth of 1791: 'The buzaars have hitherto been sufficiently well supplied to answer the immediate wants of the inhabitants; but the alarm of an approaching scarcity is now become so universal that *the poorer sort of people* will shortly experience considerable distress, as *the price of grain* and the difficulty of procuring it, *even for money*, is daily increasing.'[24] In an earlier drought (in 1774), the peasants of Burdwan complained that: 'Our condition is most miserable; for though there is grain in the hands of the Merchants . . . they have leagued together to keep the price up and we are perishing with hunger. If we *cannot procure food even with money*, how is possible for us to stay in the District?[25] These descriptions underscore the fact the money-

bazaar-market nexus determined the availability of food in the countryside as it did in the towns.

Who were the consumers? Large and intermediate towns exclusively depended on the arrival of traded rice for their sustenance. The social organization in the urban areas was complex and was based on a hierarchy of occupational groups and standards of living; therefore the structure of demand in these centres was correspondingly complex.

With regard to rural demand, the evidence is less definitive. A detailed list of houses, compiled in 1775, in *tarf* Rangamati, *chakla* Murshidabad is of interest.[26] Out of a total of 256 households, *chasis* (cultivators) accounted for 101, that is 39.45 per cent, of the total. The rest were as detailed in Table 36.

This evidence reveals a high concentration of non-agricultural occupations in rural society. Pure artisans, that is, those who were completely

TABLE 36. PROFESSIONAL GROUPS IN RANGAMATI, 1775

Profession	Number of households	% of Total
Putwa, those who breed silkworms	22	8.59
Tanti (cotton weavers)	6	2.34
Carpenters	3	1.17
Smiths	2	0.78
Barbers	1	0.39
Dokandars (shopkeepers)	2	0.78
Teli (oilpressers)	5	1.95
Widows	3	1.17
Fishmongers	7	2.73
Manjhi (boatmen)	1	0.39
Goala (milkmen)	17	6.64
Coolies	22	8.59
Chassars (silk weavers)	3	1.17
Officials	33	12.89
Mendicants	13	5.08
Bamboo cutters	3	1.17
Moochi (shoemaker)	1	0.39
Moodi (grocer)	1	0.39
Baori caste	8	3.12
Filature worker	1	0.39
Unspecified	4	1.56
Total	158	61.68

Source: WBSA, PCR Murshidabad, vol. 7, 15 February 1776.

separated from any form of agricultural production, accounted for 27 per cent of the households; officials comprised a sizeable 12.89 per cent. This suggests quite strongly that an overwhelming number of people in rural society, at least 50 per cent, depended substantially on the market for their subsistence requirements. Significantly, therefore, this *tarf* had two grocery shops plus a *haat* with thirteen shops divided as follows:

Moodis (grocers)	3
Cowri shops	3
Shroffs (moneychangers)	2
Tobacconist	1
Carpenter	1
Widows	2
Pashari (general provisioner)	1

A survey made by Robert Kyd of the villlage of Sibpur, on the western bank of the Hughli, showed that in 1791 out of a total of 419 Hindu households in that village, pure agriculturist, *chassis*, comprised 106 households thus accounting for 37.23 per cent of the residents.[27] The rest were divided in the fashion as shown in Table 37.

Thus, in Sibpur, a purely non-agricultural population of artisans constituted 15.5 per cent of the households. A further 21.72 per cent were fishermen, the nature of whose occupation made them net consumers of food. Added to this was a substantial—20.51 per cent—upper caste rural gentry who would generate a systematic demand for

TABLE 37. PROFESSIONAL GROUPS IN SIBPUR, 1791

Professionals	Number of households	% of total
Brahmins	65	15.51
Vaidya or physicians	3	0.71
Kayasthas	18	4.29
Moira or confectioners	10	2.39
Kasari or brass makers	2	0.48
Milkmen	4	0.95
Chootor or carpenter	14	3.34
Soonar or goldsmith	3	0.71
Teli or oilmen	6	1.43
Tantee or weavers	26	6.20
Bania or shopkeepers	27	6.44
Jallia or fishermen	91	21.72

Source: IOL Ms. Eur. F. 95, fol. 64.

the relatively luxury end of the market. In all, more than half of Sibpur's population depended upon the market for subsistence; of this the dependence of the artisans and the fishermen was the most critical, and it is perhaps no accident that 6.44 per cent of the population in that village was that of the *bania*.

The high concentration of a food-dependent population in rural Bengal as demonstrated by the case of Rangamati and Sibpur are not isolated cases. According to Buchanan, there were 1,00,809 houses of artisans in Rangpur. In addition were the categories shown in Table 38.

A contemporary estimate computed that the all-Bengal average of members per household was of 5.5 persons.[28] This would give us a population of 5,54,450 artisans in Rangpur alone at the turn of the century: a concentration of nearly 207 artisans per square mile of territory.[29]

The patterns of consumption are also indicative. Table 39 gives the annual consumption of various items by a 'Hindoo family in Respectable Circumstances' in Buldacaul in eastern Bengal surveyed by commissioner J. Paterson in 1789.

Paterson additionally noted that a 'Mussullman family in similar circumstances differs from the Hindoo in the addition of onions and garlick to their food and less pawn [betel-leaf] being consumed. They are less expensive in their clothes, ornaments and utensils [and] the expences of a Mussulman for religious ceremonies are not so great as those of the Hindoos.'[30]

On the other hand, the expenditure of an artisan in somewhat comfortable circumstances in maintaining a family of six persons in Rangpur is shown in Table 40. Finally, the consumption patterns of a 'common artificer' are detailed in Table 41. Buchanan's data also show that different ranks of people in Rangpur spent on food in fashion as shown in Table 42.

TABLE 38. PROFESSIONAL GROUPS IN RANGPUR, c. 1807

Professional groups	Number
Kolu or Oil Mills	3,254
Distilleries	27
Badyakar or common musicians	2,660
Notis or dancing girls	79
Kirtaniyas, those who sing the praises of different Gods	578

Source: F. Buchanan, 'Survey of Ronggoppur', Statistical Tables, Table 39.

TABLE 39. CONSUMPTION PATTERNS OF THE RURAL GENTRY
IN EASTERN BENGAL, 1789

Item	Value (Rs.)	% of Total
Rice	199.65	33.18
Salt	8.25	1.37
Oil	27.38	4.55
Turmeric	1.37	0.23
Pepper	2.81	0.47
Ginger	1.37	0.23
Dhania (coriander)	1.37	0.23
Paan (betel-leaf)	5.69	0.94
Betel-nut	5.50	0.91
Tobacco	4.56	0.76
Gur (Molasses)	22.81	3.79
Fish	17.06	2.83
Vegetables	17.06	2.83
Firewood	0.75	0.12
Chunam (lime)	0.75	0.12
Clothing	48	7.98
Bedding	29.25	4.86
Utensils and Ornaments	208	34.57
Total	601.63	

Source: IOR, BRC, P/51/40, 15 July 1789.

The fact that expenditures on food were inversely proportional to an individual's economic station in life is perhaps a strong statement in favour of the argument that overwhelming numbers in Bengal were dependent on the market for their subsistence. Coarse rice and salt were items which even people at the lowest levels of income, such as a 'man who labours for others in the field'[31] or a 'wretched boatman' in Bengal's numerous rivers and creeks[32] had to purchase on a continuous basis. In Rangpur, such people never used fish nor vegetables in their diet, but there were occasional purchases of pulses, oil, tobacco and betel-leaf.[33] The significance of such high clusters of people depending on local networks of exchange for their subsistence and other requirements is heightened when we consider the fact that the worst brunt of the famine of 1769-70 was borne by the rural artisan: 'the number of *consumers* who suffered by that calamity was greater in proportion, than that of the cultivators of grain'.[34] The mortality among this group is said to have

TABLE 40. PATTERNS OF CONSUMPTION OF AN ARTISAN IN COMFORTABLE CIRCUMSTANCES, RANGPUR, c. 1807

Item	Annual value (Rs.)	% of Total
Rice	32.25	48.59
Pulse	1.5	2.26
Salt	4.5	6.78
Mustard oil	8.25	12.43
Onions, garlic, capsicum and turmeric	1.12	1.69
Fish	5.25	7.91
Vegetables	3.00	4.52
Betel with spices	6.00	9.04
Molasses or *gur*	3.75	5.65
Tobacco	0.75	1.13
Total	66.37	

Source: F. Buchanan, 'Statistical Tables of Ronggoppur', IOL, Ms. Eur. G. 11.

TABLE 41. PATTERNS OF CONSUMPTION OF A COMMON ARTISAN: RANGPUR, c. 1807

Item	Annual value (Rs.)	% of Total
Rice	22.5	84.43
Pulse	0.22	0.82
Salt	0.75	2.81
Oil	1.01	3.79
Fish, vegetable, turmeric and capsicum	1.2	4.5
Betel-leaf and nuts	0.75	2.8
Tobacco	0.23	0.86
Total	26.65	

Source: F. Buchanan, 'Statistical Tables of Ronggoppur', IOL, Ms. Eur. G. 11.

accounted for 25 per cent (in Purnea) and 50 per cent (in Malda) of the total deaths which occurred in that catastrophe.[35]

With regard to the demand for agricultural produce in general, the fragmentary bits of evidence which are available may perhaps allow the following purely tentative reconstruction. J. Paterson, Commissioner at Commilla, calculated that in 1791 a 'Hindoo family in respectable circumstances' consisting of 16 persons, including dependents, spent

TABLE 42. PATTERNS OF FOOD CONSUMPTION
IN RANGPUR, c. 1807

Category	Total annual expenses (Rs.)	Expense on food (Rs.)	% of Total
First Class	4,095.13	658.81	16.09
Second Class	1,443.37	354.87	24.59
Third Class	447.12	195.62	43.75
Fourth Class	168.44	84.75	50.31
Fifth Class	65.01	45	69.22
Sixth Class	32	27	84.37
Seventh Class	25.75	21.37	82.99

Source: F. Buchanan, 'Statistical Tables', Table 39.

roughly Rs. 333.50 per year on food.[36] This would give us an annual per capita consumption of Rs. 20.54. This figure closely approximates to the evidence from Rangpur where F. Buchanan's estimate of the domestic consumption of 'seven ranks' of people in 1807 gives a crude per capita consumption average of Rs. 21.02 a year.[37] The population of Bengal and Bihar in 1790 was stated to be 22 million,[38] of which 12 million may have been the share of Bengal.[39] Multiplying the latter figure by the lower rate of Rs. 20.54, we arrive at a purely tentative sum of Rs. 246.8 million annually being spent on food in the province.

An estimate such as this will be open to a wide range of criticisms and all of them may perhaps be valid. This estimate does not take into account the regional variations in consumption, it assumes a fixed pattern of consumption for all social strata, it does not discuss consumption expenditures over time and fails to relate these expenditures to the price situation—these are perhaps just a few objections which can reasonably be anticipated. The estimate I have made is *purely tentative* and is an attempt to indicate the money involved in the annual demand for agricultural produce. The other reason is to draw attention to the fact that such large amounts could not have emerged from consumption in towns alone.

The Regional Food Market:
Some Evidence of Price Movements

One major indication of an integrated provincial market would be the behaviour of prices of food across the region. In this connection, M.M. Postan's study of the medieval British economy provides helpful insights into the working of a pre-modern market. It is essentially symptomised by wide fluctuations in prices and in the flow of commodities from season to season and from locality to locality, as well as by the seasonality of the consumers. 'Grains were at their cheapest in the early autumn, i.e. immediately after the harvest had been gathered, since the villagers then had some grain of their own to eat. But they almost invariably rose in the summer, since by that time many villagers had exhausted their own grain supplies and swelled the ranks of the buyers.' Incidentally, the 'narrowness' of the market 'merely widened the amplitude of fluctuations [of price]' and that, 'prices would not have risen or fallen as sharply as they did had the buyers and sellers been more numerous, the volume of commodities larger, and the access to imported food easier'.[40]

On the basis of the above typology, an integrated market, therefore, would exhibit opposite tendencies. Prices would not show such wide amplitude of fluctuations (at least not under normal climatic conditions), though a seasonal price-swing would be naturally built into the movement of prices. In other words, prices would tend to be lower immediately after the harvest and rise subsequently, not because of the impermanence of consumers but as a normal trend from season to season. In addition, a permanency of 'pure' consumers would generate a palpable demand for a basket of goods and services with a large number of buyers ensuring commodity transactions of a particular order. Further, a well-oiled machinery of trade in foodgrain would exist and importation of food both to meet exigencies and on a regular basis would be possible without any or much internal barriers. Finally, with regard to the significance of a secular price-trend, one can argue following Labrousse that 'as opposed to short term convulsive movements. . . a more extended and progressive price rise . . . spells expansion and prosperity, [while] a [sustained] decline denotes economic recession'.[41]

A recent study of the integration of the really modern grain market in Scotland[42] makes the following pertinent observations regarding the indices of market integration:

Most recent work relating price movements to market conditions has been predicated on the assumption that the degree to which prices in different markets fluctuate in unison reflects to the extent which those markets were associated or integrated. . . . As knowledge and trade flow more freely and unity of marketing is achieved over a number of previously disconnected markets, so any local imbalance of the effective supply would be transmitted throughout the wider marketing region. Grain movements would seek to address such imbalances and price movements in one area would naturally come to be reflected by price movements elsewhere in the enlarged market region.

Specific indicators of such market integration as suggested in this study are: an increased synchronicity of price movements of a particular grain in different regions and a concomitant decrease in the synchronicity of price movements of different grains within a single region. A third possible consequence of market integration 'would be a decrease in the volatility with which individual grain prices fluctuated from year to year'.[43]

While there is no doubt that market integration would mean increased synchronicity of price over different regions largely because of the nature of demand and the networks of local trade, a continuing volatility of individual grain prices would not necessarily cut into this integration. Grain prices were initially determined by the state of the current harvest which was in turn influenced by prevailing climatic conditions. As the latter variable was beyond the control of human agency, the amplitude of price fluctuation would continue notwithstanding a high degree of market integration. Under such conditions the inelasticity of demand would be the oil and local trade, the motor of this integrative process. As has been shown below, merchants would be the principal beneficiaries of the volatility of spot prices. Gibson and Smout also suggest that market integration results in a better access to food from outside, especially in case of localized crises of subsistence. This eliminates the need for substitution in grain consumption (that is switching to locally available lower grade crops) as would happen in an incompletely integrated market. 'As market integration proceeded there would be less need for such substitution, and grain prices within a single market would tend to move more independently of one another'.[44]

Figure 5 charts the movements of two basic commodities, fine and common rice, and of wheat in lower Bengal between 1700 and 1800 on the basis of the Chinsura series provided by G. Herklotts (in *Gleanings in Science*, vol. 1) in January 1829.[45] Figure 6 provides the prices of coarse rice of the *aman* and *aus* harvests in the district of Birbhum between 1784 and 1813 on the basis of the series of prices provided by W.B. Bailey in the *Asiatick Researches* (1816).[46] Both these figures are unequivocal about the synchronicity of prices.

The reason for such symmetry was to be in all likelihood the inelasticity of demand for rice in the countryside. The fact that nearly a quarter of Bengal's society, especially rural society, depended on the ebb and flow of commodities in the market is eloquent testimony of the pull of a regional demand. The impermanence or uncertainty of consumption imposing a deleterious impact on prices was thus not a problem in the economy. Naturally, demand influenced the movement of agricultural prices. In this, the demand for the most basic necessities, especially food grains, tended to exert an extremely powerful effect on the movement of prices because of its intrinsic inelasticity owing to the increase in the numbers of those who did not produce their own food, either because they lived in the cities, or belonged to the rural consumption-elite or comprised those peasants who had little or inadequate agricultural land and had to rely on wages earned as sharecroppers or agricultural labourers. Since consumption behaviour was a function of income, lower wage-earners would tend to consume coarser (that is inferior) grains within a region, thereby explaining the symmetry of price movements of rice especially as they are reflected in Figure 5.

The behaviour of the economy to such a buoyant demand for food is also apparent from the discussion of prices in Chapter 1 which shows that the steady secular increase in the prices of agricultural produce through the eighteenth century was accompanied by a positively correlated response of the production system which took the form of an expansion of the agricultural frontier.

Figure 7 compares the prices of common rice over the century from the series compiled by Akhtar Hussain for Bengal and the series provided for lower Bengal (Chinsura) by Herklotts. This figure demonstrates the striking concordance of prices between the Bengal averages, that is the prices which the Company paid in order to buy food chiefly for its garrisons and those prevailing in lower Bengal. This impression is buttressed by Figures 8 and 9 which separately show the actual price lines, quinquennial moving averages and the linear trends of the two series. The conclusion that there was a secular rise in the prices of agricultural goods is inescapable.

Prices of other agricultural products also rose in a similar fashion. Figure 10 gives the prices and their linear trends of two commodities, clarified butter and mustard oil, from the Chinsura series. Figure 11 charts the prices and linear trends of clarified butter and mustard oil in the city of Calcutta between 1754 and 1800 on the basis of the data provided by W.B. Bailey.

The price data show certain important trends. First, they reveal a generalized rise in the prices of all agricultural commodities. Second, there was a fairly continuous rise in the rural prices of rice and this feature is of great significance in understanding the developments in the provincial economy. In this context, Figure 12, which compares the prices of common rice in different districts with those prevailing at Calcutta for years when such comparative data are available, certainly allows the conclusion that it is for the first time in Bengal's history that we can speak of a provincial market integrated by uniform price-trends. That the economy was price sensitive is also highlighted by the fact that a rise in the price in some districts would immediately push up prices in far-flung areas. This aspect of price behaviour comes into sharper relief during times of dearth and famine. This I discuss in Chapter 5.

On the whole, it is the synchronicity of prices of different agricultural products which is the clearest indicator of market integration. In this connection Figure 13, which shows the movement of *aman* rice prices in Burdwan and Birbhum between 1784 and 1813, is a clear illustration of the symmetry of prices in two contiguous districts of Bengal.

Finally, agricultural prices exerted a grave influence on the prices of other commodities and on the living standards of both the agricultural and non-agricultural population, a point which requires further elucidation.

An inter-sectoral interdependency of prices, whereby the oscillation in one sector influenced prices in another seems to have been a well-established feature of Bengal's agrarian economy. Naturally, therefore, price movements influenced incomes in the different sectors of production. A rise in the price of rice or paddy meant that 'the ryotts are obliged to sell the produce of their lands dearer than formerly' while the manufacturer 'paying more than formerly for the materials & for the necessaries of life are unable to subsist without increasing the price of their goods'.[47] H.T. Colebrooke (writing in 1793) observed that the price of grain had the greatest influence on the prices of other commodities in the market.[48] The high price of cotton in 1789 was being blamed 'upon the famine [1788] which increased the value of corn'.[49] 'Should a degree of scarcity raise the price of grain above the average rate, it falls heavily on the manufacturer',[50] and an 'exorbitant encrease [sic] on the rate of the necessaries of life renders the ordinary allowances for labour insufficient'[51] are statements which clearly indicate the economic influence exerted by the state of agricultural prices on the conditions of practically every harvest-dependent social strata in Bengal. The people who were most affected by the state of agricultural prices were presumably the poorer-peasants having insufficient lands at their disposal, the rural and urban labourers and those artisans who depended exclusively on the market for their subsistence. A study of the weavers in the employment of the East India Company shows that a rise in food prices was fraught with severe consequences since these artisans had to pay more to buy food for their subsistence, but were themselves unable to raise the prices of their products.[52]

That the prices of non-agricultural products had risen prodigiously during the course of the late eighteenth century appears an undeniable fact. For instance, the market price of silk-cocoons apparently increased by more than 50 per cent within the space of one year between 1770 and 1771,[53] while there was a short-run slump in the prices of paddy and other 'inferior variety of grains' in many districts between 1771 and 1773.[54] Whatever relief this may have provided for the artisans was clearly to be short-lived, as 1773-4 and 1775 were once again years of dearth caused by bad harvests. As rice prices rose, the prices of raw materials and of the finished-good tended to rise proportionately. Thus by 1789: 'The unusual rise in the prices of the two principal articles so necessary to the weavers, Rice and Cotton, has created an encrease [*sic*] in the price of cloths at the Markets beyond what was ever known, and introduced a practice of reducing the number of threads of the warp, which has debased the cloths in their texture.'[55]

Another aspect, vital to the structure of local trade in food grains, was the nature of the price differential prevailing between town and country. Table 43 sets the widely scattered evidence regarding this in a somewhat comprehensible fashion.

Thus Bengal prices show an integrated economy undergoing a fairly

TABLE 43. PRICE DIFFERENTIALS BETWEEN TOWN AND COUNTRY (*Rs. per maund*)

Year	Place	Price in towns	Price in country
1773	Chittagong	0.62	0.55
1774	Chittagong	0.61	0.54
1775	Chittagong	0.62	0.56
1774	Burdwan	1.43	0.80
1775	Murshidabad	1.31	0.52
1788	Dinajpur	1.17	1.06
1789	Dhaka	0.68	0.53
1791	Birbhum	1.54	1.06
1792	Birbhum	1.40	1.05
1794	Dhaka	0.80	0.57
1791	Calcutta	1.38	1.14
1792	Calcutta	1.82	1.25
1794	Calcutta	0.94	0.44

Sources: IOR, BRC, P/49/44, 11 December 1773, 10 January 1774; P/49/47, 22 August 1774; P/49/50, 28 January 1775; P/51/21, 4 June 1788; P/51/50, 28 October 1789; P/52/37, 23 November 1791; P/52/42, 9 March 1792; P/52/44, 4 May 1792; P/52/45, 1 June 1792; Rajat Datta, 'Merchants and Peasants', p. 389.

noticeable rise in the rural prices of food, and this fact was important in shaping merchant-peasant linkages. In other words, the nature of price differentials between town and country, coupled with the rise in country prices had significant socio-economic implications. As will be discussed subsequently they provided a new direction to local trade in agricultural produce, not just in terms of movement of food, but also with regard to the social relationships which emerged in the countryside in this period.

Markets and the State

The importance of an orderly system of markets in the overall movement of local trade and the connection between trade and revenue was recognized both by the Nizamat and the East India Company. Murshid Quli Khan, the first really powerful Nazim of the province, seems to have made it his avowed policy to intervene in the movement of grain from the countryside to the towns only during periods of food shortages in an attempt to restrict monopolies by the grain merchants.[56] Apart from the tax motive, such intervention by the regional state was also guided by the immediate expedient of keeping supply lines open to towns under threat of scarcity. Price regulations and control by the Nizamat was an exercise in redirecting the movement of the agricultural surplus from the countryside to the towns during years of scanty harvests. In normal agricultural years, the relative balance between supply and demand was allowed to regulate both the amount of staples flowing into the towns and the prices at which these were sold. The *nirkh* (price rates) in the *bazaars* of the early eighteenth century signified 'the prices which the vendors generally regulated on their own', which were then confirmed by the official seal of the market supervisor (*daroga-i-bazaar*).[57]

The Nizamat's attitude towards the market under normal economic circumstances was guided by its own perception of mercantile activity and by the social milieu in which these markets were formed and functioned. The state's attitude to mercantile activity is perhaps evident from Alivardi Khan's assertion that the 'merchants are the Kingdom's benefactors; their imports and exports are an advantage to all men . . .'.[58] It was this attitude, held by the Nazims in general which probably resulted in their farming the trade of 'the several articles which constitute the internal commerce of Bengal and Bihar' to the local merchants: 'such commerce seems to have consisted principally, if not exclusively, in commodities of the natural produce, or manufactures, of these provinces' wrote Henry Vansittart (in 1762) about the state of internal commerce under the Nizamat.[59] Such an arrangement seems to have been mutually acceptable to both parties involved in the venture. For the state it was certainly considered financially viable;[60] so was it for the merchants.

Thus in Bishnupur, the betel-leaf (*paan*) merchants had in the past the 'exclusive privilege vested in them of vending all the betel produced in Bishnupur or brought from other districts [thereby becoming] purchasers at their own prices & vendors at what rates they please'.[61] The betel-leaf trade at Dhaka was similarly situated. The farmer of this trade (*paan mahal*) purchased zones of exclusive trade from the state and 'no other person could bring any betel leaf from the moffusul and if anyone attempted it, the farmer attached the leaf, as his property, by virtue of his engagements with the government'.[62] These exclusive rights of trade were not limited to a high-value article like betel. The Nizamat era seems to have seen the formation of such rights in practically all types of commercial activity, even at the lowest rung of the market (the *haat*) at the village level. At Nadia, petty retailers of rice and paddy paid special monetary gifts (*salami*) to the zamindar for the exclusive privilege of selling rice at the *haat*; there were *salamis* paid for acquiring similar rights over retailing salt, over gathering shells for making lime (*chunam*), for weighing and measuring commodities being brought for sale at the *haats* (*kayali*) and even for the sale of firewood to cremate the dead in the village.[63]

The other factor which influenced the state's attitude towards markets was the social origins of such places in Bengal. What mattered here were not merely the larger markets in towns like Dhaka and Murshidabad, but the whole range of small and intermediate markets (*haats* and *bazaars*) which were linked in a pervasive commercial network. As discussed in Chapter 3, some of these smaller centres of redistribution were created partly by the family members of the Nazims and by their officials. In greater part however, they were the products of zamindari initiative. A census of the number of markets in Burdwan in 1790 showed that the Rajas actually owned 7 *ganjs*, 1 *bazaar* and 16 *haats*. Of these Raja Tejchand had established 9; Trilok Chand, 11 and Kirti Chand, 4.[64] Of the 42 *haats* in pargana Buluah in 1795, 39 belonged to zamindars 'in regular succession' while the remaining three were 'dewattar'.[65] In Dhaka, several 'petty bazars' were established by holders of revenue-free lands, and the profits from such markets were 'employed in defraying the expences of different Musjids [mosques] & Takoor

Baris [temples] & for performance of religious ceremonies'.[66] It is, therefore, not surprising that the zamindars persisted in claiming from the Company 'an equal right in rents arising from gunges and bazars' as late as 1790,[67] when the entire political equation had changed in the province.

The Nizamat implicitly recognized the importance of mercantile activity and of an articulated network of markets to the regional economy. An illustrative example of how commercial considerations shaped the attitudes of the potentates of the state can be given from the case of liquor manufacturing and retailing in the environs of Murshidabad. Here 38 out of 80 shops selling high grade liquor, 28 out of 53 shops retailing toddy, and 62 out of the 139 shops dealing in opium in 1790 were owned by Nawab Mubarak-ud-daulah and his family.[68] Given the fact that the 'vend of liquor is extensive in towns',[69] ownership of retailing outlets provided supplementary income for the town elites, thereby showing the extent to which the upper crust had become integrated with the commercial network. Further down the social scale we find merchants, even at the *haat* level, willing to pay a tax (even a bribe) in exchange of acquiring exclusive privileges of trading even in basic staples like rice.

But there were still major barriers in the way of a regionally integrated market in the first-half of the century. The fact that merchants, and even petty vendors at the *haat* level, were able to carve out petty domains of privileged trade, and that zamindars and other landed proprietors were the prime agents for the establishment of these markets jointly militated against the development of an unfettered system of markets in the province. The reason for this zamindar/*byapari* combination was largely due to the state's internal need to balance the two social strata in order to ensure its own stability[70] in the midst of a prosperous economic situation. However, the overall outcome of such an arrangement seems to have been a combination of two developments: (*a*) a proliferation of zamindari *chowkis* (outposts) to collect tolls at various rates dictated by the financial predilections of an individual zamindar;[71] and (*b*) continuous conflicts between merchants and zamindars, and between zamindars and other landed proprietors, over the rate of tolls, over market jurisdictions and the movement of commodities. These conflicts often assumed violent proportions and could even disrupt marketing networks in the short run.[72]

The post-1757 era saw the state and markets interacting along significantly restructured lines because of the changing political and commercial situations in Bengal which necessitated a closer control over internal markets than what had prevailed under the previous regime. After 1757, state intervention assumed the apparently contradictory forms of rigorous control in the marketing of some commodities and a relatively striking non-interference in the movement of others. The pressures of a world market meant that commodities like textiles and opium (and later indigo) were to be rigorously controlled at all levels including production. After 1772, salt was added to the list of official mono-polies, not so much for its overseas value as for its being lucrative in the internal trade of the province.[73]

For trade in other commodities, especially that in agricultural produce, the Company's attitude was one of ensuring 'fair trade',[74] which entailed the dissolution of the restraints not only of the type inherited from the Nizamat, but also of the type fostered by the Company's own officials in the form of the Society of Trade. The latter task was tackled by the administrative exercise of prohibiting the private trade of its officials from 1771 onwards, and by attempting to place severe restrictions on profiteering and hoarding of grain during times of scarcity.[75] The former job occupied a major portion of its official business, especially after the disastrous famine of 1769-70. With regard to the internal barriers to the movement of trade, the Company's attitude was that *chowkis* were inimical to fair trade and to the honest trader as these, and the tolls levied there, meant that merchants were 'too frequently and . . . unnecessarily subjected to the exercise of authority' other than the Company's and that too many transit duties tended to push up prices beyond any reasonably accepted standards. The logical step from this type of reasoning was that the *chowkis* had to be abolished and duties had to be streamlined so that prices 'of manufactures and of the necessaries of life' could be brought down.[76] Implicit also was the need to display the Company's newly acquired political power. 'A market is a place', wrote Vansittart in 1778, 'where authority must be exercised to regulate the weights and scales, to preserve order and to afford protection to the persons who frequent it. . . .'[77]

The first major step in the realization of the Company's aims came in 1773 when:

(1) all duties levied upon grain 'in its transportation from the country' were abolished;
(2) duties on trade in agricultural produce were henceforth to be collected 'only at the capital towns whither it is brought for

consumption', and the management of such duties was to be under five customs houses to be established and stationed at Calcutta, Hughli, Murshidabad, Dhaka and Patna;

(3) 'all road duties [*rahdari*] whether by land or water exacted antecedent' to the regulations of 1773 were to be made null and void;

(4) 'all the inferior types of chokies [*chowkis*]' over all types of trade routes were to be dismantled in a phased manner; and

(5) the right of the local merchant to be 'at liberty to carry his merchandize where ever he thinks proper for sale' was to be ensured. [78]

The thrust of these regulations seems to have had an immediate impact, at least on the movement of trade to the towns. For example, the general levies made on zamindari piers (*ghat chowkis*) between Murshidabad and Calcutta declined from a previous (in 1756) range of Rs. 4 to Re. 1 to a high of Rs. 3 and a low of Re. 0.12 in 1774.[79]

The control exercised by the zamindars and talluqdars over markets was not easy to tackle. The main reason for this was the continuing resistance of these people to any state interference in what they considered to be their hereditary and ancient rights.[80] The problems in this regard were many, thereby showing the complexities involved of state intervention in a sphere it considered a pure economic institution but which was in reality a distinct form of agrarian property. Markets were established by 'zemindars, talookdars and every denomination of rent free holders'[81] with a view to their income, but like all things privately owned, these were often sold to a host of buyers which created major problems in the way of the state's plans for outright dispossession. For instance, a detailed survey of markets situated around the town of Murshidabad in 1793 showed that 23 of the 32 markets situated in its proximity had changed hands within the past fifty years.[82] The buyers of such places, according to John Shore, were 'proprietors' and dispossessing such a person would amount to taking 'away his whole property from him; and this in Bengal would excite clamour and discontent in the proprietors'.[83] The apparent explosiveness of the situation forced the Company to drag its feet till about 1790 when the first major steps were taken to bring the landed proprietors to heel. The option of outright dispossession was hotly debated in the Board of Revenue, but was finally shelved. In order to circumvent the 'clamour and discontent', the state took a close look at the pattern of ownership and the structure of taxes and duties (*sair jihat*) in these market places, established a major difference between the two and decided to continue one and abolish another.

What the Board of Revenue did was to make a separation between rents collected in these markets and taxes collected on trade. Rents were designated as 'any collections made . . . as a consideration for the use of grounds, shops and other buildings belonging to landholder', whereas taxes were deemed to be those levied as a 'duty on commodities' and on transit of goods. A landholder's right to collect rent was considered his 'private right' and therefore not to be interfered with, but the right to tax was construed as the 'exclusive right of government'.[84] Thus on 20 July 1790, landholders were prohibited from all involvement in the collection of *sair jihat* in lieu of a fixed compensation, and henceforth 'no proprietor of land [would] be admitted to any participation thereof, or be entitled to make any claims on that count'.[85]

The Company's intrusion had a major impact on the state of Bengal's internal market. It led to that crucial bit of state intervention necessary for the final crystallization of an integrated market for agricultural produce in Bengal, thereby bringing to a culmination the processes already set in motion under the Nizamat.[86] The Regulations of 1773 were able to free the merchant from the clutches of zamindari outposts and toll-stations, at least in the commercially important areas of the province. The merchants were quick to react. Almost immediately they resorted to establishing '*private gunges* for the reception of all goods brought to the market by themselves and others' and were reportedly busy turning out from 'their golahs [granaries] and landing places the sircars [officials] and kyalls [weighmen] who are employed by the [Company's] Customs House'.[87] For the merchants this jurisdictional redistribution actually served to expand their direct control over the internal market, both over the networks in them and over their spatial distribution.

The countryside of Bengal underwent a proliferation of such 'private gunges'. By the 1770s Calcutta was definitely undergoing an increase in 'various kinds of hauts and bazars',[88] which, in some quarters (like the police) was seen as a 'great public nuisance . . . by the general exposure of provisions of all kinds in the highways and the innumerable shops and sheds erected thereon'.[89] In the 1780s, the 24-Parganas, situated south of Calcutta, was seeing an 'extension of hauts and bazars . . . by which old hauts are destroyed and new ones constructed'[90] and by 1792 this district had 144 markets of which *haats* numbered 100.[91] In 1790 Jessore was served by a chain of 225 *haats*,[92] and Burdwan had 17 *ganjs* and 345 *haats* in the same year.[93] The figures in Table 44 give some

TABLE 44. NUMBER OF MARKET-PLACES IN SELECTED DISTRICTS

District	Year	No. of markets	Year	No. of markets
Dhaka	c. 1765	536	1791	650
Jessore	c. 1778	69	1790	225
Rangpur	c. 1770	321	c. 1807	591
Dinajpur	c. 1770	206	c. 1807	635

Sources: Taylor, *A Sketch of the Topography and Statistics of Dacca*, p. 203; BR Misc., IOR, P/89/37, 27 July 1791; BM, Add. Ms. 29088, fol. 150.

indication of the increase in the numbers of market places in some districts of Bengal during our period.

The scale with which markets in Bengal appear to have expanded in this period was almost spectacular. A comparison with China is revealing. The Ch'ing period which was characterized by an expansion of trade and the growth of 'periodic markets',[94] was undergoing a general 'accretion of markets'. For instance in Szechwan the number of standard markets increased from 4 in 1622 to 13 in 1875.[95] Compare this to the fact that in Dhaka district alone there were 536 markets (*haats, bazaars,* and *ganjs*) in 1765; these had increased to 650 in 1791 (see Table 44). Obviously, areas previously deficient in markets were now brought under their purview. This facilitated peasant-market integration which, as is being argued here, was one of the distinctive developments in the eighteenth century. Skinner sees the proliferation of marketing centres in Ch'ing China as the 'intensification of the rural landscape'. In China this meant an increase in the volume of trade in an average market day at already existing markets and/or an absolute increase in the number of fixed market days.[96] Considering the phenomenal expansion in Bengal, it seems likely that these features would be present here; and indeed they were. Apart from the *ganjs,* which were fixed daily markets, the *haats*— the periodic (usually weekly) village market—in rural Bengal were beginning to be assembled every alternate day in certain localities.[97] Even in low-lying areas prone to inundation and floods by the seasonal overflowing of its rivers, markets were held 'for four & sometimes in some places for six months of the year on board of boats'.[98]

Commercial Profile of Local Marketing Networks

One aspect, crucial to the movement of local trade in Bengal was that many of the local marketing systems (*haat, bazaar* and *ganj*) had become specialized agencies for the circulation of food and other items of immediate consumption. Bhagwangola, the main centre feeding Murshidabad handled about 18 million maunds of rice per year in the 1760s.[99] Large markets (*ganjs*) in Calcutta, like Baitakhana and Sovabazar, derived their tax revenue chiefly from the rice and paddy brought there for sale.[100] Similar patterns can be seen even in those areas where large towns did not exist. Purnea's annual trade in grain was worth 2 million maunds; single parganas in Dhaka circulated about 1.8 million maunds of paddy and rice a year; and individual *haats* (village markets) in Bakarganj handled on an average 3,50,000 maunds of rice and 1,20,000 maunds of paddy a year.[101]

Estimates of the share realized by agricultural produce in the total amount of local trade are few, but those available for rice and paddy do strongly suggest that these were remarkably high (see Table 45).

Commercial dealings in these markets seem to have been quite buoyant in the period between 1760 and 1800, both in terms of the frequency of transactions and in terms of the social participation in them. In Jessore, 'every pergunnah and village have established bazars and hauts. Several of the villagers keep shops in them, while others hold them at their houses'.[102] The *moodies* (grocers) of Calcutta traded in *bazaars* and from their own houses, while the *tahbazaris*[103] went to markets during the day 'exposed their goods on stalls, or in temporary shops outside the established market' and returned home at night, only to arrive the next day 'with no other intention than to vend their articles which are usually of a perishable nature and must be sold within the day . . .'.[104] In Dhaka, the *haats* assembled twice or thrice a week, those in the city were open daily, and the main items of trade consisted of 'agricultural produce and of native manufactures'.[105]

These markets catered for a variety of local needs, both of commodities and services. A purely rural market supplying a few villages (*tarf*) of Rangamati (with a total of 256 houses in 1776) had 13 shops which

TABLE 45. VOLUME OF LOCAL TRADE IN FOOD GRAINS

Year	District	Trade in grain (% of Total)
1771	Jalalpur	56.22
1779	Pagladanga[106]	62.37
c. 1807	Dinajpur	67.9
c. 1807	Rangpur	49.64

Source: Rajat Datta, 'Merchants and Peasants', p. 145; IOR, CCR, P/67/76, 12 March 1779; Martin, *Eastern India*, vol. 3, Appendix F; F. Buchanan, 'Statistical Tables of Ronggoppur', IOL, Ms. Eur. G. 11, Table 38.

included 4 grocers, 1 tobacconist, 3 dealers in cowries and 2 money-changers (*sarrafs*).[107] A market established in early 1778, about '760 cubits away' from the big Sovabazar of Calcutta, had, by 1779, 54 shops comprising 39 *tahbazaris*, 9 *moodies*, 3 fish-mongers and 3 sellers of 'threads and blankets'.[108] A *bazaar* in Sutanuti (established in 1777) had, by December 1778, managed to attract a sizeable number of 101 permanent shops and 731 *tahbazaris* vending their goods in the open;[109] 6 new shops were added to this market in 1779, of which one was that of a *sarraf*.[110]

The apparent vibrancy in these markets can be explained partly by the exercise of the Company's political will in freeing the markets from the traditional social control of the landholders and partly by the significance of town demand. The *haats* of Jessore, for instance, seem to have proliferated in the 1780s, precisely at a time when its estuarine marshes were being reclaimed; and these markets functioned as a chain of feeder lines between the Sunderbans and Calcutta, which along with Katwa and Kalna formed the three consumption and redistribution centres in western Bengal.[111] But equal emphasis must also be given to the place occupied by rural demand in this situation. It is perhaps significant that Tippera, commonly recognized as having one of the most inhospitable terrains in its hilly parts, had 'upwards of 300 hauts' in 1790,[112] and Sylhet had 'no fewer than six hundred gunges and bazars' in the same year.[113] The flood prone area of Contai had *haats* where, apart from the usual trade in agricultural produce, 'small quantities of thread, coarse weaving cloth, mats made of split bamboos, brass and tutenag plates, koddalies [spades], ploughs, plough shares and ruts for winding thread' were sold at regular intervals.[114] In Rajshahi *haats* took to boats during peaks of monsoon flooding.[115] These examples reinforce the argument that there was a widely pervasive structure of rural demand. It is no accident, therefore, that the grain *byaparis* based in towns would often be found sending their surplus stocks, ostensibly kept to meet town demand, back into the countryside at the slightest available opportunity.

A Social Profile of the Byapari

Trade in agricultural produce was the task of a specialized community of merchants at all levels of the tiered markets in Bengal. At the *ganjs* and *bazaars*, there existed the 'principal' or 'capital' merchant, variously called the *goldar* (*golahdar* or owner of granaries), *aratadar* (wholesaler) and the *bhusi mahajan* (wealthy in grain) depending on the colloquialisms of different areas.[116] Below them existed a wide range of petty traders who usually traded with their stocks, but also functioned as part of an extensive network of commercial dealings created by the town based merchants. The *farias* (pedlars) of Bakarganj[117] and the *paikar* and *baladiya* of Purnea[118] are examples of such lower groups of traders who, while functioning independently at one level also combined the role of middlemen for some other, and obviously bigger, merchant. This would make the structure of local trading networks more of a multiform set of exchange relationships between a number of differentiated partners rather than a 'pyramidal structure' with an easily identifiable base and apex.[119] Multifarious relations of exchange would explain the function of the itinerant traders, like the ones in Birbhum who were merchants 'not residing but trading in the *zillah* [district] through agents in grain' in 1796.[120] Such relationships would also explain the existence of *bara byaparis* of Rangpur who came on boats 'partly loaded with salt and other commodities and partly with cash' and made their purchases of 'grain, tobacco, oil and sugar', which were in turn sold to other merchants trading in the bigger centres.[121] In Dinajpur, the principal grain *byaparis* were stationed in Rajnagar (the main grain mart of the district) and traded through a wide range of commissioned agents (*gomasthas*) who made spot purchases and then transported the grain to Rajnagar for onward distribution.[122]

Such mercantile linkages also gave rise to various circuits of exchange between a whole set of merchants depending upon the nature of an individual's operations. In Jessore, for instance, *golahdars* sold to *bhashaneah* (river based) traders, who in turn dealt with petty retailers making spot purchases, but these retailers would often make spot purchases directly from the *golahdars* as well from the 'occasional vendors in the Bazars'.[123] We also come across a whole range of shopkeepers (*dokandars*) of Birbhum who were 'all retailers of grain' and dealt with the wholesale merchant who resided in the *zilla* and the itinerant trader who was usually non-resident.[124] Then there were the *tahbazaris* who, as in Calcutta, traded in open spaces outside an established *bazaar* mostly in items of immediate consumption. In Rangpur, Buchanan provides evidence for the existence of 17 specialized retailing outlets ranging from the grocer (*moodi*) to the dealer in unbleached cloths (*kaporiya*).[125] The existence and the apparent proliferation of such trading networks indicate the speed with which the circuits of exchange were completed in the local markets.

Who were the traders in agricultural produce? Data for reconstructing the social profile of the entire range of merchants in this sector are scarce. There is some evidence, however, for analysing the social origins of the 'principal' traders in the following tentative fashion. Buchanan lists merchants from Benares, presumably of the Khatri caste, and Gosains of north India as the richest merchants in Rangpur at the turn of the century.[126] In Dinajpur, the principal trader was the family-based concern of Bhoj Raj, an Oswal merchant from Rajasthan.[127] The single longest piece of evidence so far available of these merchants comes from the letters of R.P. Pott, the comptroller of government customs at Murshidabad, who (in 1787) described 'the whole body of grain beparis' in Murshidabad as being composed of 'four tribes', these being 'coyer, buccali, ouzineah and moorchak'.[128]

D.H. Curly makes an orthographical study of these names and then suggests that 'coyer' (*kaya*) refers to the merchants from Rajasthan, 'buccali' (*baqqali*) points to a traditional group of grain merchants common in all parts of north India, 'ouzineah' (*ujjaini*) alludes to merchants native to Ujjain and 'moorchak' (*murcha*) was the name of a merchant group trading with Murshidabad on the Jalangi river. Curly further suggests that these names do not refer to specific caste groups but rather to a set of regional identities and occupational positions; and that the *kaya* and *baqqal* probably represent a split between the Marwari and Bengali merchants.[129] A list of merchants trading in Rangpur in 1770, shows a dominance of the *banias* or *baniks* (*saha, pal, seth* and *poddar*),[130] while similar caste-names (*pal, saha, sheel* and *addi*) figure prominently among the principal grain traders in Calcutta's Sovabazar in 1787.[131] Of a total of 45,835 maunds of grain allegedly withheld from the markets of Midnapur during the famine of 1788, grain *mahajans*, or *baniks* figure prominently as responsible for controlling 37,710 maunds; the rest (8,125 maunds) was held back by the zamindars and chaudhuris.[132]

The overall weight of the evidence shows that wholesale trading in food-stuffs was the function of a specialized social community, the *banik* or *bania*,[133] and it is quite likely that it was in a process of transition from previously occupational groups into a distinct social caste in the eighteenth century. H. Sanyal's study of the Sadgops (an agricultural caste in south-west Bengal) and the *telis* (oil-pressers) documents the social movement of groups from occupational categories to specific castes or sub-castes in agrarian society from the seventeenth century.[134] His study perhaps lends credibility to the view I propose of the grain merchants coalescing into a specific caste during the period under review.

Mercantile Strategies: Creation of Intermediaries and Circumventing State Control

Crucial to the principal traders were the links they could establish between their trading headquarters and the supply bases, and the manner in which they did so is of central importance to the ways in which they exercised their control over markets. In essence what the merchants did was to establish a chain of intermediate dealing agents, the *gomastha* and the *paikar*. They traded with loaned capital for a commission, buying up from the *haats* as well as directly from the peasants.[135] The reason why these agents had easier access to the peasants' threshing floor was due to their ownership of pack oxen which facilitated their movements in the *muffassal*.[136] These agents were also instrumental in forwarding seed advances to the needy *raiyat* on behalf of the superior trader.[137]

Gomasthas and *paikars* purchased directly from the cultivators, and these purchases were made both at the latter's house and at the *haat*.[138] The *paikars* took small advances from the non-resident merchant (in Purnea these advances ranged from anything between Re. 1 and Rs. 30 at a time),[139] gave 'ample security for the money' and made their purchases 'at a rate sufficiently moderate to admit of the pykars selling of it at the gunge price without loss to himself'.[140] The *kaya* and *baqqali* merchants of Murshidabad had their kinsmen stationed in the principal markets of Dinajpur;[141] so did the non-resident merchants purchasing their grain in Rangpur[142] and in Birbhum.[143] But the networks did not stop at this. Big traders also entered into ties with other social groups not belonging to the same caste or kin. The *baladiyas* considered to be of a lower caste were the *paikars* of the grain merchants of Rangpur[144] and Purnea.[145]

The establishment of these trading intermediaries represents the growth of an interlinked capital in Bengal's countryside, a type of capital whose main concentration was in towns, and which, by a process of internal division became fragmented into numerous clusters. Yet, in essence, these clusters were only the other face of a centralized deployment of capital. That the management of capital was centralized is apparent from the way in which these different merchants functioned. The Murshidabad-based merchants, divided as they were into four major groups, managed their affairs under the leadership of a head who was appointed by the group. Each 'tribe', writes Comptroller Pott, 'has a head who manages the business in Murshidabad for the collective body'.[146] What these so-called 'tribes' did was to form tightly knit groups (*dala*) under the leadership of a *paramanik* or *dalapati* (the 'head' in Pott's language) who regulated the activities of its members spread out over an extensive catchment area. The Dinajpur merchants did so by having regular meetings (*baithak*) with their *dalapati* in Murshidabad where they formulated general principles and regulations in mutual consultation.[147] The importance of their trading concerns meant that restrictions on commercial and social dealings on the basis of caste could not be rigidly enforced. Other castes had to be incorporated in the wider network of the *dala*; even those considered low in the caste hierarchy were brought into its ambit as, for example, the *baladiyas* of Purnea and the *sahu* or *teli* merchants in Rangpur.[148]

All these factors resulted in a honeycomb of intermediate agents who were crucial to the control these merchants could exercise over markets. Close control over supply lines from the *muffassal* and an overriding profit motive were two inter-related considerations for any trader and these intermediaries were one way in which such aspirations could be realized.

Another crucial factor in merchant operations was the manipulation of agricultural prices. The price mechanism comprised two elements: the prices in the urban centres or in towns and the prices in the countryside.[149] The latter sphere determined what the merchants considered their price of procurement. The other aspect of the price situation (to which I have already drawn attention) was its apparent volatile nature, even during relatively normal agricultural years. Famines, or near-famines, and harvest failures exerted further destabilizing effects on prices.

To circumvent the problem of oscillating prices and to ensure remunerative returns at both ends of the trading scale were some of the principal concerns of these merchants. Any interference in this sphere was viewed with immediate hostility which even the Company realized much to its chagrin, especially during times of scarcity. Thus, in August 1774, when an 'unusual drought' was looming large and prices had risen to 'an alarming height' in the city of Murshidabad, boats loaded with rice at Bhagwangola would not proceed to the city as the *byaparis* 'were in hopes that the price would still rise, particularly if the unfavourable weather should continue . . .'.[150] During the widespread flood-induced famine of 1787, merchants continued to send rice to Calcutta from the already deficient places like Sylhet, Dinajpur, Dhaka and Rajshahi: 'the famine has already raised the price of rice considerably and the merchants continue daily increasing it by exportation'.[151] An attempt was made to place an embargo on exports from Rangpur by the Collector D.H. MacDowall, but the *byaparis* continued sending their stocks to Murshidabad, to be re-routed to Calcutta, saying that 'we have never sold grain at Rungpore, and from selling here great loss will accrue to us . . .'.[152] In fact, even Warren Hastings was forced to deregulate prices and suspend *ganj* duties to coax merchants to bring rice to Calcutta in 1784 (a year of a partial famine) as 'they were deter'd from bringing it to market because they were obliged to sell it at an arbitrary valuation'.[153]

Governmental efforts to meet this problem head-on met with less success than had been anticipated. They simply could not break through the mercantile web and deal directly with the cultivators. Merchants bought all the surplus grain available at the slightest suspicion of state interference. 'The mahajans having by some means obtained information of Government's intention are endeavouring to purchase up all the grain of the country in the expectation of making their own terms' wrote the collector of Tirhut in February 1795.[154] In Sylhet, attempts to keep the state's involvement in the purchases of grain under wraps failed in October 1794 as 'the rice merchants forming an idea that the government are in want of grain store it up to enhance the price and thereby distress the District'.[155] State buying in Purnea was similarly troubled. The collector there wrote in dismay that his attempts to purchase 10,000 maunds had increased the spot price of rice 'by one-fourth the next market day', whereas 50 merchants could purchase 20,000 maunds in one day '*without any enhancement of price*'.[156]

This covert resistance by the grain merchants made the Company reverse its policy initiated in 1794, to buy directly from the producers at regulated prices and to stockpile in 'public granaries'.[157] This resistance forced them to make purchases 'by private arrangements with grain merchants and other persons in such manner as may appear best calculated to procure grain at the cheapest rate',[158] even though it was universally known that the 'mohajons always sell at a higher price than what can be purchased from the ryotts'.[159]

Merchants, Price Control and Profits

Having circumscribed the interference of the state in agricultural prices the merchants set about determining these themselves. It was commonly accepted that the sale prices in towns and *ganjs* were determined by the merchants themselves,[160] presumably in their periodic *baithaks*. The merchants of Calcutta who purchased grain from the ones trading from Murshidabad and Dhaka frequently complained that coming to agreeable sale prices between these groups was difficult because of the interference of the Company in Calcutta's *bazaars*; their logic clearly being that prices had to be negotiated between the traders themselves, and that the merchants would trade wherever they received a better price: 'more will be brought to the market if the merchants are permitted to sell [at] what price they please' said the *byaparis* of Calcutta to John Shore in February 1788.[161]

Apart from manipulating prices during times of scarcity (discussed later), these merchants also took advantage of the seasonal flow of the principal riverain systems upon which the larger towns depended on their supplies of food from long distances. The following description pertaining to the supplies of food to Calcutta in 1791 is extremely revealing:

The times when rice becomes most scarce and dear are in a month or two after the communication between the Great River [i.e. the Ganges] by means of the Gelingee, Cossembuzar and Sooty Rivers, with the river Hoogly are stopt [*sic*], and become unnavigable, even for small craft, which usually happens by the end of November, or early in December, and continues to the end of May, or beginning of June, when these channels of conveyance again are opened, and become navigable; from which it appears that all communication with the upper provinces, and the importation of grain from the these parts are shut up, for about six months in a year. This therefore is the season when rice usually becomes dear, not because a real dearth or scarcity reigns, but that here in Calcutta, and the province of Bengal, the corn-merchants in wholesale and retail raise the price of their grain; and this they always do from December to June when the communication with the upper country is open again.[162]

It was precisely the determination of 'the price they please' which made the merchants hostile to state intervention, as I have just shown. At another level, it caused them to chart out numerous strategies in order to control prices and supplies as well as to ride the numerous price crests and troughs. These strategies were designed to control the source of all trade: the peasant. One such agency was the chain of intermediaries who made spot purchases from the *raiyat* houses as well from the *haats*. This was complemented by the ownership of storehouses (*golahs*) at various places capable of storing up to 5,000 maunds of rice or paddy for more than five years without damage, and by the possession of those crucial modes of transportation which the *raiyat* lacked.[163]

The *gomastha* and *paikar* both purchased whatever surplus was available and forestalled competition, thus enabling particular merchants to emerge as pre-emptive buyers in their own trading regions. The *golahs* enabled the merchants to tide over seasonal variations in agricultural prices between the time of ploughing and sowing (when prices were at their highest and grain was sold or advanced as seed loans) and harvests (when prices ebbed and grain was purchased). These *golahs* also enabled them to even out any disadvantages which may otherwise have arisen from unseasonal price oscillations created by an intermixture of good and bad agricultural years. In good years, when grain prices dropped sharply, the merchants would buy up, store in their *golahs* and release in both directions (town and country) in bad years brought about by crop failures.[164] The *gomasthas* were under strict instructions to cease making spot purchases (usually done immediately after the harvest) if they felt that the prices were not low enough, or that the peasants were bargaining for better terms.[165] The producers faced major problems if this ever materialized as it would immediately jeopardize the payment of revenue to the state in cash since 'without the assistance of the merchants, the ryotts suffer the greatest distress to liquidate the demand for rent upon them'.[166] This pressure forced the cultivators to suffer an enforced reduction of price in order to appease the merchant. It is therefore hardly surprising that the *raiyats* were often 'obliged to dispose of their grain on any terms, for one third, often for half less, than the customary market price'.[167]

Profits from trade seem to have emerged from three levels. At the first level were those profits which accrued from the act of selling in the towns. At the second level we can see those profits which emerged while purchasing grain from peasants, and at the third level were the super profits which merchants made during times of scarcity. The first two levels were ever present in any type of transaction,[168] whereas the third was periodic but immense nevertheless.

Contemporary notions that average rates of profit from selling grain in towns ranged from 15 to 20 per cent under normal circumstances[169] may have been a conservative estimate as the following figures shows.[170]

Year	Sale of rice in	% profit
1775	Murshidabad	52.6
1794	Dhaka	42.8
1794	Calcutta	77.7

Trade with the towns was not a profitable venture in grain alone. Betel-leaf merchants of Dhaka, who had carved petty monopolies of their own had an extremely lucrative trade going precisely because of the price differentials which prevailed between the town and the *muffassal*.

'the difference between the price at which *paun* [betel-leaf] is sold in the city of Dacca, and the rate at which it is purchased, probably within three miles of the town, is frequently one thousand per cent' was how the profit from this trade was described by W. Douglas, the collector of Dhaka in October 1789.[171]

Such profits could emerge because of the prevailing modes of procurement which forced the peasant to remain at the lowest rung of the price mechanism prevailing in the countryside,[172] and it was here that we see the second level of profit in operation. The fact that peasants were coerced into selling at prices below a third or even half the 'customary market price' perhaps indicates the techniques of pricing strategies adopted by the merchants. In fact the data available shows that the extent of under-pricing the producer could be higher than half the prevailing market prices at the village level.[173]

Year	Locality	Produce	% profit in procurement
1777	Burdwan	Rice	88.6
1777	Dinajpur	Rice	66.6
1794	Murshidabad	Paddy	75
1794	Murshidabad	Rice	44.4
1794	Purnea	Paddy	33.3

At the third level (during years of scarcity) the merchants made profits ranging from 150 to 200 per cent by manipulating the prices between town and country. Famines were boon years for these merchants from another point: these were also the years when villages became gross importers of food and profits ranging from 40 to 20 per cent were easily made by sending grain to the *muffassal*,[174] but these were much below the advantages arising from trading with towns during such years.[175]

Merchants and the Agrarian Economy

As it is quite evident, the connections between the grain merchants and agricultural production were caused by a combination of factors in the late eighteenth century. Briefly stated, the linkages seem to have arisen out of a buoyant demand for food both in the towns and in the countryside, rising agricultural prices and a recurrence of famines and semi-famine situations which plagued the province from 1769 onwards (discussed in Chapter 5). These situations made it imperative for traders

THE AGRARIAN ECONOMY AND THE COMMERCIAL TRANSACTIONS   217

to keep a tight control over lines of supply which they did in two ways: (*a*) by the creation of a wide network of trading intermediaries (*paikar* and *gomastha*) scattered over an extensive catchment area; and (*b*) by the formulation of strategies designed to control the peasants freedom of choice in the market. An analysis of the latter provides a picture of mercantile penetration in agricultural production.

There is some evidence to suggest that the grain *byapari* had started getting involved in the process of agricultural reclamation as a device to ensure steady supplies in the long-run. Evidence for merchants providing advances of money, through their agents, to cultivators wanting to reclaim wastes or cultivate their own lands 'to be repaid in kind at the time of cutting the crops' is available from Jessore. Such advances seem to have become endemic here as can be seen from the description of such annual advances as 'the *invariable mode*, and the Ryotts having no other means of obtaining seed, rely upon their respective merchants, the whole of their profits arising from their lands, having been applied solely to the discharge of other debts, and their own subsistence'.[176]

In Rangpur, merchants could also be seen participating directly in production, that is, organizing production with the use of sharecroppers, apparently in a widely prevalent fashion.[177]

The merchants' participation in agricultural production was guided not only by their long-term interests as merchants, but also by certain constraints on Bengal's peasant economy and society which forced the producers to reach out to external agencies to fulfil crucial production requirements. One such constraint was the shortage of material resources in the hands of the 'inferior ryott' or the 'poorer class of ryotts'. Lack of resources made small peasant production vulnerable to any unbalancing forces, and the periodic incursions of famines and partial crop failures made this vulnerability chronic. It was precisely this susceptibility, caused by persistent shortage of working capital and the uncertainties of production, which enabled merchants to enter the realm of production by a combination of subsistence and production loans.

Consumption loans were taken by the peasants for a variety of reasons, the chief of which appears to have been to tide over particularly bad agricultural years. In Birbhum, 'money borrowed by the ryotts assumes such a variety of shapes that I [the collector] am at a loss what term to give it. Generally however [a loan] corresponds with the exigencies of a borrower'.[178] Interest on such loans ranged from 24 to 36 per cent per annum depending on the dictates of the creditor.[179] In the context of a persistent shortage of the means of livelihood and productive resources, these subsistence loans were the first steps in what subsequently would become a vicious cycle.[180] Once a loan was incurred, a series of good harvests in continuous succession would be the only way in which households could circumvent the prospects of prolonged indebtedness. Alternating cycles of good and bad agricultural years, because of the impact these had on the prices of their produce and their incomes, tended to drive peasant households further into arrears so that finally they were confronted by a seemingly insurmountable wall of debt servicing as well as a range of creditors. 'His debts annually accumulating, the ryott becomes enslaved to his creditor', wrote Henry Colebrooke about the situation in Purnea.[181] Subsistence loans also provided one lever by which merchants could get a grip on the end produce antecedent to any interference in production.

Consumption loans were often accompanied by production loans, and it seems probable that loans of the latter type tended to predominate. The 'annual practice of the beparris to advance to the poorer class of ryotts a sufficient quantity of grain to sow their lands to be repaid in kind at the time of cutting their crops' had become the 'invariable mode' in Jessore by the late 1780s.[182] Half the standing crop of Burdwan in October 1794[183] and 'one half of the whole cultivation' of Dinajpur in 1807[184] were estimated to the product of production loans. The cultivators of Purnea were 'accustomed to loans from the principal merchants which, however oppressive are absolutely necessary to them as they are unable to maintain cultivation unless they receive such assistance'.[185]

In Rangpur, where the general level of indebtedness was said to be lower than in its neighbouring district of Dinajpur, loans for the production of grain had nevertheless grown into a 'ruinous system' by the turn of the century.[186]

Production loans were given as advance payments much before the commencement of the agricultural season, or as our sources describe 'long before the crops of the poorer ryotts are fit to gather'.[187] Writing in 1772, James Stuart commented that 'large sums of money are yearly lent out to the ocupyers [sic] of the lands in order to advance the improvements of the soil. The interest exacted for such loans is exorbitant because the repayment of capital is precarious.'[188]

Bengal had two major agricultural seasons, the *aman* (winter) and the *aus* (spring), and loans were contracted on both occasions. The winter rice harvest was considered of greater market value 'bearing a higher price and sought after by all'[189] and was, therefore, a prime area of merchant intervention. The seedlings for this harvest were universally sown in the Bengali month of *Assar* (June-July) and reaped in *Agrahan* (November-December). Advances on this crop were made in the months of *Pous* or *Magh* (December to February) and the repayments were made in the subsequent month of *Agrahan*, thereby completing a yearly cycle of loans and repayments.[190] For the *aus* (spring) crop, merchant strategies were two-fold: *gomasthas* (commissioned agents) would be sent to the villages '*to purchase it from the ryotts before they could dispose of their surplus crops*';[191] alternatively, advances to the cultivator were made in the month of *Assin* (September-October), six months before sowing in the month of *Baisakh* (April-May), to be repaid next *Assin*, immediately after the harvest in the month of *Bhadro* (August-September).[192] The timing of these loans was critical as any delay, even by a month, could lessen the amount of interest in the annual cycle of advances and repayments. Thus a 'man who shall make advances in Bhadoon expects and obtains a greater increase than the man who makes his advance in Assin and in this manner the increase is lessened as the time approaches for cutting the grain', wrote a contemporary observer in October 1794 in relation to the system of advances for the *aus* crop.[193]

Such loans were not limited to the cultivation of rice alone. Almost all major agricultural products were tilled under varying degrees of advance contracts. As discussed in Chapter 1, betel-leaf in eastern Bengal, and sugarcane in Birbhum were based on such arrangements with merchants who then sent refined sugar to Calcutta. Tobacco grown in Nadia and Rangpur was partly financed by merchants of Calcutta, Dhaka and Murshidabad, and the rest was purchased by their agents on the spot. Ginger produced in Rangpur was sold immediately by the farmers to merchants as 'the whole is paid for in advance'.[194]

How did these loans operate? The terms on which such loans were given were elaborately laid down. First, the peasant contracting for these loans was not given any written document; the whole episode was conducted on a 'verbal basis'.[195] Second, he was to receive a maximum of two-thirds of the value contracted for as advance and the 'balance at the delivery of grain and at the rate that may first be established in the pergunnah after the reaping is over';[196] but the proportion of the total loan being given in advance depended upon the type of rice being cultivated. Peasants cultivating *aman* rice could hope to get up to half the amount in advance while those cultivating the intrinsically inferior *aus* grain had to be satisfied with a quarter.[197] Third, the terms of the agreement had to be faithfully 'observed and abided' by the borrower as he was 'under penalty of making good every loss that may occur [to the lender] from the non-payment, in addition to the amount being returned with interest'.[198]

Regarding modes of repayment, it appears that the interest charged on such loans was higher than what was demanded on loans of immediate subsistence. The available evidence suggests that these were to be repaid in kind at rates ranging from 38 to 50 per cent,[199] which would make the interest on such loans one and a half times greater than on loans for consumption. Moreover, the price mechanism under which such loans were given was crucial as well as additionally profitable for the merchants. These advances were made six months before the commencement of the sowing season when prices were at their highest in the seasonal swing. Repayments had to be made immediately after the harvest— 'settlement of account takes place as soon as the crops come in',[200] at a time when prices were at their lowest, thereby making the cultivator part with a larger portion of the produce while making adjustments for the seasonal price variation. The loans 'are made in the season when grain is dearest and repaid when the price is lowest'.[201] Peasants could stand to lose between '2 annas per rupee'[202] (or 12.5 per cent) and 25 per cent in real terms while making these adjustments.[203] 'The cultivators, of necessity, have to bear with all this', remarked an Indian observer of agrarian matters in this period.[204]

These loans symbolize the intrusion of merchant capital into the very core of Bengal's economy; they also provide the point of reference for studying the dynamics of social domination over the processes of agricultural production. These loans were the 'usual and long standing custom'[205] between the merchant and the peasant all over Bengal, which meant that, apart from debt servicing, the cultivators were tied into straight to the merchant 'in preference to bringing grain to the market'.[206] It is, therefore, hardly surprising that the trader could often procure supplies 'without making purchases with ready money'.[207] The fact that the peasants of pargana Burdwan could declare unequivocally in the midst of a bad agricultural season that 'it is only because of the merchants that we have the means of purchasing our subsistence and preserving our lives' clearly shows the grip of merchant capital in rural Bengal.[208] The power inherent in a relationship of this kind is apparent from the fact that when merchants decided to call in, or discontinue such loans, the peasants were immediately 'obliged to dispose of their grain on any terms'[209] to appease the merchants as without their financial assistance 'the Ryotts suffer the greatest distress'.[210]

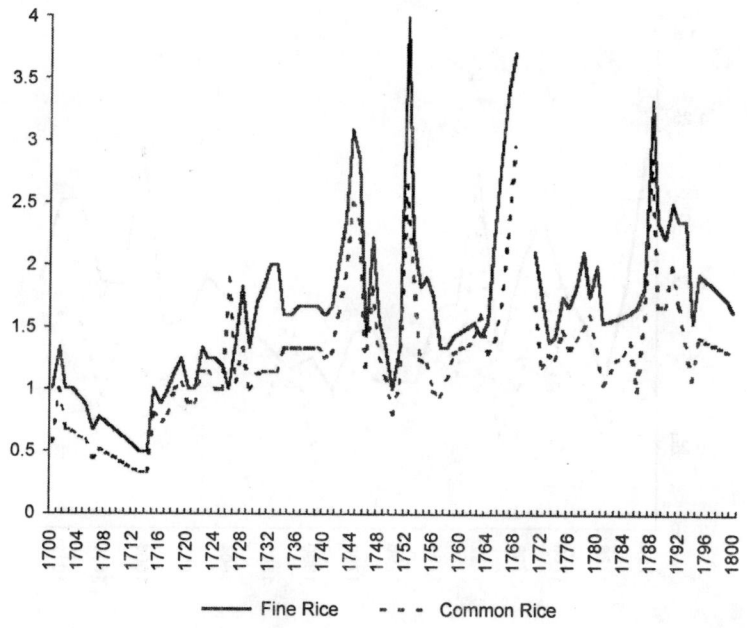

Note: Famine Prices Excluded.

Source: G. Herklotts, 'Prices in Lower Bengal', *Gleanings in Science*, vol. 1, January 1829.

FIGURE 5. FOOD GRAIN PRICES IN LOWER BENGAL, 1700-1800
(*Rs. per maund*).

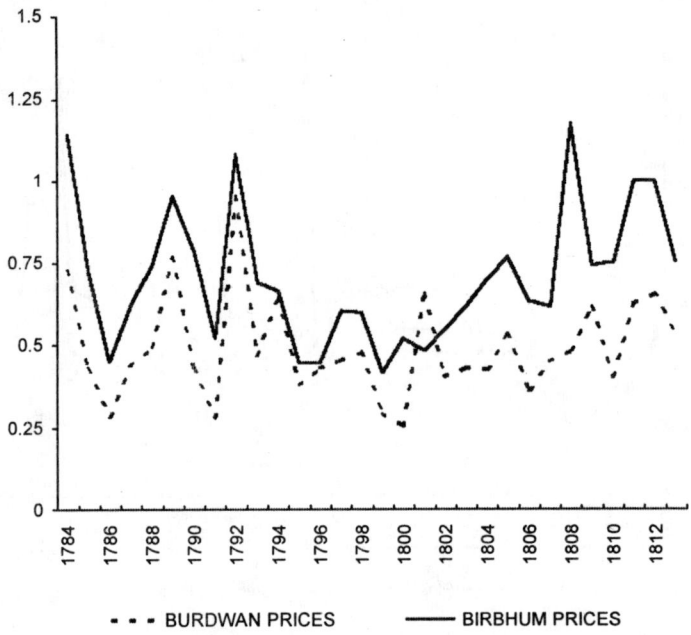

Source: W.B. Bailey, 'Statistical View', *Asiatic Researches*, vol. 12, 1816.

FIGURE 6. COMPARATIVE RICE PRICES IN BIRBHUM, 1784-1813, COARSE RICE (*Rs. per maund*).

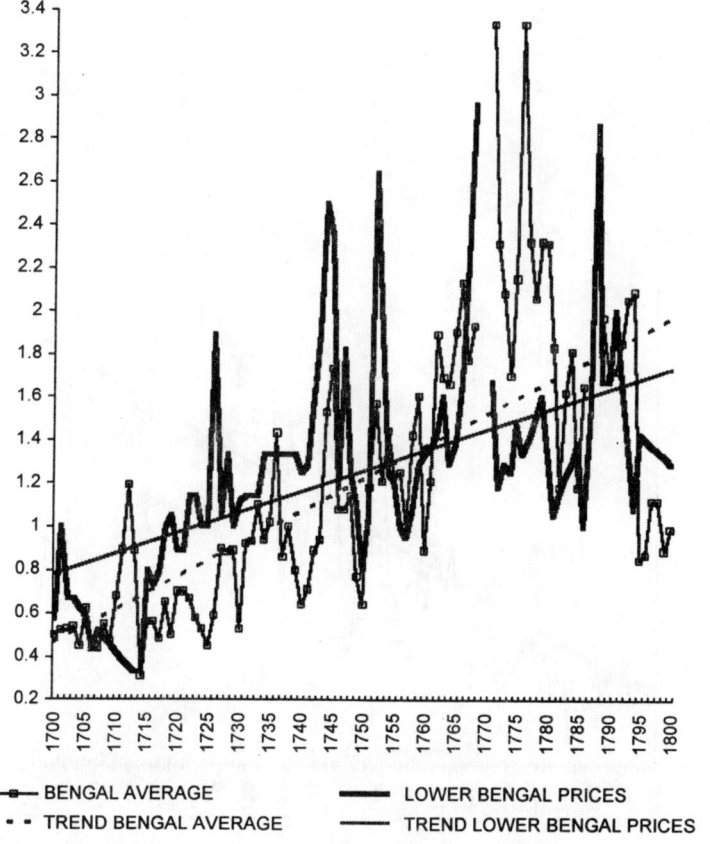

Note: Famine Prices Excluded.

Sources: G. Herklotts, 'Prices in Lower Bengal'; Hussain, 'A Quantitative Study', pp. 368-9.

FIGURE 7. COMPARATIVE RICE PRICES IN BENGAL, 1700-1800: COMMON RICE (*Rs. per maund*).

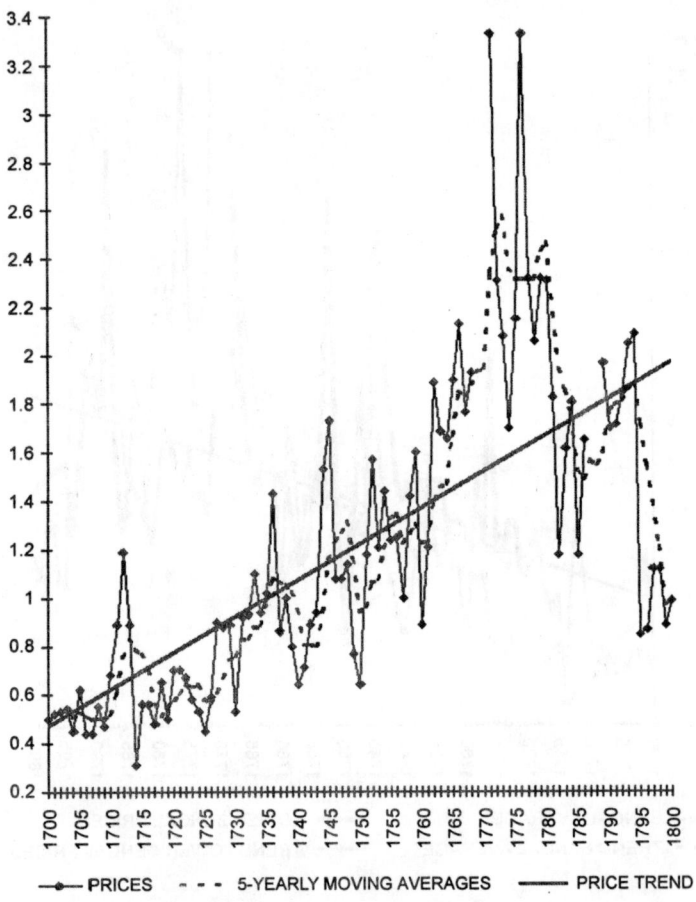

Sources: G. Herklotts, 'Prices in Lower Bengal'; Hussain, 'A Quantitative Study', pp. 368-9.

FIGURE 8. BENGAL RICE: PRICES, 1700-1800, PRICE LINE, TREND AND MOVING AVERAGES OF COMMON RICE
(*Rs. per maund*).

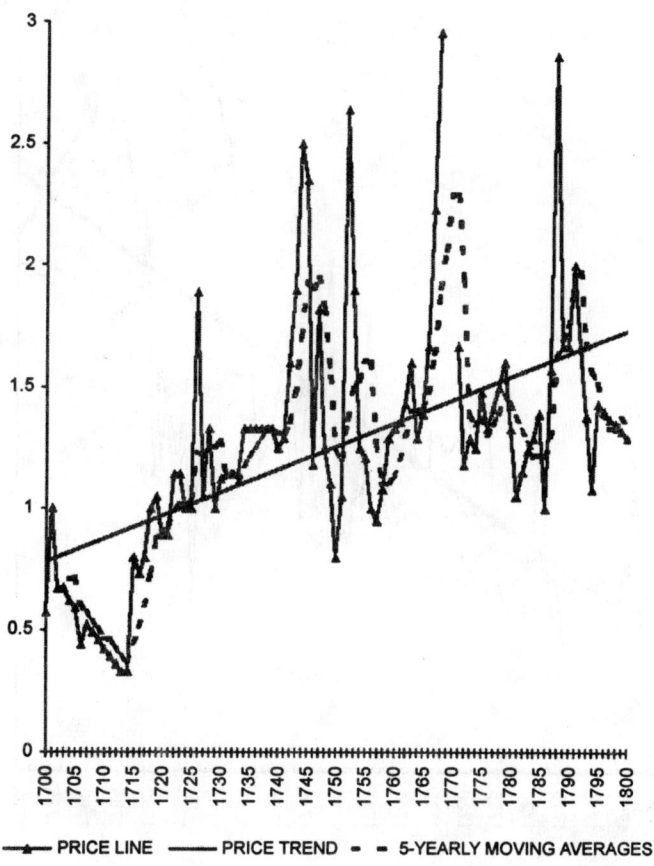

Sources: G. Herklotts, 'Prices in Lower Bengal'; Hussain, 'A Quantitative Study', pp. 368-9.

FIGURE 9. LOWER BENGAL RICE: PRICES, 1700-1800, PRICE LINE, TREND AND MOVING AVERAGES OF COMMON RICE (Rs. per maund).

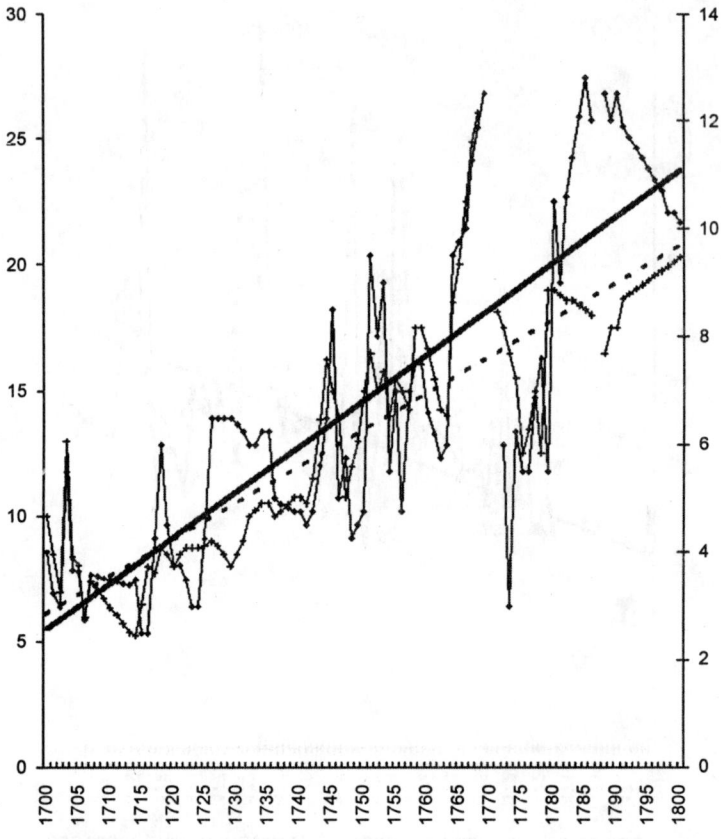

Note: Famine Prices Excluded.

Source: G. Herklotts, 'Price in Lower Bengal'.

FIGURE 10. PRICES IN LOWER BENGAL: OIL AND *GHEE*, 1700-1800 (*Rs. per maund*).

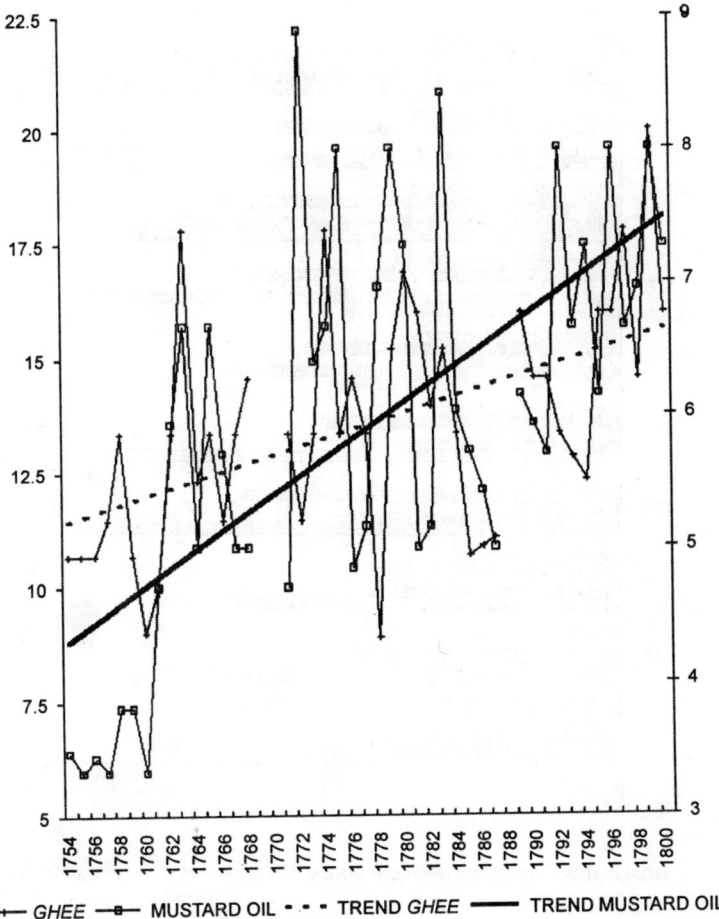

Note: Famine Prices Excluded.

Source: Bailey, 'Statistical View', *Asiatick Researches*, vol. 12, 1816, pp. 560-1.

FIGURE 11. CALCUTTA: MUSTARD OIL AND *GHEE* PRICES, 1754-1800
(*Rs. per maund*).

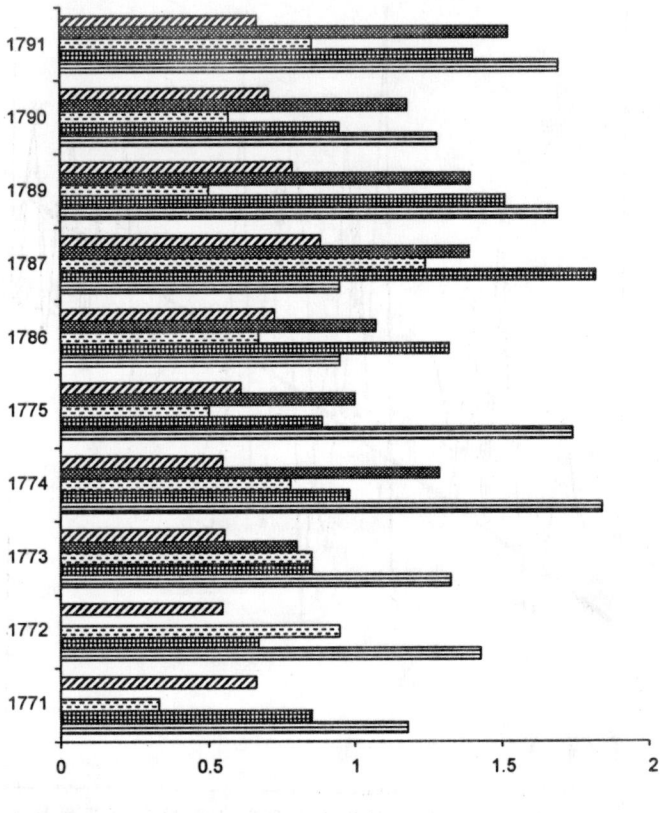

Note: Famine Prices Excluded.

Sources: Bailey, 'Statistical View', *Asiatick Researches*, vol. 12, 1816, pp. 560-1; IOR, BRC, P/59/42 to P/52/40; WBSA, Grain, vol. 1, 17 October 1794 and WBSA, PCR, Dinajpur, vols. 4 and 6, Appendix to Proceedings.

FIGURE 12. RICE PRICES IN BENGAL DISTRICTS: ORDINARY RICE
(*Rs. per maund*).

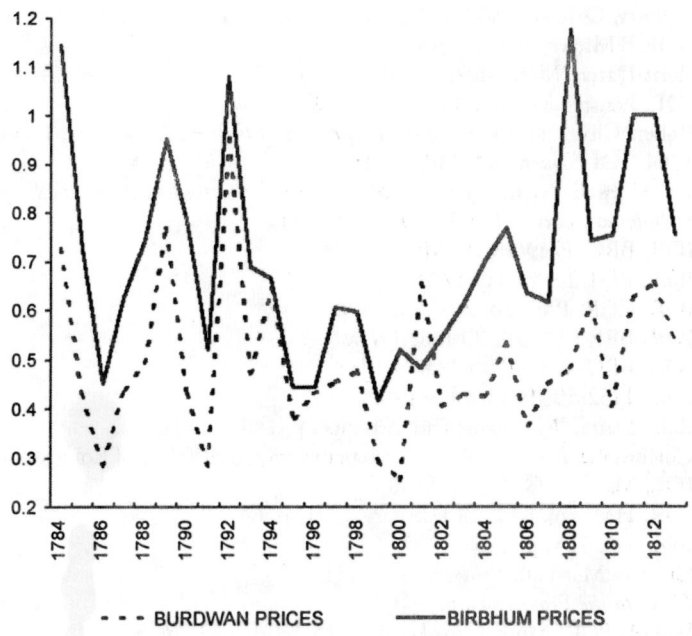

Source: Bailey, 'Statistical View', *Asiatick Researches*, vol. 12, 1816, pp. 516-17.

FIGURE 13. *AMAN* RICE PRICES, BURDWAN AND BIRBHUM, 1784-1813 (*Rs. per maund*).

NOTES AND REFERENCES

1. This question is particularly important in the context of eighteenth century Bengal because the writing of economic history has hitherto been overwhelmingly concerned with Bengal's relations with the world market, or with the marketing structures and strategies for the high-profile commodities like textiles, opium and indigo.
2. P. J. Marshall, *East Indian Fortunes: the English in Bengal in the Eighteenth Century*, Oxford, 1977, p. 24.
3. IOR, HM, vol. 465, p. 146.
4. Rajat Datta, 'Merchants and Peasants', p. 287.
5. IOL, Francis, Ms. Eur. E. 12, fol. 250.
6. Robert Clive, *An Address to the Proprietors of the East India Stock*, London 1764 (IOL, *Tracts*, vol. 113), p. 19.
7. J.Z. Holwell, *Interesting Historical Events Relative to Bengal and the Empire of Indostan*, part 1, London, 1765, p. 194.
8. IOR, BRC, P/49/38, 23 March 1773.
9. Ibid., P/51/20, 7 May 1788.
10. IOR, CCR, P/67/70, 30 April 1778.
11. IOR, BRP, P/71/70, 29 June 1789.
12. IOR, BRC, P/51/17, 4 March 1788.
13. Ibid., P/52/50, 19 October 1792.
14. Rajat Datta, 'Merchants and Peasants', p. 148.
15. Colebrooke, *Remarks*; also F. Buchanan, 'Statistical Tables of Ronggoppur', IOL, Ms. Eur. G. 11, Table 40.
16. IOR, HM, vol. 434, 28 February 1791, p. 647.
17. Ibid., vol. 765, p. 146.
18. Cited in Marshall, *Bridgehead*, p. 151.
19. *Letter to the Proprietors*, p. 6.
20. WBSA, PCR, Murshidabad, vol. 8, 15 February 1776.
21. IOR, BRP, P/70/48, 7 November 1788.
22. WBSA, Grain, vol. 1, 17 October 1794.
23. Ibid., 29 October 1794.
24. IOR, BRC, P/52/27, 23 November 1791; emphasis added.
25. Ibid., P/49/47, 30 August 1774; emphasis added.
26. WBSA, PCR, Murshidabad, vol. 7, 15 February 1776.
27. IOL, Ms. Eur. F. 95, fol. 64.
28. W.B. Bailey, 'Statistical View of the Population of Burdwan & Ca.', *Asiatick Researches*, vol. 12, 1816, pp. 557-8.
29. Rangpur's area was 2,679 square miles according to James Rennell (*Bengal Atlas*, p. 53).
30. IOR, BRC, P/51/40, 15 July 1789.
31. IOR, BRC, P/51/40, 3 April 1789.

32. See Indrani Ray, 'Journey to Kasimbazar and Murshidabad: Observations of a French Visitor to Bengal in 1743', *Bengal Past and Present*, vol. C, part ii, no. 191, July-December 1981, p. 56.
33. F. Buchanan, 'Statistical Tables of Ronggoppur'.
34. 'Memoir by Sir George Campbell on the Famines which affected Bengal in the Last Century', in J.C. Geddes, *Administrative Experience Recorded in Former Famines*, Calcutta, 1874, p. 421; emphasis added.
35. See Chapter 5.
36. IOR, BRC, P/51/12, 19 October 1787.
37. In Rangpur the 'seven ranks' of people in society having a total of 66 persons, including dependents and servants, spent a combined sum of Rs. 1,387.43 a year on food (F. Buchanan, 'Statistical Tables of Ronggoppur', IOL, Ms. Eur. G. 11, Tables 38, 39; hereafter, 'Ronggoppur'). These figures pertain to predominantly rural populations, patterns of consumption in cities would differ as large numbers of wealthy people residing there would tend to spend more on different types of food.
38. W. Tennant, *Indian Recreations*, 2, p. 2.
39. IOR, HM, vol. 434, George Smith to Henry Dundas, 28 February 1791, p. 647.
40. M.M. Postan, *The Medieval Economy and Society: An Economic History of Britain in the Middle Ages*, 1975, p. 265.
41. C.E. Labrousse, 'Economic Fluctuations and the Individual', in M. Aymard and H. Mukhia (eds.), *French Studies in History, vol. 1: The Inheritance*, Delhi, 1989, p. 271.
42. A.J.S. Gibson and T.C. Smout, 'Regional prices and market regions: the evolution of the early modern Scottish grain market', *Economic History Review*, vol. 48, no. 2, pp. 258ff.
43. Ibid., pp. 260-1.
44. Ibid., p. 261.
45. I have used Akhtar Hussain's interpolations for the gaps in Heklotts's data. A part of this series is reproduced by Brij Narain, *Indian Economic Life: Past and Present*, Lahore, 1929. Brij Narain has given the details only of rice prices whereas Herklotts gives the prices of a basket of major agricultural products.
46. Figure 5 also highlights the fact that notwithstanding the long-term provincial trend of a secular rise in price, there was a short-lived slump in prices just prior to the Permanent Settlement. This price trend is also revealed in Figure 13 which compares the prices of *aman* rice in the districts of Burdwan and Birbhum between 1784 and 1813. Though prices seem to have recovered around the turn of the century, the price slump was the unfavourable conjuncture amidst which the Permanent Settlement was introduced, the implications of which are beyond the purview of this discussion. The synchronicity of the inter-district and inter-harvest price lines despite the

temporary sharpness of the slump clearly indicates the relative inelasticity of rural demand.
47. IOR, HM, vol. 206, p. 149.
48. *Remarks*, p. 67.
49. IOR, BRP, P/71/9, 15 May 1789.
50. IOR, BRC, P/51/22, 15 August 1788.
51. IOR, HM, vol. 393, 6 November 1788, p. 121.
52. Hamida Hossain, *The Company Weavers*, pp. 9-11.
53. IOR, HM, vol. 769, 22 August 1771, p. 361.
54. See WBSA, CCRM, vol. 3, 25 January 1771; WBSA, Circuit at Purnea, 2 February to 9 February 1771, p. 8; IOR, BRC, P/49/40, 26 June 1773; ibid., P/49/46, 31 May 1774.
55. 'Report of Commercial Occurrences', 6 August 1789, IOR, HM, vol. 393, pp. 254-5.
56. Francis Gladwin, *A Narrative of the Transactions in Bengal*, Calcutta, 1788, p. 122; also cf. IOR, BPC, P/1/10, 15 January 1733 and ibid., P/1/26, 19 November 1757.
57. IOR, BRP, P/70/44, 11 July 1788; emphasis added.
58. Alivardi Khan to Barwell, 9 January 1749, IOR, HM, vol. 804, p. 61.
59. Ibid., vol. 92, 15 December 1762, p. 32; emphasis Vansittart's.
60. Ibid., p. 30.
61. IOR, BRP, P/71/23, 25 March 1790.
62. IOR, BRC, P/51/50, 24 October 1789; emphasis added.
63. IOR, BR Misc., P/89/36, 13 May 1790.
64. IOR, BRC, P/52/16, 9 July 1790.
65. IOR, BR Misc., P/89/42, 29 May 1795.
66. Ibid., P/81/36, 13 May 1790.
67. Ibid., P/52/9, 9 April 1790.
68. IOR, BR Misc., P/89/36, 29 October 1790.
69. Ibid., 7 May 1790.
70. See Philip Calkins, 'The Formation of Regionally Oriented Ruling Group in Bengal', *Journal of Asian Studies*, vol. 29, no. 4, August 1970 pp. 799-806; Marshall, *Bridgehead*, pp. 63-4 and J. MacLane, *Land and Local Kingship*, for the political necessities facing the Nizamat.
71. For example in the crucial trade route between Calcutta and Murshidabad there were 18 *chowkis* (K.M. Mohsin, *A Bengal District in Transition: Murshidabad, 1765-1794*, Dhaka, 1973, p. 104).
72. Rajat Datta, 'Merchants and Peasants', p. 395.
73. See A.M. Serajuddin, 'The Salt Monopoly of the East India Company's Government in Bengal', *Journal of the Economic and Social History of the Orient*, vol. 21, part 3, October 1978; Balai Barui, *The Salt Industry of Bengal, 1757-1800: A Study in the Interaction of British Monopoly Control and Indigenous Enterprise*, Calcutta, 1985; P.J. Marshall, *Bridgehead*, pp. 111-12.

THE AGRARIAN ECONOMY AND THE COMMERCIAL TRANSACTIONS   233

74. Minute of Warren Hastings, 9 March 1773, IOR, HM, vol. 217, p. 29.
75. Marshall, *East Indian Fortunes*, p. 243; Rajat Datta, 'Merchants and Peasants', p. 380.
76. Ibid., pp. 28-9; also cf. IOR, BRC, P/49/38, p. 836, 1032.
77. Minute of Henry Vansittart, 16 December 1778, IOR, CCR, P/67/72; emphasis added.
78. 'Regulations for the Future Establishments and Regulation of Duties of the Country Government, March 1773', IOR, HM, vol. 217, pp. 44-9; also BRC, IOR, P/49/38, 23 March 1773.
79. Compare K.K. Datta, *Studies in the History of the Bengal Suba*, Calcutta, 1936, pp. 159-60; and WBSA, PCR, Murshidabad, vol. 2, 20 June 1774.
80. For the opposition of the zamindars in Burdwan, IOR, BRC, P/52/9, 9 April 1790; in Jessore, ibid., P/52/14, 24 June 1790; and in Nadia, IOR, BRP, P/70/32, 28 August 1787.
81. IOR, BRC, P/52/12, 28 May 1790.
82. Ibid., 15 May 1793.
83. John Shore in *FR 2*, p. 493.
84. IOR, BRC, P/52/13, 11 June 1790.
85. Ibid., P/53/1, p. 475.
86. This is in complete contrast to the assertion made by A.K. Bagchi that British rule completely 'disrupted' the 'network of markets' which were 'oriented towards domestic exchange' and 'failed often to compensate for this disruption' ('Markets, Market Failures and Transformation' in Stein and Subrahmanyam (eds.), *Institutions and Economic Change*, pp. 51-2).
87. IOR, BRC, P/49/38, 19 February 1773; emphasis added.
88. IOR, CCR, P/67/72, 16 December 1778.
89. Ibid., P/67/62, 13 May 1776.
90. IOR, BRP, P/70/27, 8 May 1787.
91. IOR, BR, Misc., P/89/37, 27 June 1792.
92. Ibid., P/89/36, 5 July 1790.
93. IOR, BRC, P/52/16, 9 July 1790.
94. Rawski, *Agricultural Change*, p. 69.
95. W.G. Skinner, 'Marketing and Social Structure in Rural China', *Journal of Asian Studies*, vol. 25, no. 2, part 2, February 1965, p. 195.
96. Skinner, 'Marketing and Social Structure', p. 196.
97. John Taylor, *Dacca*, p. 294.
98. IOR, BR, Misc., P/89/36, 15 September 1776; emphasis added.
99. Rajat Datta, 'Merchants and Peasants', p. 145.
100. IOR, CCR, P/68/8, 3 September 1781.
101. Rajat Datta, 'Merchants and Peasants', p. 145.
102. IOR, BRP, P/71/25, 28 May 1790.
103. Literally below the market: a name given to the itinerant petty-pedlars in the *bazaars*.
104. IOR, CCR, P/67/72, 16 September 1778; emphasis added.

234 SOCIETY, ECONOMY AND THE MARKET

105. Taylor, *Dacca*, p. 294.
106. This was a market in close proximity to Calcutta and handled the grain arriving from the 24-Parganas.
107. WBSA, PCR, Murshidabad, vol. 8, 15 February 1776.
108. IOR, CCR, P/67/65, 24 January 1779.
109. IOR, BRC, P/50/13, 22 December 1778.
110. Ibid., P/50/18, 27 July 1779.
111. Ibid., P/51/7, 26 February 1788.
112. Ibid., P/52/19, 7 September 1790.
113. Ibid., P/52/13, 8 May 1790.
114. IOR, BR Misc., P/89/41, 15 April 1794.
115. See n. 100.
116. For a discussion of the hierarchy of markets and merchants in neighbouring Bihar, see Kumkum Banerjee, 'Grain Traders and the East India Company: Patna and its Hinterland in the Late Eighteenth and Early Nineteenth Centuries', *IESHR*, vol. 23, no. 4, October-December 1986.
117. Beveridge, *Bakarganj*, p. 282.
118. WBSA, Grain, vol. 1, 17 October 1794; also Buchanan-Hamilton, *Purnea*, pp. 695-6.
119. K. Banerjee ('Grain Traders and the East India Company') views local trade as a pyramidal structure.
120. *WBDR, ns, Birbhum*, p. 64.
121. F. Buchanan, 'Survey of Ronggoppur', IOL, Ms. Eur. D. 75, fol. 85.
122. WBSA, Board of Revenue, Judicial Branch, 2 June to 30 August 1790, 14 July 1790; WBSA, Grain, vol. 1, 15 October 1794.
123. IOR, BR Misc., P/89/40, 20 January 1795.
124. *WBDR, ns, Birbhum*, 4 June 1796, p. 64.
125. IOL, Ms. Eur. D. 75, fols. 83-4.
126. Ibid., fol. 87.
127. Martin, *Eastern India*, vol. 2, p. 759.
128. IOR, BRP, P/70/35, 27 September 1787, letter dated 19 November 1787; emphasis added.
129. 'Rulers and Merchants in Late Eighteenth Century Bengal', unpublished D. Phil. thesis, University of Chicago, 1980, pp. 24-5.
130. K.M. Mohsin, *Murshidabad*, pp. 24-5.
131. IOR, BRP, P/70/40, 18 April 1787.
132. IOR, BRC, P/51/21, 2 July 1788.
133. 'Dealers in the necessaries of life are of the same cast or tribe and connected with each other', and this was seen as an important reason for enabling 'them to form combinations against the public with greater facility than in other countries' (IOR, BRC, P/52/26, 21 October 1791).
134. H. Sanyal, 'Continuities of Social Mobility'.
135. See WBSA, Grain, vol. 1, 17 October 1794 for Burdwan; 18 October 1794 for Jessore; 10 October 1794 for Bakarganj; 15 October 1794 for

Dinajpur; 17 October 1794 for Purnea; 10 November 1794 for Tippera.
136. Kumkum Banerjee ('Grain Traders and the East India Company', p. 408), following Buchanan-Hamilton, says that the *paikars* of Purnea did not use oxen, considered sacred beasts, as they were men of pure birth. Evidence to the contrary is however available. The *paikars* of Purnea 'from keeping a number of cattle are enabled to go some distance into the country and purchase the grain immediately as it is harvested from the cultivator . . .' says a report from the collector of Purnea in October 1794 (WBSA, Grain, vol. 1, 17 October 1794).
137. Rajat Datta, 'Merchants and Peasants', pp. 396-7, 399.
138. Beveridge, *Bakarganj*, p. 282; BDR, *Dinajpur, vol.* 2, 3 June 1788, p. 231.
139. WBSA, Grain, vol. 1, 17 October 1794.
140. Ibid.
141. WBSA, Board of Revenue: Judicial Branch, 2 June-30 August 1790, 14 July 1790.
142. IOL, Ms. Eur. D. 75, fol. 84.
143. *WBDR, ns, Birbhum*, p. 64.
144. Ibid., fol. 88.
145. Buchanan-Hamilton, *An account of the District of Purnea*, Calcutta, 1928, p. 528.
146. IOR, BRP, P/70/35, 27 November 1787; emphasis added.
147. Martin, *Eastern India*, vol. 2, p. 758.
148. IOL, Ms. Eur. D. 75, fol. 84.
149. 'In distant parganas, far removed from proper means of transport and market towns, grain sells considerably cheaper, the price rising in places more contiguous to marts of grain' (WBSA, Grain, vol. 1, 31 October 1794).
150. IOR, BRC, P/49/51, 13 February 1775.
151. Ibid., 9 August 1787.
152. Ibid., P/51/9, 21 July 1787.
153. 'Abstract of an Examination of Several Grain merchants in Calcutta', Sir John Shore to Council, 1 February 1788, in ibid., P/51/7.
154. WBSA, Grain, vol. 3, 14 February 1795.
155. Ibid., vol. 1, 19 October 1794.
156. Ibid., 31 October 1794; emphasis added.
157. Ibid., p. 2.
158. Ibid., p. 147.
159. Ibid., 29 October 1794.
160. IOR, BRC, P/50/30, 5 January 1781; IOR, BRP, P/70/40, 1 April 1788; ibid., P/70/44, 20 August 1788; Taylor, *Dacca*, p. 296.
161. IOR, BRC, P/51/17, 1 February 1788.
162. Ibid., P/52/28, 1 March 1791; emphasis added.
163. Rajat Datta, 'Merchants and Peasants', pp. 400-1.
164. *Ibid., pp. 395, 399-400.

165. WBSA, Grain, vol. 1, 17 October 1794.
166. IOR, BRP, P/71/30, *arzi* of the Raja of Burdwan, 31 July 1790.
167. IOR, CCR, P/68/7, 4 May 1781; emphasis added.
168. In fact the distinction between the first and second levels of profits is largely formal in an effort to show the various channels which contributed to the making of mercantile profit. In actual operation these two levels were intertwined in a complex fashion, and it was level two (profits made from underpricing the peasant) which actually determined the range of profits which could be made in level one.
169. *CPC, vol.* 8, letter no. 158, 12 February 1788.
170. Calculated from Rajat Datta, 'Merchants and Peasants', p. 389.
171. IOR, BRC, P/57/50, 24 October 1789; emphasis added.
172. Also cf. Rajat Datta, 'Merchants and Peasants', pp. 397-8.
173. Ibid., p. 398.
174. Ibid, pp. 389-90.
175. This relative disadvantage would perhaps explain the higher relative mortality in primarily rural areas (like Purnea, for instance) during the famine of 1769-70.
176. IOR, BRC, P/51/20, T. Henckell to Board, 7 May 1788; emphasis added.
177. IOL, Ms. Eur. D. 75, fol. 103.
178. IOR, BRP, P/70/35, 6 November 1787.
169. IOR, BRC, P/49/38, 15 December 1772; IOR, BRP, P/70/35, 13 November 1787.
180. 'Once begun, a chronic cycle of indebtedness tends to reproduce itself. Prior debt prevents saving after the harvest, because the creditor calls in his loan, and means of personal and productive consumption will consequently be likely to run short again before the next harvest' (H. Friedmann, 'Household Production and the National Economy: Concepts for the Analysis of Agrarian Formations', *JPS*, vol. 7, no. 2, February 1980, p. 172).
181. IOR, BRP, P/71/26, 18 June 1790.
182. IOR, BRC, P/51/20, 7 May 1788.
183. WBSA, Grain, vol. 1, 17 October 1794.
184. M. Martin, *Eastern India*, vol. 3, p. 906.
185. IOR, BRP, P/71/25, 26 May 1790.
186. IOL, Ms. Eur. D. 75, vol. 2, book 4, fol. 84.
187. WBSA, Grain, vol. 1, 17 October 1794.
188. 'Memoirs of the Coinage in Bengal', IOR, HM, vol. 62, p. 46.
189. WBSA, Grain, vol. 1, 17 October 1794.
190. Ibid., 29 October 1794.
191. Ibid; emphasis added.
192. Ibid., 17 October 1794; also IOR, BRC, P/51/21, 29 January 1788.
193. WBSA, Grain, vol. 1, 17 October 1794.
194. IOL, Ms. Eur. D. 75, vol. 2, book 2, fol. 16.

195. IOR, BRP, P/71/26, 18 June 1790.
196. WBSA, Grain, vol. 1, 22 October 1794.
197. IOR, BRP, P/71/26, 18 October 1790.
198. WBSA, Grain, vol. 1, 22 October 1794.
199. Rajat Datta, 'Merchants and Peasants', p. 159.
200. IOR, BRC, P/51/29, 29 January 1788.
201. IOR, BRP, P/71/26, 26 April 1790.
202. IOR, BRC, P/51/21, 29 January 1788.
203. IOL, Ms. Eur. D. 75, vol. 2, book 4, fol. 84.
204. *Risala-i-Zira'at* (tr. Harbans Mukhia, *Perspectives on Medieval History*, Delhi, 1993), section vii.
205. IOR, BRC, P/51/21, 14 June 1788.
206. Ibid.
207. IOL, Ms. Eur. D. 75, vol. 2, book 4, fol. 84.
208. IOR, BRC, P/49/37, 30 August 1774; emphasis added.
209. IOR, CCR, P/68/7, 4 May 1781.
210. IOR, BRP, P/71/20, 11 October 1790.

CHAPTER 5

Dearth and Famine: Ecological Dislocation, Subsistence and Crises

F AMINE AND DEARTH denote two existential crises of survival. Famine typifies a situation when a subsistence and mortality crisis become combined in a critical conjuncture. There is both excessive starvation and excessive mortality in a famine. It is, therefore, a one-off event of calamitous proportions. Dearth, on the other hand, represents a more continuous saga of small scarcities. Although they have a lesser impact than famines, they are nevertheless important since their periodicity creates major problems of subsistence among the harvest sensitive strata (the artisans, labourers and the town poor), women and children in the household, as well as among the peasants, especially those who work their lands with inadequate resources.

That there has been a connection between ecological dislocations—an intense drought or a severe flood—and subsistence-crises in the past is almost axiomatic in any analysis of famine-causation. Ecology is the systemic interface between human and natural conditions of existence in which the rhythms of the nature often tend to break loose into ecological disaster points. A certain fragility of the ecosystem is an inbuilt precondition, underpinned as it is by its vulnerability to climatic conditions. This periodically unleashes cycles of crises, which are then followed by phases of recovery. The length, duration or effectiveness of this recuperation is dependent on the next ecological downturn.

Naturally, other factors contribute to convert the rumblings of a crisis into a full-scale eruption. Climatic conditions have been assigned an 'instrumental' and not 'a causative role' in engendering crises of subsistence in order to avoid the pitfalls of 'climatic determinism',[1] and as I have also argued below, the movement of agricultural prices was the key in transforming a famine-warning into a famine-point in eighteenth century Bengal. On the other hand, the specificity of the ecological factor, and the differential nature of its impact are often underplayed, sometimes in the most outstanding analyses of such crises of subsistence. Thus, Amartya Sen's approach to famine-causation relegates the climatic variable to the fringes. Though he accepts that in the case of famine in the Sahel 'it would be stupid to pretend that the drought was not seriously destructive',[2] he makes no attempt to chart its ecological implications and its role in intensifying the breakdown of entitlement. The marginality of the ecological variable is evident in another important contribution to the literature on famine-studies in which Dreze and Sen pose a leading question: 'Are Famines Natural Phenomena?', dismissively relegate their answer to two pages ('blaming nature can, of course, be very consoling and comforting' they tell us), and devote the remaining three hundred pages of the book to the question of entitlement and changes in policy.[3]

To what extent have natural phenomena been responsible for famines? This is a question which can allow a nuanced perspective on the proximate cause of a subsistence-crisis, and help in avoiding the danger of ecological-determinism. In other words, we need to study the extent to which an ecological disturbance has to reach before it can trigger a crisis because different societies have different thresholds of withstanding such climatic aberrations. It is in this perspective that this essay attempts to study the impact of short-run ecological dislocations caused by a drought or a flood on subsistence and to assess the wider implications of coping.

Food Shortages in the Eighteenth Century: An Overview

As Table 46 shows, Bengal was no stranger to acute food shortages, caused either by insufficient or excessive rains.

An agricultural economy, like the one in Bengal, which depended almost exclusively on the monsoons for irrigating its fields, or for nourishing the soil by the silt deposits left by the seasonal flooding of its rivers, was naturally vulnerable to periodic cycles of droughts or floods. It is also significant that out of the fifteen documented cases of scarcity listed in Table 46, there were only three years (1711, 1752 and 1763) in which people actually died, presumably due to starvation. In 1711 'several thousands have famished for want of rice'.[4] The year 1752 was described as the worst famine in the past 60 years and 'many inhabitants have perished in the town [Calcutta], a truth well known to every one'.[5] In 1763 'many thousands are continually perishing thro' [*sic*] want'[6] and

TABLE 46. FOOD SHORTAGES AND NATURAL CALAMITIES
IN BENGAL, 1700-60

Year	Type of calamity	Spatial location	Symptoms
1711	Drought	West Bengal	Scarcity and deaths
1718	Drought	All-Bengal	Scarcity
1727	Drought	Unspecified	Price increase
1728	Drought	Unspecified	Scarcity
1732	Unspecified	Unspecified	Dearth
1734	Drought	All-Bengal	Scarcity
1737	Flood	West Bengal	Scarcity
1741	Flood	West Bengal	Mulberry crop ruined
1742	Flood	Unspecified	Scarcity
1752-3	Flood	All-Bengal	Scarcity and death
1755	Flood	West Bengal	High price and scarcity

Sources: Relevant volumes of BPC, IOR, P/1/2 to P/1/49; Letters Received, IOR, E/4/20 to E/4/30; Letters from Coast and Bay, IOR, E/4/1 to E/4/6; IOR, HM, vol. 804, 20 November 1752, p. 237; *BDR, Midnapur*, vol. 1, 18 July 1767 and 10 November 1767, pp. 171, 191; ibid., vol. 4, 30 January 1771, p. 47; Robert Orme, *Historical Fragments of the Mogul Empire*, London, 1805 (original 1782), p. 405.

'many thousands of walking skeletons' flooded on to the streets of Calcutta.[7] The year of famine mortality may have been 1738 since in 1769 the people of Purnea apparently considered it to be the worst calamity to have befallen that district 'for thirty years past'.[8]

In a large sense, therefore, the scarcities and famines which struck the province in the late eighteenth century would appear to be part of a saga of recurring food shortages. Yet, there appears to be a strong case for arguing that for some unknown climatological reasons, the ones in the late eighteenth century were more intense and therefore had more pervasive social consequences. It is certainly undeniable that the scale of mortality in 1769-70 far exceeded the deaths described in the worst scarcities in the early part of the century. Even if one questions, as indeed I do, the general consensus that 10 million people actually died in that famine (discussed later), that fact, nevertheless, remains that nothing like 1769-70 had ever occurred in the province. Similarly, the cyclone and flood of 1787 was particularly disastrous in eastern Bengal since they apparently coincided with the changes in the course of the river Teesta. Therefore, the effects of the famine in 1788 were quite disastrous even though the mortality was considerably less than what seems to have occurred in 1769-70. All these issues will be discussed at length during the course of this chapter.

Ecology and Crises of Subsistence

Under normal circumstances, Bengal produced three rice harvests, *aman, boro* and *aus*. *Aman* was the winter harvest which was universally sown in the Bengali month of *Assar* (June-July) and harvested in *Agrahan* (November-December). This harvest was generally considered of great market value 'bearing a higher price and sought after by all'.[9] The other major crop was that of *aus* (spring), sown in *Baisakh* (April-May) and reaped in *Bhadro* (August-September). Compared to the winter crop, the spring harvest was intrinsically inferior, being consumed overwhelmingly by the 'lowest and poorest classes'.[10] *Boro* rice was an intermediate crop producing the coarsest quality rice which was sown in *Chaitra* (March-April) on extremely low lands[11] and reaped in either *Assar* (June-July) or *Srawan* (July-August). Therefore unlike the two major crops which required a gestation period of six months, the *boro* was a quick-ripening crop, capable of providing a harvest in four months.[12] Moreover, this crop was not produced in all areas of Bengal. For instance, it was not grown in Burdwan, Purnea and Midnapur,[13] which indicates its relative insignificance in the overall supply of food in the province.

Then there were those crops which were grown by the peasants purely for sale in the market, or for their conversion into artisanal produce. Table 47 illustrates the cropping patterns of some the principal cash crops.

Bengal also depended almost exclusively on the monsoons for irrigating its fields, or for nourishing the soil by the silt deposits left by the seasonal flooding of its rivers. The silt, 'which on the waters subsiding settles on the lands, fertilizes them, and affords the Ryots [cultivators] a

TABLE 47. CROPPING PATTERNS OF SOME CASH CROPS

Crop	Month sown	Month harvested
Cotton	October-November	May-June
Betel	All year through	
Mulberry	April-May[14]	August-September
Tobacco	July-August	January-February or February-March
Various types of lentils	Between April-May and August-September	January-February and March-April

Source: IOR, BRP, P/70/40, Appendix to Proceedings of April 1788.

very great relief for their loss of crops; their lands thus circumstanced yields them a produce of sursoo [mustard-seed], tobacco, daun [paddy], turcaree [vegetables] and pulse.'[15]

For obvious reasons droughts caused immediate problems for agriculture[16] but the scale of the subsequent scarcity depended upon the intensity of the drought. A temporary delay in the monsoon was easily accommodated with the minimum of difficulty. Each village had its own tanks and ponds, dug by the local landed-proprietors or by individual peasant families, 'from religious or other motives' which were used to irrigate the adjoining fields in the 'event of the rains ceasing before the crops approach maturity'.[17] These tanks were used to irrigate those fields which were sown with crops considered most valuable to the peasants. Such fields were known as the *colla* (high) lands which were utilized to cultivate the commercially valuable food crops (lentils, onions, mustard) along with mulberry for silk worms, tobacco and raw cotton.[18] Relatively lower lands were meant for the cultivation of *aman* rice. The *jala* lands (literally water prone), so-called because of their situation in close proximity to river banks or streams, were primarily used for the cultivation of the lower grade grains of the spring (*aus*) and *boro* crops of rice, and for the cultivation of various edible farinaceous-roots (*kochu*) which were used as essential food-supplements especially by the poor.[19]

Everyone lived under the unrelenting tyranny of nature. For local landed proprietors natural catastrophes were inevitable evils which they had to endure, even though these meant falling incomes and desertion by their peasants. For the producers, the peasants and the artisans, the fickleness of weather was a bitter fact of life. For them such events were nothing short of God's own scourge on earth: not only were their enterprises severely shaken; not only did they die in large numbers; they were also left with the crippling burdens of making good the losses suffered by the state and by their landed-superiors. For the producers, dearth and famines were more than just 'bad' agricultural years, they were times of sheer crises of survival. Finally, the state, and especially Company's state in the late eighteenth century was a veritable barometer, registering each and every little change in the economic atmosphere caused by scarcities or bad harvests as they jeopardized the collection of revenue.

The question is: how do we differentiate between dearth and famine in the context of eighteenth century Bengal? Obviously, social perceptions classified various food-crises according to their scale and magnitude. Paul Greenough is correct in stating that the cultural perception of prosperity, and conversely of adversity, in traditional Bengali society depended on the nature of the current paddy harvest and the ways in which different social groups were allowed access to it.[20] Yet, dearth and famines were more than rude shocks to established cultural constructs of plenitude ('society's conception of the good life')[21] and scarcity. They caused severe economic dislocations. Production faltered, prices soared and people died. 'Traditional' means of support within the community no longer sufficed to maintain social bonds, or to alleviate the sufferings of the famine stricken. Table 48 provides a brief outline of the number of times Bengal suffered from such occurrences between 1761 and 1800.

The significance of Table 48 lies in the fact that it shows the recurrence of severe bouts of dearth in between two major famines. Taken together, these events show an economy particularly vulnerable to periodic cycles of scarcity occasioned by adverse cycles of weather. The connection between weather and subsistence crises, therefore, requires some explanation.

When a drought hit, the *colla* land was immediately affected. This had two consequences on the life of the inhabitants. First, some of the finest quality rice grown in the *aman* harvest was immediately threatened.

TABLE 48. DEARTH AND FAMINE IN BENGAL, 1761-1800

Year	Occurrence	Spatial location	Symptoms
1761	Unspecified	West Bengal	High prices
1763	Flood	East Bengal	Scarcity and deaths
1767	Flood	East Bengal	Partial loss to crops
1767	Drought	West Bengal	Scarcity
1769-70	Drought	Regional	Famine, dearth and high mortality in western Bengal
1773-4	Drought/Flood	Regional	Dearth, high prices
1775	Drought	West Bengal	Partial scarcity, high prices
1777	Drought	West Bengal	Harvest failure, dearth and high prices
1779	Drought	West Bengal	Crops destroyed, high prices and a famine panic
1783	Drought/Flood	Regional	Crop failure and 'public scarcity'
1787-8	Flood	Southern and eastern Bengal	Famine and mortality in in east, dearth elsewhere
1791	Drought	West Bengal	Food shortage and high prices
1793	Drought	West Bengal	Similar symptoms

Sources: IOR, SCC, P/A/10; relevant volumes of IOR, BRC, P/49/38 to P/54/6; relevant volumes of IOR, BRP, P/70/1 to P/72/37; S. Islam (ed.), *Bangladesh District Records, Dacca District*, vol. 1, 1784-87 for floods in 1783.

Next in line were the 'cash' crops of tobacco, mulberry and cotton. A shortfall in both the sectors meant that both peasant and artisanal incomes were jeopardized. Second, droughts also impinged on the water stocks necessary for the physical survival of the rural population. Ponds (*dighi*) dried-up and the water levels in wells receded. The former affected the amount of water which cultivators could give to their standing crops, whereas the latter reduced the availability of potable water. Thus during the drought of 1791, the Board of Revenue instructed the collectors of revenue 'to enjoin their officers to be careful to leave in the tanks and reservoirs in question a quantity of water sufficient for the consumption of the inhabitants of those villages which have usually been supplied from them'.[22]

Obviously, the magnitude of the crisis depended upon the intensity of the drought. The worst recorded case of this type was in 1769, when a drought, lasting for 6 months from August 1769 to January 1770, created havoc in the economy of western Bengal. This drought was unprecedented 'insomuch as the oldest inhabitants never remember anything like it'.[23] The rice lands had 'so harden'd for want of water, that the ryotts have found a difficulty in ploughing and preparing it for the next crop',[24] and the fields of rice 'parched by the heat of the sun are become like fields of straw'.[25] Water resources were critically hit. The intensity of the drought meant that the cultivators desperately tapped their ponds, etc., for supplies which soon dried up: 'water in lakes & reservoirs . . . was soon exhausted by the Ryotts in watering their rice grounds. Yet notwithstanding their industrious endeavours, no crop was produced but the water was expended', and these tanks 'from the extreme drought have become parched and barren'.[26]

The drought also damaged those crops which were crucial to Bengal's artisanal production. Thus by November 1769: 'The Ryotts find themselves totally incapacitated to cultivate the valuable crops of Cotton and Mulberries and the inferior ones of Gram, Pease, Barley, Tobacco, Betel leaf & ca. *which succeed the rice harvest.*'[27]

It is quite obvious from Table 47 that the drought of 1769 ruined the produce from the highly-valued sectors of Bengal's economy[28] and this perhaps explains the subsequent misery faced by the artisans.

Fortunately, Bengal did not face another drought of a similar magnitude during the remaining years of the eighteenth century. Yet droughts did occur and exerted their influence on the state of harvests. There was a threat of a scarcity in 1773, but it was apparently short-lived. According to Warren Hastings, 'vigorous and early measures were taken for its removal and succeeded'.[29] The monsoon of 1774 failed, and by August the Company's administrators had begun worrying about 'the late unusual drought [and] the damage which the grain crops have already sustained by the want of rain'.[30] Parallels with the 1769 situation were raised when the rice withered and peasants 'turned their cows into the rice fields to graze upon it'.[31] The winter and spring rice harvests of 1775 were said to be in a 'sorry state' owing to another drought, and the peasants 'find it very difficult to fulfil their *kists* [payment of revenue]'.[32] In 1777, western Bengal 'hath so little rain fallen here this season, that there are the most alarming appearances . . . and both farmers and inhabitants of all ranks are under great apprehensions'.[33] A drought in 1779 had jeopardized the cultivation of the winter rice: 'the sowing lands are waste, the arable not ploughed' in western Bengal, whereas in the east crops of cotton, sugarcane and rice had suffered extensive damage.[34] The drought in 1783 had raised 'the present apprehension of a public scarcity arising from the deficiencies of a late harvest and the appearances of a similar failure in the present crop'.[35] In 1791, the 'unpromising appearance of crops, and little prospect of a speedy fall of rain' had raised the spectre of 'an impending calamity' in the province.[36]

Droughts, therefore, took their toll of harvests and seriously affected food supplies; but the degree to which both were affected depended upon the intensity and the duration of these phenomena. Thus 1769 became a year of a major famine whereas the other years of drought were limited to being years of dearth. Another feature of droughts, perhaps peculiar to Bengal, was the geographical asymmetry of such occurrences. For example, the intense drought of 1769 ravaged with its greatest severity in seven districts of west and north-east Bengal,[37] the rest of the province (especially the south-western and south-eastern parts) were relatively unaffected.[38] The latter areas did receive copious amounts of rain and 'the rivers have overflown & fertilized the lands, *even in this remarkably dry season*'.[39] A study of later droughts also confirms this regional picture.

The drought of 1773 was characterized by a similar geographical asymmetry. The overall worry was of an 'approaching scarcity', and some districts did report an estimated shortfall of 'one third of the usual produce',[40] but the food and price situation in other areas was strikingly different. The districts of Bishnupur and Burdwan (in west Bengal) were said to be in the throes of a 'great cheapness of grain' owing to good harvests.[41] In the east, the districts of Rangpur and Dinajpur had such plentiful harvests that in 1773 there was an 'extraordinary cheapness of the necessaries of life [which] has been of great disservice to the collections [of revenue]'.[42] The cheapness (caused by available stocks) of grain in these districts carried on into 1774,[43] at a time when western Bengal was said to be suffering a 'late unusual drought'.[44] Here, the crops had withered to such an extent that the cultivators, for instance of Purnea, 'daily complain loudly . . . that great part of their crop is scorched up, and is daily spoiling more and more';[45] others (in Birbhum) had given up their hopes of reaping the winter harvest of rice and had 'turned their cows into their fields to graze upon them'.[46] The situation in 1775 was characterized by a delayed monsoon. The rains not having arrived up to the end of July caused 'the most alarming appearances' of scarcity, but by early August 'a very plentiful fall of rain' had 'relieved the minds of the people'.[47] Finally the failure of the rains in 1779, which jeopardized the winter harvest and ruined the cash crops in some regions of west and east Bengal paradoxically occurred, amidst 'a great plenty of grain' in southern Bengal and a price situation so low that the 'Ryotts cannot dispose of their crops'.[48]

The role of the *jala* lands was also of particular importance during droughts. Situated in close proximity to water-channels, these lands retained their intrinsic moisture for a longer duration than the *colla* lands, and were therefore capable of producing the quick-ripening crop of *boro* paddy even in the middle of a drought. Eastern and southern Bengal were the most favoured in this respect. Such areas, being prone to seasonal water-logging, managed to retain crucial supplies of water during a dry season in the rest of the province.[49] Thus during the drought-induced famine of 1769-70, the low-lying districts of Dhaka and Bakarganj continued to furnish the 'greatest supplies of grain to the [Bengal] Presidency'.[50] Further west, rice was produced in adequate quantities in the *jala* lands of Midnapur[51] during that famine, so did the district of 24-Parganas which was situated in close proximity to Calcutta: 'owing entirely to [its] situation, which being low [it] retained what rainfall and enabled the people to water and preserve their crops [and] this aided so much to moderating the price of grain'.[52]

Unlike western Bengal, which was prone to periodic incursions of droughts, floods were an endemic problem in eastern and southern Bengal.[53] Here one has to distinguish between seasonal overflowing of its rivers and floods. Normally floods between three and six feet on low lands did not endanger the standing winter crop of rice as the grain naturally grew to keep its ear above water; but floods above 6 feet in height destroyed the crop.[54] Such seasonal floods were a part of the agricultural life in eastern Bengal, as James Rennell's description shows:

'By the latter end of July, all the lower parts of Bengal, contiguous to the Ganges and the Berrampooter are overflowed and form an inundation of more than a hundred miles in width'.[55]

The natural ingredients of a favourable agricultural season in the district of Dhaka were described by James Taylor as 'A high inundation in the preceding year, followed in the cold season by a moderate fall of rain which enables the husbandman to plough his land. This succeeded by frequent, but not heavy showers, in the spring months, and subsequently by a gradual rise of the rivers, constitutes the most favourable weather for the growth of rice'.[56]

Flood waters had to rise above 6 feet to be able to damage standing crops. A flood of this magnitude occurred in 1786-7 (discussed later). There are other recorded cases of floods but these were more localized in nature, caused by rivers which either overflowed their embankments or breached them in places. Such flood seems to have occurred in the districts of Jessore and Rajshahi in 1770 and 1771, when some crops were reportedly destroyed;[57] but the damage actually caused appears to have been quickly recouped as the month of September 1771 did see an 'extraordinary cheapness of grain' in Rajshahi.[58] Some damage to crops were also reported by such inundations in 1773, when the river Damodar overflowed its banks in some low-lying areas in the district of Burdwan.[59] In 1773, the district of Laskarpur suffered from the floods of the Padma and Burril rivers which: 'not only operated to this pergunnahs prejudice by destroying the harvest on the ground and sweeping away whole villages, ruining the Ryotts & occasioning great desertion; but when the water subsided, they left large tracts of land which was before cultivated, entirely choked up with sand & so impoverished as to preclude all hopes of bringing it again into cultivation'.[60]

Crops were also partially damaged in Midnapur because 'of the inundations in the beginning of September 1773'.[61] It is certainly true that the inundations of 1773-4 destroyed the standing crops in some areas and therefore 'impoverished the Ryotts'.[62] Yet, the impression which also appears inescapable is that the damage done was partial and limited to those lands having the lowest topographical situation. In Burdwan, the worst affected were 'those pergunnahs situated on the banks of the Damoodah [Damodar]', other parganas were only marginally affected.[63] The fact that the 'extraordinary cheapness of grain' said to be prevailing in 1773 (discussed earlier) would strongly suggest that the floods were relatively localized.

The flood of 1787 was entirely different from these instances. Two reasons made this event so different. First, there was a freak monsoon in that year which began from the month of March 1787. The rains continued without let-up till July, by which time 'nothing but a sheet of water' was to be seen in most areas of eastern Bengal.[64] Second, there was a major cyclone in the east which commenced from 30 October 1787 and lasted with undiminished intensity till 2 November. The cyclone was the proverbial last straw on the province's back. Practically, the whole of eastern Bengal was devastated.[65] Villages were under '7, 8 & 9 feet under water'.[66] A flood of such magnitude was accompanied by 'destruction of cattle, the blowing down of vast number of trees and levelling the houses of the poorer people and unthaching and unroofing of others. . . .'[67] People drowned and died as entire villages were swept away,[68] standing crops were destroyed by as much as 50 per cent in some areas,[69] and the flood-afflicted streamed in large numbers to Calcutta and Murshidabad.[70] The spectre of a full-blown famine once again stalked the province. Yet it must be emphasized that the combined effects of the unusually heavy monsoon and the cyclone were largely concentrated in eastern and south-western Bengal.[71] Elsewhere, for instance in Burdwan and Nadia, floods caused by rivers overflowing their banks (or breaching embankments) destroyed habitations and damaged the standing crop.[72] Therefore the famine which followed was concentrated primarily in those areas where floods and the cyclone had wreaked the worst damage. Elsewhere there were problems created by the spillover effects from the famine hit districts which manifested in high prices and dearth, but not in any excess mortality.

Ecological Crises and Vulnerability

Droughts meant that either standing crops withered, or that the lands were rendered unfit for sowing the next crop. It is a fact that the droughts which came after the one of 1769 were not particularly severe as to cause a famine on a provincial scale. Yet they were serious enough to cause food shortages at localized levels, and such shortages characterize a situation of dearth as short-run and localized crises of subsistence. But the events of 1769-70 had deeply influenced the province's psyche, and this fact meant that even partial scarcities raised fears of an impending famine. Thus the drought in 1773 was enough to 'alarm the inhabitants with apprehension of a renewal of their recent sufferings from famine'.[73]

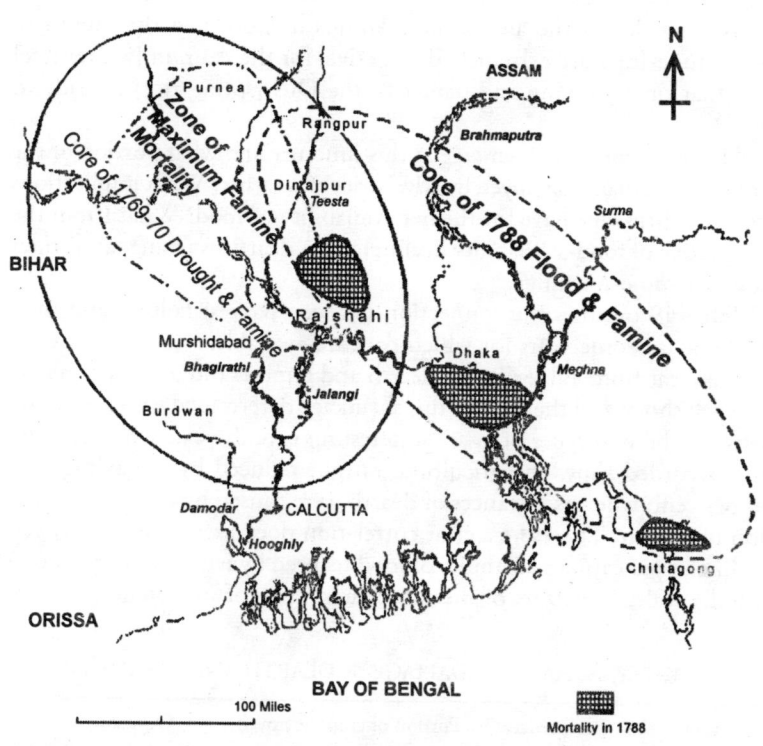

MAP 3. THE SPATIAL SWEEP OF FAMINE: 1769-70 AND 1788.

Floods also caused similar situations and fears. Here the important distinction was between a seasonal inundation and a flood which destroyed crops and livestock, thereby creating dearth or, as in 1788, a famine. It seems that a proper monsoon, characterized by regular rainfall between the months of July and September, was ideally suited for such inundations,[74] but unseasonal (post-monsoon) rains tended to cause floods and destroyed crops. Even a severe monsoon could cause a devastating flood: 'the annual inundations if sudden in the month of Assar [June-July] are extremely destructive, for the Aumun [winter rice] is at that time growing and tender & the Ous [spring rice] is ready to cut'.[75]

The questions which emerge at this juncture are: what portion of the crops were actually destroyed by adversities of weather? What implications did such shrinkage have for the net availability of food? What (from the perspective of food-availability decline) was a dearth-warning, as distinct from a famine-warning?

Table 49 outlines the connection between harvest failures and their outcome for some years for which data are available.

It is clear from Table 49 that dearth and famines did occur when harvests fell short; and the scale of this shrinkage determined the boundaries between the two types of crises. Interestingly, both years when famines were recorded show an agricultural output reduced by a maximum of 50 per cent; whereas instances of dearth were caused by a smaller reduction of output. Therefore a clear correlation does seem to exist between decline in per capita availability of food (caused by harvests falling short) and the scale, or nature, of the subsequent crisis of subsistence.

TABLE 49. HARVEST DAMAGE IN DEARTH AND FAMINE

Year	Occurrence	Portion of crop destroyed	Outcome
1769-70	Drought	28 to 50 per cent	Famine
1775	Drought	33.3 to 37.5 per cent	Dearth
1779	Drought	5 per cent	Dearth
1786-7	Flood	37.5 to 50 per cent	Famine
1791	Drought	33 per cent	Dearth

Sources: WBSA, Proceedings of the Select Committee, vol. 4, Proceedings of July 1770, p. 444; BDR, Chittagong, vol. 1, 30 November 1769; WBSA, Murshidabad, vol. 6, 11 September 1775; ibid., vol. 11, 25 May 1779, 30 May 1779 and 7 June 1779; IOR, BRC, P/51/22, 25 June 1788; IOR, BRP, P/71/45, 16 November 1791.

The case for the real decline in the availability of food comes into sharper relief from the state's perception of what happened to food-supplies during years of scarcity. In a Minute, dated 21 October 1791, the Board of Revenue noted that in a famine, or even in case of a serious dearth, it was very difficult to meet the shortfall by imports since 'the Bengal monsoon prevailing throughout the inland countries with which we have any communication, they are generally involved in the same calamity, and instead of being able to afford assistance, depend upon us for the supplies of their own wants'. Therefore overland supplies of food were obviously inadequate in meeting any part of the provincial demand. Supplies by sea was seen as a feasible, though expensive, option but the difficulty with this was that 'no judgment can be formed of the harvests until the month of October'[76] and the possibility of a scarcity at that stage meant that 'a considerable time must elapse before the ships which may happen to be lying in the river unemployed can be fitted out, as the owners ascertain from what countries the rice can be procured at a price that will render it worth their while to import it'. The other problem with the importation of rice was the apparently fluctuating prices of agricultural produce in the province: 'as the failure of the crops in one season occasions the price of grain to rise to an exorbitant height, so an abundant harvest in the next reduces it to a very low value; unless therefore the rice is imported before the next crops come to maturity the importer is liable to sustain a considerable loss'.

Provision of food at affordable prices by imports was not feasible under such conditions, and as the Board noted any relief by this mechanism was insignificant, and that the province had 'only its internal resources to depend upon'.[77]

Also crucial to the question of food-availability decline was the manner of its internal distribution during seasons of scarcity. Two features were of central importance here: (i) the spatial asymmetry of adverse weather meant that stocks of food could be procured elsewhere in the province, albeit at high prices, and (ii) the entire range of local trade in agricultural produce was handled by a specialized community of grain-merchants.[78] As soon as there was even a semblance of scarcity, these merchants purchased grain from areas with relatively greater surpluses and moved their stocks to the deficient places, especially to the cities, to be sold at high prices. Their strategies were facilitated because they owned, or controlled the means of bulk transportation, and they possessed storehouses (*golahs*) where rice and paddy could be stored for up to ten years without damage. Producers in relatively unaffected districts had no choice but to sell their surpluses as they were almost, without exception, tied to these merchants through an elaborate system of production and consumption loans, which meant that their end produce was already hypothecated to the grain trader even before they actually sowed their lands. The net results of all these schemes were that spot prices in less deficient places also rose, thereby hitting the harvest-dependent non-agricultural population there, and the food which came in to the deficit regions was sold at such high prices that the most vulnerable could barely purchase it. The following description of the dearth of 1791 shows the connection between harvest failures and a noticeable decline in the availability of food: 'The bazars have hitherto been sufficiently well supplied to answer the immediate wants of the inhabitants, but the alarm of an approaching scarcity is now become so universal that the poorer sort of people will shortly experience considerable distress, as the price of grain and the difficulty of procuring it, even for money, is daily increasing'.[79] Starvation by the poor was clearly the inevitable outcome of any kind of harvest failure. They were made 'at once bereft of present subsistence and future occupation'.[80]

How many, or what kind of people, did such shortages immediately affect? The people most severely affected during the famine of 1769-70 were said to be 'the workmen, manufactures and people employed in the river [boatmen]', and this was so because they 'were without the same means of laying by stores of grain as the husbandmen'.[81] The highest concentration of famine-deaths in 1769-70 were recorded amongst the rural artisans, the urban poor and among those who had migrated to the towns after being uprooted from their lands (presumably the poorer peasants). Contemporary descriptions suggest that nearly 50 per cent of rural artisans perished in the worst affected districts,[82] and more than 500 uprooted migrants died daily in the city of Murshidabad during the peak of the famine.[83] That famine had caused unprecedented 'distress to the poor' and threw up 'large numbers of deplorable objects [*sic*]' at the mercy of whatever relief was available in the towns.[84] The flood of 1786-7 took a heavy toll of the poorest in the villages. Nearly 7,000 people died and 12,000 people were forced to migrate from Chittagong within a fortnight of August 1786.[85] As the water-level increased, even the *colla* lands became submerged[86] and the rural poor of southern Bengal 'either wandered to Calcutta to receive support from the charity, or went into other districts in hopes of employment, or simply resigned their lives'.[87]

In eastern Bengal 'great number of ryotts [cultivators] have been drowned by the inundations, and others being destitute of food for their support, and reduced to the greatest distress, have deserted from the districts'.[88]

A dearth did not occasion misery of similar magnitude as famines did. Nevertheless the first to be affected were invariably the 'poor class of natives'[89] who comprised the 'large proportion of the lower and most useful class of people'.[90] The essential problem during a dearth was not so much an absolute reduction in the quantity of food at the provincial level, but the severe food-availability decline at local levels which consequently raised the spot-prices of rice and paddy up to a point where the local poor were unable to purchase their subsistence. The tenuous balance between weather, output, prices and incomes meant that even the slightest variation in the first two variables (weather and output) immediately affected the state of prices, which, in turn, put severe strains on the income of producers. The reasons that they did not die in large numbers during dearth were mainly because of the short duration of such phenomenon and that food-supplies came in from the unaffected areas, albeit at high prices. But dearth were nevertheless serious crises of subsistence. The vulnerable sections of society were exposed to the problems of not having enough to eat despite their already low-levels of subsistence.[91] Almost invariably the producers complained that a dearth meant that 'we cannot procure food even with money' though 'there is grain in the hands of the merchants'.[92] Dearth meant that the producers were 'at once bereft of present subsistence and future occupation' which left them with no choice but to 'abandon their habitations to escape the horrors of a famine'.[93] It was this constriction in the availability of food under a dearth situation which explains why 'the bulk of the people' at Rajshahi,[94] in 1788 'have barely fed themselves and *not* with their usual food'.[95]

It is, therefore, possible to posit the existence of fairly numerous harvest-sensitive people in the province of Bengal whose very lives were at risk during famines and whose subsistence was jeopardized in a dearth. These people were obviously those who were 'without the same means of laying by stores of grain as the husbandmen'—people who suffered the greatest number of deaths during the famine of 1769-70. These were the artisans, poor peasants and labourers who either possessed insufficient land or depended exclusively on the current harvest for their sustenance and did not have alternative or stored-up supplies of food. In the towns, these were the labourers and the town-poor who depended entirely on the wages they could earn and whose existence was severely threatened by any reduction in food-supplies from centres of supply as any such contraction forced prices to spiral, while their wages remained constant.[96]

Price as a Factor in Famine and Dearth

Central to this discussion is the notion that both famines and dearth were crises of subsistence of differing intensities whose initial conditions were created by a shortage in the quantity of the harvest. The other factor which was equally crucial in both cases (famine or dearth) was the state of agricultural prices because food shortages are occasions of high prices, and the extent to which prices climb, and the damage they cause, depends on the temporal spread of such shortages.

In a later study of famines in Bengal W.W. Hunter pointed to the critically small dividing line between a famine warning and a famine point in the province, the central indicator of which was the price situation. 'In Bengal', wrote Hunter, 'even the slightest rise in the prices of agricultural produce makes a whole difference between a famine warning and famine point.'[97] This statement suggests a clear connection between supplies of food and prices and between purchasing power and subsistence, and there is ample evidence to show the existence of such connections on a widespread basis in Bengal's economy of the eighteenth century.[98]

At what price-point did a dearth arise, and at what point did a famine occur? Figures 14 and 15 show the extent to which winter rice prices rose during the famines of 1769-70 and 1788. Figure 16 reflects the state of prices in Midnapur in the former famine whereas Figure 17 reveals a comparative picture of rice prices in five Bengal districts during the famine of 1788. Some other evidence for 1769-70 also suggest an extremely steep rise in price. For instance rice prices had gone up 500 per cent in a season in Purnea during that famine. In Murshidabad prices climbed from Rs. 1.14 a maund (in July 1770) to Rs. 2.5 in August. At Calcutta prices ranged from Re. 1 a maund in 1769 to Re. 0.77 in 1770.[99] The relative cheapness of rice in Calcutta was mainly due to the Company's strategy of stocking provisions to feed the city, notwithstanding which 76,000 people reportedly died on its streets between July and September 1770.[100] But the price rise in 1788 suggests that the Company failed to take adequate measures to alleviate the crisis in the city.

Table 50 makes an attempt to present the largely scattered price data of dearth in a somewhat comprehensible fashion.

TABLE 50. RICE PRICES UNDER DEARTH IN THE
LATE EIGHTEENTH CENTURY (Rs. per maund)

Month/Year	District	Price
1773	Burdwan	0.49
1774	Burdwan	0.57
1773	Chittagong	0.54
1774	Chittagong	0.61
1774	Rajshahi	0.60
1775	Rajshahi	0.80
1774	Laskarpur	0.60
1775	Laskarpur	0.82
1784	Dhaka	2.22*
March 1791	Rangpur	0.66
June 1791	Rangpur	1.05
January 1791	Murshidabad	1.14
February 1792	Murshidabad	1.49

Note: *The previous price is not known, but 1784 was a localized famine in Dhaka.

Sources: For prices between 1773 and 1775, IOR, BRC, P/49/42, P/49/44, P/49/47, P/49/54; WBSA, PCR, Murshidabad, vol. 3, 18 August 1774 and vol. 6, 11 September 1775; S. Islam, *Bangladesh District Records*, op. cit., p. 98; IOR, BRC, P/52/32, 30 May 1791; ibid., P/52/40, 6 January 1791 and 24 February 1792.

These price data show that typically famine prices were characterized by a sudden steep rise; prices in dearth also rose sharply but not to the same extent. The other important aspect of price behaviour during a famine is the uniformity with which prices rose throughout the province. Thus in the famine of 1788, the average selling price of rice in the cities of Calcutta, Murshidabad and Dhaka was Rs. 1.84 per maund; in Bakarganj (which was recognized as one of Bengal's granaries and had continued to supply rice to Calcutta during the height of the 1769-70 famine) rice prices had shot up to an all-time high of Rs. 1.05.[101] The intensity with which prices rose during a famine provides an important insight into the probable cause of the starvation deaths which occurred. Obviously when the price of basic staples rose by as much as 500 per cent (for instance in Purnea in 1769-70), deaths by starvation were bound to follow. Moreover high prices in the epicentre of the famine had a spillover impact on the prices in relatively unaffected areas. Thus during the famine of 1769-70, grain-merchants flooded to Midnapur, where some stocks of rice were still available, and forced an escalation in price.[102] Similarly, Mymensing which had food-stocks during the crisis of 1787-8 was invaded by merchants eager to purchase rice and paddy;

consequently prices rose sharply in April 1788.[103] What we are not sure is whether these spillover effects actually caused starvation deaths in these latter areas.

Certainly, therefore, the price situation was an important factor in causing famine deaths. People died because they could not purchase their subsistence at the prevailing prices. Yet, foodgrain shortage did not automatically lead to excess mortality. A dearth was relatively free from such deaths. 'Foodgrain shortage', says Alamgir, 'only when translated into *prolonged* foodgrain *intake* decline, causes death',[104] and it is this phenomenon which perhaps explains the massive deaths reported in western Bengal in 1769-70, when the drought, which lasted for six months, presumably forced the famine-vulnerable to starve for the same period. Many, apparently, also died in 1787-8, but there are no contemporary estimates to go by. One estimate however says that 152 households, out of a total number of 287 households in a village of Rangpur perished during the flood of 1787 and the famine which followed.[105] The question of famine mortality will be discussed shortly.

Famine and dearth prices thus provide an important insight into the nature of subsistence crises in a commercialized economy. Harvest failures were only proximate causes for triggering these events. Their real magnitude can be comprehended only by looking at them as severe dislocations in the food-market caused by sharp rise in prices. Subsistence was obviously now a function of an individual's purchasing power, and this fact must clearly indicate the virtual absence of 'traditional' forms of subsistence 'security'.

Famine and the State: Revenue and Relief in 1769-70

The other factor often ascribed a crucial responsibility in causing famine misery was the tax burden. The famine was a critical crisis of subsistence, and the amount of money which ordinary people could spare for purchasing essential items of survival obviously depended on the amount of money they had after paying the state's revenue.

The Company's revenue policies in Bengal were guided by a mixture of commercial and military considerations. From 1760, and especially after 1765 (when the Company was granted the Diwani of Bengal and Bihar), Bengal's territorial revenue was used to finance its commercial 'investments' in the province, as well as to finance its military ventures in the rest of India. Obviously, these considerations meant that collections of revenue had to be strictly enforced, and the fact that this indeed was

TABLE 51. REVENUE COLLECTED DURING 1765-72

Year	Collections (Rs. in million)*
1765-6	13.59
1766-6	16.69
1767-6	16.87
1768-6	15.81
1769-70	14.74
1770-1	13.82
1771-2	15.73

Note: *The collections between 1765 and 1772 have been converted from, £ sterling into *sicca* rupees at the rate of Rs. 8.88 = £1, which seems to have been the prevailing rate of exchange in the mid-eighteenth century.[107]

Source: *Fourth Report*, 1783, pp. 99-101; G.W. Forrest, *Selections from the State-Papers*, pp. 264-5.

the case can be seen from the statement of the collections made between 1765-6 and 1771-2 (see Table 51).[106]

The table shows that there was a reduction in the amount collected but from all the accounts of the famine one gets the impression that this reduction was inadequate in ameliorating distress. Perhaps, too little came much too late.

On the other hand. we have the details of revenue actually collected from different districts during the famine which allow a more differentiated picture of the magnitude of the tax burden imposed by the Company. These figures are given in Table 52.

The figures in Table 52 show that more revenue was collected during the height of the famine than in the subsequent year, and this partially confirms Hastings' statement that there was greater misery during the famine of 1769-70 than ever before because the pitch of collection was 'violently kept up to its former standard'.[108]

However, we get a different picture if we look at these figures in a comparative perspective keeping in mind the geographically asymmetric nature of the famine. It is extremely significant that collections from three of the most devastated parganas, Birbhum, Nadia and Hughli, were less during the famine than in some others which were only marginally or not in the least affected (for instance, Dhaka or Jessore). In fact, the combined collections from the 7 worst-hit districts, (excluding Burdwan for which figures are not given), are shown in Table 53.

On the whole, those areas situated in the epicentre of the famine paid Rs. 5,53,000 less as revenue in 1769-70 than what they did in 1770-1,

TABLE 52. COLLECTIONS FROM DISTRICTS,
1769-70 AND 1770-1 (Rs. in 10 thousand)

District	Collections 1769-70	Collections 1770-1	Difference
Rajshahi	8.10	7.96	+ 0.14
Laskarpur	2.23	2.05	+ 0.18
Rokunpur	2.27	2.86	-0.59
Fatehsing	1.43	1.32	+ 0.11
Chunakali	1.44	0.85	+ 0.59
Jahangirnagar	1.74	1.74	Nil
Silberis	0.50	0.49	+ 0.01
Kasimpur	0.46	0.51	- 0.05
Baharband	1.13	1.11	+ 0.02
Bhetoria	15.47	10.85	+ 4.62
Purnea	9.77	9.25	+ 0.52
Birbhum	7.25	8.49	- 1.24
Dhaka	18.07	17.84	+ 0.23
Sylhet	0.33	0.27	+ 0.06
Tipperah	0.71	1.02	- 0.31
Nadia	3.64	7.16	- 3.52
Dinajpur	17.42	14.51	+ 2.91
Rajmahal	1.16	0.66	+ 0.5
Rangpur	6.65	6.76	- 0.11
Jessore	6.17	5.47	+ 0.7
Hughli	3.73	7.96	- 4.23
Total	109.67	109.23	+ 0.44

Source: FR, Murshidabad, IOR, G/27/1, 4 April 1771, p. 672.

TABLE 53. REVENUE COLLECTED FROM THE WORST AFFECTED
DISTRICTS IN 1769-71 (Rs. in 10 thousand)

District	Collections in 1769-70	Collections in 1770-1
Rajshahi	8.10	7.96
Birbhum	7.25	8.49
Purnea	9.77	9.25
Nadia	3.64	7.16
Dinajpur	17.42	14.51
Rangpur	6.65	6.76
Hughli	3.73	7.96
Total	56.56	62.09

Source: FR, Murshidabad, IOR, G/27/1, 4 April 1771, p. 672.

and this indicates that the relationship between famine-misery and taxation was not as direct and certainly more complex than what has been made out to be. The severity of the drought and the steep escalation in the price of food grains perhaps contributed more to causing famine-misery than the burden of taxation which was higher in those areas where the famine had made only a marginal difference.

What the Company's state failed to do, and which surely increased the misery of the people, was to provide any form of institutional relief even when it had the cash reserves to do so. Insignificant cash was spent in providing immediate assistance to the uprooted and hungry who flooded in large numbers to the city of Murshidabad. Up to April 1770 the Company had advanced only Rs. 1,00,000 'for the purchasing of rice on account of a charitable distribution made to the poor in and around Murshidabad'.[109] This amount compares unfavourably with the cash relief of Rs. 47,250 provided by Nawab Mubarak-ud-daulah, Mohammad Reza Khan, Roy Durlabh and Jagat Seth to run gruel kitchens in the city.[110] In Patna, Shitab Roy spent Rs. 30,000 ferrying rice from Banaras to feed the hungry who were apparently flocking into the capital and 'in this manner an immense multitude came to be rescued from the jaws of imminent death'.[111] Elsewhere, the state's performance was more dismal. In Rangpur, John Grose (later accused of profiteering during the famine)[112] could only bring himself to provide a daily distribution of Rs. 5 worth of rice amongst the indigent even when 'the distress of the poor at this place still continuing very great from the scarcity of grain, and such a large number of deplorable objects [sic] daily applying for relief'.[113]

However, the most important dimension of institutional relief is the material help given to support production and the producers both of which are caught in the middle of a severe short-run crisis of survival under famine duress. Here, the Company performed miserably. A clear indication that some steps to support production were necessary was given by Thomas Rumbold from Bihar as early as November 1768 when he wrote to say that the inhabitants had to be encouraged and relieved from oppressions 'and it will be necessary hereafter to make some allowances to farmers in those places which have suffer'd most from want of rain. . . .'[114] Yet very little was actually done.

There was an insignificant amount of financial aid (*taqavi*) provided to help the cultivators to begin production for the next agricultural season. The drought and high prices had destroyed the productive resources of the majority of producers and it was universally acknowledged that they needed cash help on easy terms of repayment (*taqavi*) in order to sustain their future economies. Yet only Rs. 4,07,055.50 were advanced from the Murshidabad treasury as *taqavi* between 11 April 1769 and 31 May 1770; of this amount, Rs. 3,68,793.62 (nearly 91 per cent) were distributed in the worst affected districts of Nadia, Birbhum, Purnea (in western Bengal) and Rajshahi, Rangpur and Dinajpur (in the east).[115]

The question is: could this *taqavi* have alleviated the plight of the producers, especially in the worst affected districts? The answer is negative when we compare the actual collections (*hasil*) in 1770-1 with *taqavi* given in 1769-70 (see Table 54) in districts for which comparable data are available.

Famine Mortality: A Reappraisal

Obviously the scale of the misery of these harvest-sensitive strata depended on the magnitude, measured in terms of the spatial and temporal spread, of these subsistence crises. Famines were occasions when a subsistence and mortality crisis came together in a critical conjuncture. This happened in 1769-70 and to a relatively lesser extent in 1788 as a carry-over from the massive floods of 1786-7.

The most written about catastrophe in the history of eighteenth century Bengal is the famine of 1769-70, an event whose demographic consequences have almost become an orthodoxy in provincial historiography. Ever since Warren Hastings wrote about this aspect in 1772, subsequent most historians have accepted his estimate of 10 million deaths causing a decimation of one-third of the population without any

TABLE 54. *TAQAVI* AND *HASIL* IN SOME BENGAL PARGANAS, 1769-70 AND 1770-1 (*Rs. in 10 thousand*)

Parganas	Taqavi in 1769-70	Hasil in 1770-1
Rajshahi	0.50	7.96
Rangpur	0.35	6.76
Dinajpur	0.45	14.51
Rokunpur	0.15	2.86
Purnea	1.25	9.25
Nadia	0.30	7.16
Birbhum	0.84	8.49
Hughli	0.02	7.96

Source: Taqavi figures are from SCC, IOR, P/A/10, 31 May 1770; *hasil* statistics are from Table 53.

questions.[116] Others who have questioned these mortality estimates have, nevertheless, left the issue unresolved owing to the paucity of evidence.[117]

Having arrived in Bengal only in 1772, Hastings' estimate could only have been a rough all-Bengal assessment made without any quantitative information at his disposal. Except in a few exceptional cases, the Company's beleaguered administration had comprehensively failed to collect any reliable information about famine mortality. Therefore Hastings' estimate was at best an inspired guess he made from the descriptions he read from some of the most devastated parganas. Other contemporary estimates place the death toll from 80 per cent (Shitab Roy in 1771)[118] to 20 per cent (John Shore in 1789).[119] Of these, Shitab Roy's would be the most reliable as he was an eye-witness, but his account was limited to Bihar, especially to Patna, where he was during the entire duration of the famine and where famine mortality was indeed very high.[120]

The question is: how many people actually died during the course of the famine? All answers must remain provisional and absolutely tentative because of the paucity of quantitative data.

The descriptions contained in the sources suggest a famine misery on a calamitous scale. Deserted villages stretching over vast areas, large tracts of uncultivated land turning into jungles and abandoned homes both in the towns and in the villages—these are some of the main features of the devastation in rural Bengal recorded by eye-witnesses during and after the famine. In 1783, Ghulam Hussain spoke of 'vast multitudes' being swept away in the famine: 'whole villages and whole towns were swept away'.[121] Contemporary evidence available from 9 villages 'contiguous to Nattore' in Rajshahi suggests that these were reduced to 212 'dilapidated houses' in 1771 from 1267 households in 1769.[122] Rangpur suffered 'a great scarcity of inhabitants'.[123] In Birbhum 'many hundreds of villages are entirely depopulated & even in the large towns here are not a fourth part of the houses inhabited'.[124] Later inquiries revealed that Pachet and 'northern parts of Burdwan suffered a great loss of inhabitants during the famine'.[125] The famine had taken a daily toll of 'more than five hundred people' in the city of Murshidabad and 'in the villages and country adjacent the numbers said to have perished exceed all belief'.[126] By February 1770, it was feared that 'one half of those who were about to pay revenues & cultivate the lands will unavoidably perish' in Purnea.[127] It was later estimated that one-half of 'the labouring and working people' had actually died in Purnea between 1769 and 1770.[128] The villages of Rajmahal were 'for the most part totally abandoned [because] of a want of hands to follow the employment of tillage'.[129] The number of looms 'employed in the Company's investment' at its factory in Malda 'decreased during that calamity by near one half'[130] and there was 'a universal scarcity of Ryotts' in Hughli in the immediate aftermath of the famine.[131] What we cannot be sure is whether the descriptions given in the preceding paragraph refer to depopulation actually caused by mortality or to desertion caused by the flight of peasants and artisans. The fact that the producers actually migrated during this event, and also during subsequent episodes of dearth and famine, is indicated in the sources, whose implications will be examined shortly.

Regarding deaths, the situation in Purnea and Murshidabad may perhaps permit the following reconstruction, keeping in mind that the peak period of famine devastation was a maximum of six months and that the worst affected were the 6 districts of western and north-eastern Bengal (Map 3).[132] Purnea seems to have lost a total of 2,00,000 people during the course of the famine.[133] The estimate of 500 deaths per day in Murshidabad gives us a total of 90,000 dead during the peak of famine mortality in that city. Even if we take the death-toll in Purnea as the representative figure of famine fatalities in the affected districts and apply that to the six districts which are described as having the highest rates of famine deaths, we get a figure of 1.2 million dead. This would be a reasonable estimate because (i) south-western and south-eastern Bengal had practically no excess deaths and (ii) the famine had a clearly asymmetrical geographical sweep. The latter aspect has been discussed at various places in this chapter. The former point (that of the absence of excess deaths in south-eastern Bengal) is apparent from the figures regarding deaths and desertions in four contiguous villages of Rajshahi and Chittagong between 1769 and 1772 which are set out in Table 55.

Nevertheless, the depopulation in the districts most affected was enormous as contemporary estimates in the following districts show:[134]

Purnea	50 per cent
Birbhum	25 per cent
Rajshahi	33 to 50 per cent
Malda	50 per cent

There is, therefore, some justification in the statements of Sinha and Chaudhuri that the famine caused unprecedented deaths. But these deaths were limited only to the worst affected districts and the previously accepted estimate of 10 million deaths now appears a largely inflated figure for a number of reasons.

TABLE 55. MORTALITY AND DESERTIONS IN 1770:
RAJSHAHI AND CHITTAGONG

Year	Deaths: Rajshahi	Desertions: Rajshahi	Deaths: Chitagong	Desertions: Chitagong
1769	7	40	112	13
1770	548	153	54	Nil
1771	9	31	55	Nil
1772	Nil	Nil	59	Nil

Sources: BM, Add. Ms. 29076, fol. 140; *BDR, Chittagong,* vol. 1, no. 202.

First, the peak intensity of the famine was located between the months of January and August 1770 when the highest concentration of deaths and desertions were reported from the most affected districts in western and north-eastern Bengal. In effect, this gives us a peak mortality period of six months. In the space of six months 10 million deaths would make the case of Bengal unique in the history of global demographic catastrophes.[135] The implausibility of so many deaths happening in such a short time becomes more apparent when it is remembered that the famine was not accompanied by a general epidemic. The only recorded case of an epidemic comes from Purnea, where in April 1770 we are told of the 'Horror of Pestilence being added to those of Famine'.[136] It is interesting that a 'great fever epidemic' (presumably malaria) was recorded in 1762 leading to the deaths of 50,000 people,[137] but no such descriptions exist for the 1769-70 situation.[138]

Second, there is an unresolved problem with the death toll itself. It tells us nothing about *excess mortality* during the famine. It is generally recognized that practically, in all pre-modern societies there was a clear distinction, though often rendered tenuous, between normal and excess mortality. Normal mortality were regular deaths which occurred in the absence of a mortality crisis occasioned by war, famine or an epidemic. Excess mortality were the *additional* deaths caused in a crisis year and the case of pre-industrial Europe shows that these deaths could rise up to 6 or 10 per cent over and above the 'normal' deaths of roughly 3 per cent a year.[139] The estimate of 10 million dead for Bengal in 1769-70 conceals the proportion between normal and excess deaths, though the scale of the crisis in that year would probably point to a fairly high concentration in the latter category. Once again however one has to bear in mind that such a concentration would apply in the *worst* affected districts. Elsewhere the proportion between normal and excess mortality was much smaller, precisely because the catastrophe was less severe in those parts.

Most of our views of mortality during this famine are based, not on any quantitative evidence (which are not available), but on descriptions of supposedly vast stretches of land falling out of cultivation (*pateet*) or deserted (*palataka*) or simply lost (*lokshan*) to cultivation (*jote*). The most common understanding is that the famine forced one-third of the land out of cultivation.[140] The Amini Commission was charged with investigating the state of Bengal's agrarian economy and its report (1778) is used to support such notions of devastation.[141] A recent, though synoptic, examination of the evidence suggests a depopulation in excess of one-third on the basis of some evidence of *palataka* contained in this Report.[142] Finally, Cornwallis' statement (in 1789) that 'one-third of the Company's territory in Hindostan is now a jungle inhabited only by wild beasts'[143] is invoked as testimony of the after-effects of the famine.[144]

The evidence therefore needs to be re-examined. To begin, Cornwallis' assertion makes no reference to the famine as his point of departure. He was situating himself in the ongoing debate on the need to formulate a policy of settling Bengal's revenue on a permanent basis, and his specific comments were made in response to a Minute tabled by John Shore on this problem in June 1789. The substance of his argument was the problems associated with getting people to cultivate *new lands*.[145] Also, the evidence which is available to show the extension of cultivation in different areas of the province (Chapter 1) does not square with a notions of vast abandoned or deserted stretches of land undergoing reforestation because of a severe depopulation in these parts.

With regard to the evidence of deserted lands, the Amini Commission's evidence presents a more differentiated picture. The Commission recorded the assessed revenue (*hast-o-bud jama*) under two heads, *palataka* (literally, absconded) and *hazera* (literally, present/occupied). *Palataka* was 'rated revenue of land unoccupied [i.e. deserted] but still kept in the rent-roll', whereas *hazera* was 'revenue of occupied lands and other actual sources'.[146] Table 56 reproduces the data of the Amini report under these three heads.

The evidence is unequivocal. Barring Birbhum, and to some extent Rangpur and Purnea, the arable in most other areas of Bengal where the Amini Commission went was under optimum utilization. Particularly significant is the case of Rajshahi, whose revenue-paying lands were fully occupied a mere 8 years after having witnessed a mortality from 33 to 50 per cent of its population.

TABLE 56. EXTENT OF CULTIVATION IN BENGAL: THE EVIDENCE OF THE AMINI COMMISSION, 1778

Pargana	1 Hast-o-bud jama (Rs.)	2 Palataka jama (Rs.)	3 Hazera jama (Rs.)	3 as % of 1
24-Parganas	13,56,461	62,688	12,93,773	95.38
Town of Calcutta	1,10,299	Nil	1,10,299	100
Hughli or Muhammad Aminpur	2,76,062	13,582	2,62,480	95.08
Nadia	15,85,498	2,42,842	13,42,656	84.68
Muhammadshahi	3,80,409	38,744	3,41,665	89.81
Jessore	4,83,388	87,807	3,95,581	81.83
Saidpur	1,19,580	19,191	1,00,389	83.95
Hijli	2,94,945	28,275	2,66,670	90.41
Shuja Mutah	65,291	5,919	59,372	90.93
Alisoda	1,91,647	25,527	1,66,120	86.68
Midnapur	9,94,757	Nil	9,94,757	100
Birbhum	11,44,825	4,11,613	7,33,212	64.04
Bishnupur	5,18,731	17,918	5,00,813	96.54
Pachete	1,54,423	49,673	1,04,750	67.83
Rajshahi	29,64,631	Nil	29,64,631	100
Rokunpur	3,65,090	26,773	3,38,317	92.67
Khas Taluks	1,25,519	12,941	1,12,578	89.69
Fateh Singh	1,62,633	21,651	1,40,982	86.69
Jahangirpur	3,63,570	1,04,629	2,58,941	71.22
Laskarpur	2,97,846	30,338	2,67,508	89.81
Dhaka ('exclusive of Sylhet')	43,63,561	Nil	43,63,561	100
Chittagong ('exclusive of Tippera')	6,68,529	Nil	6,68,529	100
Rajmahal	2,94,541	38,096	2,56,445	87.06
Hatinda	77,788	4,317	73,471	94.45
Bhagalpur	4,66,329	2,518	4,63,811	99.46
Purnea	19,09,214	4,96,198	14,13,016	74.01
Rangpur	16,50,655	3,71,695	12,78,960	77.48
Total	21,389,216	21,12,215	19,276,521	90.12
Average	7,92,193.19	78,230.18	713,945.22	90.12

This picture is also confirmed by demographic data. That Bengal's population was steadily growing was 'obvious to [the] common observation' of John Shore in 1789.[147] It is undeniable that population estimates for this period are both woefully inadequate and hopelessly unreliable. Contemporary estimates are shown in Table 57.

Notwithstanding the crudeness of these estimates, these figures tell us that in 34 years between 1789 and 1822, Bengal's population increased by 15.6 million, giving us an augmentation of 0.473 million people (a phenomenal growth rate of 2.15 per cent) per annum. Therefore, there is very little substance in the argument put forward by a number of historians that the famine of 1769-70 created a long-term imbalance between land and labour in rural Bengal as 'now there was more land awaiting cultivation than there were tenants to till it'.[148] There was no shortage of people in rural Bengal despite the severity of the famine of 1769-70.

The death-toll during the flood-induced famine of 1788 is less detailed though one still gets the distinct impression that it was quite high in eastern Bengal. Nearly 7,000 people died in Chittagong in a fortnight in August 1786.[149] Massive deaths were reported from all over Bengal. At Burdwan, 'almost every house in the town and every village contiguous to it fell down' by the onslaught of the flood waters and 'large number of people died and a prodigious quantity of cattle destroyed'.[150] The flood and the havoc it caused to the standing crops had killed a 'great number of ryotts' in Rangpur while the survivors had been rendered 'destitute of food for their support'.[151] Dhaka also suffered large-scale mortality and desertion by the surviving rural population.[152] The district of the 24-Parganas, which had largely escaped from the devastating drought of 1769-70,[153] seems to have been rudely shaken by the flood. Here the problem caused by the flood was aggravated by an outbreak of small-pox causing multitudes of people to 'resign their lives'.[154] The crisis in north Bengal (Malda, Rangpur and Dinajpur) was compounded by a change in the course of the Teesta river in the 1780s. The Teesta deserted its previous course and 'now [1788] runs through . . . an infinite variety of different streams, many of which are entirely new, & have been formed

TABLE 57. ESTIMATES OF POPULATION IN BENGAL

Year	Population	% growth since 1789
1789	22 million	
1790	24 million	9.09
1793	25 million	13.64
1800	27 million	22.73
1822	37.6 million	70.91

Sources: Hunter, Annals of Rural Bengal, pp.32-4; Colebrooke, Remarks, p. xii; Sugata Bose, Peasant Labour, p. 20.

by the water penetrating thro' the cultivated fields; whilst others which before were insignificant Nullahs [streams] have from the above event become considerable rivers'.[155] When the flood struck it damaged lives, livestock and property to such an extent that 'there does not remain a vestige of the Ryotts houses nor even a mark of the places where they had formerly been'.[156] Flood-waters destroyed standing crops in Purnea and Nadia and caused peasants to flee from their lands, but there were relatively fewer deaths in these places,[157] which indicates the lesser severity of the subsistence crisis in western Bengal.

Worst conditions were still to follow in eastern Bengal. The first wave of mortality was caused by people drowning during the course of the floods in August-September 1787 and then by the cyclone which struck immediately afterwards. But it was the famine in the east, which lasted during the major part of 1788, which caused the greatest misery. Between 37.5 and 50 per cent of the standing crops were destroyed in 1787 and the harvests of 1788 were affected because the lands had soaked up moisture in excess of what was considered good for proper cultivation and so could not be ploughed.[158] Like the previous famine in 1769-70, supplies of food were reduced among the harvest sensitive strata of the population, prices rose sharply and remained so 'for a period of above eight months' in 1788.[159] Starvation deaths were reported from all over Bengal, 'many destitute and miserable' looking for some form of relief entered the 'skirts of the town' of Calcutta, thereby causing a disruption in the supplies of rice there,[160] and the general debilitation of the people elsewhere was 'too lamentable a proof of the excessive dearness of the provisions and the distressed situation of the inhabitants, who sell their children to enable them to purchase a few meals to prolong their miserable existence'.[161] The surviving population: 'Are greatly reduced in bodily strength from want of food, and have neither cattle nor other effects. The few who have anything left cannot cultivate their old possessions for want of seed and enhanced price of cattle; hence much of the old cultivated lands are lying fallow. . . .'[162]

Unfortunately we are not in a position to account for the number of people who actually died during the flood of 1786-7 and the famine which followed. Some evidence from Jessore and Dhaka suggests that this catastrophe had resulted in as much as 75 per cent of the cultivated land in some parganas falling out of cultivation as 'many of the Ryotts are dead from the famine that has prevailed the whole year while others have left their habitations and gone to other parts of the country'.[163] Another estimate of a village in Rangpur tells us that 152 out of a total number of 287 households (recorded in 1786) perished during the flood of 1787 and the famine which followed.[164] What we cannot be sure is whether the estimates of lands falling into decay were caused by the physical decimation of the producers or by the temporary out-migration of a harassed peasantry and artisans under extremely trying circumstances. There is a strong possibility that both may have occurred, though the proportion of one to the other must remain uncertain. Moreover, the fact that the famine (which followed the flood and cyclone of 1786-7) was concentrated primarily in eastern and south-western Bengal must (like the famine of 1769-70) point to a spatial asymmetry in the crisis of mortality; but the paucity of evidence compels the issue of deaths during 1787-8 to be left wide open.

Famine, Dearth and Migration

The preceding sections have indicated that people migrated from their established bases of production during famines, or even dearth. In 1769-70, producers did desert in large numbers as 'under this hardship, to forsake his [the producer's] profession and country has been the easiest and speediest means of relief',[165] which must indirectly reflect the fact that the such people actually viewed migration as a strategy of survival under crises.

There is some indication in our sources that migration for survival indeed happened in Bengal during 1769-70. Around 1770, Dhaka division was described as having a 'remarkably luxuriant' soil but thinly populated owing to its 'low situation'; floods were a problem in a topography of this type but droughts were relatively unknown, these being 'pernicious only to the crops of the upper or northern pergunnahs'.[166] Circumstantial evidence indicates that the famine of 1770 saw a movement of people from the west to the east as by 1775 'the Balance of trade is in favour of this province [as] its southern districts furnish the greatest supplies of grain to the presidency'.[167] Midnapur certainly received large numbers of uprooted migrants from Burdwan, Birbhum and 'other parts of Bengal' who came 'in a very starving condition in hopes of finding the scarcity less here than in their own country'; many travelled to Orissa.[168] The district of the 24-Parganas was not only free from the demographic consequences of the famine, it also received an influx of peasants and artisans from other districts which explains its relative prosperity during the height of the famine and later.[169] Sinha states that Rangpur 'gained an accession of population by the calamity'[170] but this is in the teeth of overwhelmingly contradictory evidence which is available to show a substantial depopulation in that district by *flight* of the peasantry even though actual deaths here were relatively insignificant.[171]

Both the facts, that able-bodied peasants survived the famine and that they migrated in large numbers, are also supported by the statements of peasants being asked to return to their abandoned homesteads by various zamindars,[172] by the reports of a fairly rapid reclamation of deserted (*palataka*) land, even in a badly affected district like Birbhum after 1771[173] and by the accounts of petty-landholders (talluqdars) in Rangpur 'enticing the ryotts from other parts' to settle in their lands on concessional terms of revenue.[174] By 1773, those who had fled from their lands in Purnea had started to return 'to their fields, except a few who have no calling' who had chosen to remain in Murshidabad 'to seek employment'.[175] Thus, the famine of 1769-70 appears to have caused a cross-migration of people from north and west Bengal to the districts situated to the south-west and the south-east.

The famine in 1787-8 saw another large-scale movement of the uprooted people all over southern-western, south-eastern and northern Bengal from those places which had been worst affected by the floods and the cyclone. The first reaction of the afflicted was to move to higher grounds and wait for the flood-waters to subside. Thus, in Burdwan, where the rivers Damodar and Ajoy breached embankments, 'nothing but the banks of tanks remained for the reception of every living animal';[176] and in Rangpur, villagers fled to 'high spots of land, and lived with those, whose houses were above the reaches of the waters', others 'took up their abodes on trees and so saved their lives'.[177] But the waters did not recede and the cyclone in the east added to their problems. Thus began the long, and debilitating, process of out-migration as the survivors were forced to 'quit their habitations and [go] to other parts of the country'.[178]

Where precisely were these 'other parts'? There is some evidence to show that some people moved to the district headquarters in the hope of some relief.[179] The uprooted from the district of 24-Parganas moved in large numbers to Calcutta 'in hopes of employment'.[180] Migrants from Rajshahi, Dinajpur and Rangpur also entered the town of Murshidabad, thereby causing immense strains on the stocks of food already under pressure and forcing a further escalation of price.[181] An unspecified number of people of Burdwan, in south-western Bengal, who seem to have been displaced by the floods of 1787 (though not to the same extent or numbers as the people in the east) were nevertheless forced to move to districts situated further westward (for instance to Birbhum and Bishnupur) where the floods had caused practically no damage and where grain, and presumably work, was available.[182] Cultivators from Rajshahi also trekked to Birbhum, where they were able to acquire land for cultivation as non-resident (*pahikashta*) peasants.[183] Some inter-district migration seems to have occurred in eastern Bengal too, where people moved from the worst affected areas to those which were somewhat better situated, though the difference between better and worse was only a matter of degree given the massive problems caused by floods and cyclone all over eastern Bengal in 1787. Thus people from Tipperah and Sylhet, where large-scale devastation had taken place,[184] appear to have moved to Mymensing where crops had only been partially damaged.[185] In Dhaka, the uprooted 'deserted their old habitations, and [have] gone to different places; and many are in the city of Dacca, [others] to Gunges [market towns], Golahs [granaries] & ca. begging their bread with sighs and tears'.[186]

Even cases of dearth could cause a temporary out-migration of people so affected. Thus the partial drought in 1773-4 caused the cultivators to desert. In Birbhum such desertions had reduced the number of cultivators in some parganas 'upwards of a third & in many places a much larger proportion'.[187] Similar flights were reported in 1784 when Tippera was apparently flooded; an estimated 12,000 'souls have left this country' since 'the labour of the cultivators was totally lost and the calamities of famine succeeded their disappointment'.[188] Dhaka district also suffered a flood in 1784, 'causing a real and great scarcity of Grain';[189] some parganas were almost entirely washed away 'by the inundation from the northern mountains', forcing the producers to leave their habitations; in pargana Dawoodpur 'there are only 11 Families remaining... and these from the difficulty in procuring provisions and the inhanced [*sic*] price of these are in the greatest distress'.[190] The Damodar river flooded Burdwan twice more after 1787, in 1791 and 1794, causing extensive damage to crops, cattle and houses.[191] In 1791 Midnapur was also visited by another flood which destroyed between 31 per cent and 37.5 per cent of the harvest ('between five-sixteenths and six-sixteenths of the total'), caused a dearth-like situation and forced peasants to flee their lands.[192] In 1791, 1,005 out of a total of 8,049 peasant house-holds in Mymensing were said to have 'deserted their lands & taken refuge elsewhere'.[193] In 1794, the peasants in some parganas of Nadia 'have deserted owing to the water overflowing their lands'.[194]

It must, however, be pointed out that the extent, direction and nature of such migration differed in accordance to the magnitude of the subsistence-crisis. Famines were occasions when the displaced were forced to migrate over long distance in search of survival. The most obvious places were the towns. Large numbers migrated to the town of Murshidabad during the famine of 1769-70 and to Calcutta during the famine of 1788. Those who survived such arduous journeys under conditions of abject debilitation returned to their villages once a semblance of normality reappeared, but there were also those (said to have 'no calling') who stayed back 'to seek employment'.[195] The proportion between those who returned and those who stayed back is unclear, because of the paucity of information. P. J. Marshall suggests a high rate of return to land in the late eighteenth century for two reasons: (i) even the poorest migrants had tiny plots of land in the village to which they naturally reverted, and (ii) prices of food were generally lower in the countryside than in towns.[196] Moreover, given the low potential of towns in the eighteenth century (including Calcutta) to absorb surplus labour on a continuous basis, the possibilities of long-term productive employment for such people, on wages high enough to beat the higher costs of living in towns, appears largely improbable. Famines, therefore were occurrences when migrations caused a re-distribution of population *within* the countryside in the midst of a severe crisis of subsistence and of mortality.

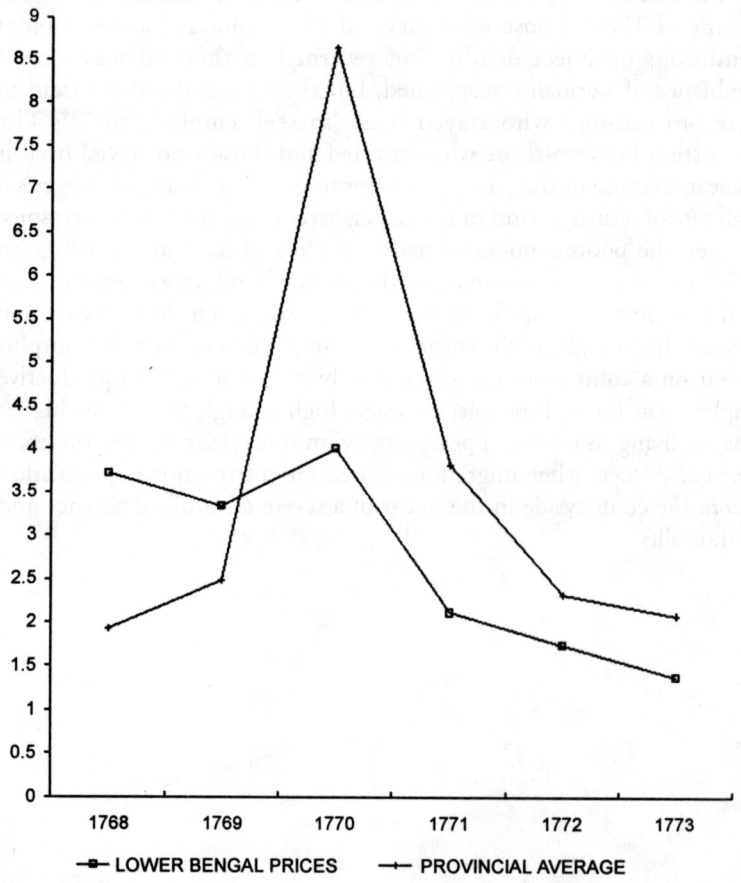

Sources: G. Herklotts, 'Prices in Lower Bengal', *Gleanings in Science*, vol. 1, January 1829; Hussain, 'A Quantitative Study', pp. 271-2.

FIGURE 14. 1770 FAMINE PRICES: ORDINARY RICE (*Rs. per maund*).

Sources: G. Herklotts, 'Prices in Lower Bengal', *Gleanings in Science*, vol. 1, January 1829; Hussain, 'A Quantitative Study', pp. 271-2.

FIGURE 15. 1788 FAMINE PRICES: ORDINARY RICE (*Rs. per maund*).

Sources: J.C. Price, *Notes on the History of Midnapur: As Contained in the Records of the Collectors Office*, Calcutta, 1876, p. 83.

FIGURE 16. MONTHLY RICE PRICES IN MIDNAPUR, OCTOBER 1769 TO SEPTEMBER 1770 (*Rs. per maund*).

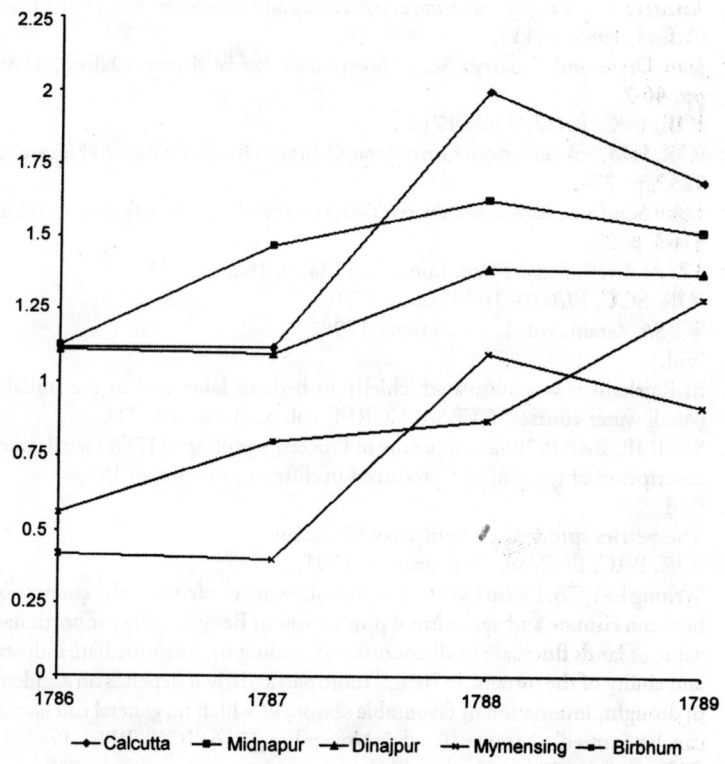

Sources: Calcutta: IOR, BRC, P/51/17 and P/51/34; Midnapur, Dinajpur and Mymensing: IOR, BRC, P/51/12, P/51/16, P/51/17, P/51/19, P/51/34; Birbhum: see W.B. Bailey, 'Statistical View', p. 516.

FIGURE 17. RICE PRICES IN 1788 FAMINE: ORDINARY RICE
(*Rs. per maund*).

NOTES AND REFERENCES

1. David Arnold, *Famine: Social Crisis and Historical Change*, Oxford, 1988, p. 31.
2. Amartya Sen, *Poverty and Famines: An Essay on Entitlement and Deprivation*, Oxford, 1988, p. 115.
3. Jean Dreze and Amartya Sen, *Hunger and Public Action*, Oxford, 1989, pp. 46-7.
4. IOR, BPC, P/1/2, 9 July 1711.
5. IOR, HM, vol. 804, from Govindaram Mitra to Roger Drake, 20 November 1752, p. 237.
6. Luke Scrafton, *Reflections on the Government & ca. of Indostan*, London, 1763, p. 27.
7. J.Z. Holwell, *India Tracts*, London, 1774, p. 165.
8. IOR, SCC, P/A/10, 16 February 1770.
9. WBSA, Grain, vol. 1, 17 October 1794.
10. Ibid.
11. In Rajshahi it was cultivated 'chiefly in beds of lakes and in the nullahs' (small water courses) (WBSA, CCRM, vol. 5, 30 April 1771).
12. See, IOR, BRP, P/70/40, Appendix to Proceedings of April 1788 for a detailed description of types of rice produced in different districts of Bengal.
13. Ibid.
14. The berries sprouted in April-May (*Baisakh*).
15. IOR, BRC, P/52/38, 19 November 1791.
16. Writing in 1776, Philip Francis had the following to say about the connection between climate and agricultural production in Bengal: 'The proportionate value of lands fluctuate in all countries according to the immediate industry and ability of the owners. In Bengal more particularly it depends on accidents of drought, inundation or favourable season, of which no general calculation can be formed' (Minute dated 5 November 1776, IOR, BRC, P/49/65, 5 November 1776).
17. IOR, BRC, P/52/36, 21 October 1791.
18. BM, Add. Ms. 19286, fol. 2. *Colla* lands were usually divided into three categories, (i) *colla basat bari* (homesteads); (ii) *colla do-fasli* (land producing two crops a year); and (iii) *colla sih-fasli* (land producing three crops a year), WBSA, PCR, Burdwan, vol.1, 23 May 1774.
19. Ibid. In Jessore and Rajshahi, part of the *aus* and the whole of the *boro* paddy was cultivated 'chiefly in beds of lakes and in the nullahs and jheels' (WBSA, CCRM, vol. 5, 30 April 1771).
20. Paul Greenough, *Prosperity and Misery in Modern Bengal, The Famine of 1943-44*, Oxford, 1982, pp. 42-52. Perhaps such constructs were not unique to traditional Bengali society alone. Did traditional European societies differ substantially in their conceptions of dearth and plenty? It would be difficult

to argue that they did on two grounds. First, these societies were predominantly agricultural and therefore survival was essentially harvest-dependent [see, Hoskins, 'Harvest Fluctuations and English Economic History', in W. E. Minchinton (ed.), *Essays in Agrarian History*, vol.1, Devon, 1968; J. Meuvret, 'Food Crises and Demography in France in the Ancien Regime', in Harbans Mukhia and M. Aymard, *French Studies in History, vol. 1: The Inheritance*, New Delhi, 1988]. Second, each society has its own hierarchical arrangements which arranged society into prescribed slots (for instance, lord-vassal-serf) whose interdependencies were regulated by 'traditional' customs and mores. The elites were culturally enjoined to be just and benign; the lower orders obedient. A 'good king' who ruled with even-handedness was as much a part of the European ethos as it was Asian (or Bengali). The Bengali peasants expected to see their superiors come to their aid during times of crisis; so did their European counterparts: 'it was of crucial importance for the maintenance of the social order that dearth was not only met, but was seen to be met by action on the part of the authorities' [J. Walter and K. Wrightson, 'Dearth and Social Order in Early Modern England', *Past & Present*, no. 71, May 1976, p. 41]. It is also true that such received notions seldom worked in real life in Europe; they also failed in Bengal.

21. Greenough, *Prosperity and Misery*, p. 12.
22. IOR, BRC, P/52/36, Minute of the Board of Revenue, 21 October 1791.
23. IOR, HM, vol. 102, p. 94.
24. IOR, SCC, P/A/8, 29 November 1769.
25. Ibid., P/A/10, 28 April 1770.
26. Ibid., 28 April 1770, pp. 192, 197.
27. IOR, BPC, P/1/44, 20 November 1769; emphasis added.
28. Thus in April 1770, Asad-ul-Zaman Khan, the zamindar of Birbhum wrote to the Select Committee that 'the cultivation is render'd fruitless by the dryness of the season and no Hope is left of the Cotton Harvest' (IOR, SCC, P/A/10, 28 April 1770).
29. IOR, BRC, P/50/68, 20 November 1783.
30. Ibid., P/49/46, 23 August 1774.
31. Ibid., P/49/47, 30 August 1774.
32. WBSA, PCR, Murshidabad, vol. 7, 4 December 1775.
33. IOR, BRC, P/49/48, 22 December 1775.
34. Ibid., vol. 21, 23 May 1779, 25 May 1779 and 30 May 1779.
35. IOR, BRC, P/50/68, 20 November 1783; also see ibid., P/50/46, 1 July 1783 and P/50/51, 26 March 1784 for statements of crop failures in Midnapur and 24-Parganas respectively.
36. Ibid., P/52/36, 21 October 1791.
37. These were the districts of Rangpur, Birbhum, Nadia, Murshidabad, Dinajpur, Jessore and Rajshahi.

38. IOR, BRC, P/49/52, 7 April 1775; *BDR, Midnapur*, vol. 4, no. 91, 7 March 1771; IOR, FR, Murshidabad, G/27/1, 28 September 1770 and G/27/2, 22 February 1771 and 25 March 1771; WBSA, CCRM, vol. 6, 29 September 1771; BPC, IOR P/1/44, 14 November 1769 and 20 November 1769; IOR, SCC, P/A/10, 30 March 1770 for statements of the spatial asymmetry of the drought of 1769.
39. IOR, SCC, P/A/10, 30 March 1770; emphasis added. In general droughts were never considered a major barrier to agricultural production in east Bengal, 'these being pernicious to the crops of the upper or northern pergunnahs [districts]' of the province (IOR, HM, vol. 206, pp. 212-13).
40. IOR, BRC, P/49/42, 21 December 1773.
41. Ibid., P/49/39, 27 March 1773 for Bishnupur, and 6 April 1773 for Burdwan.
42. Ibid., P/49/40, 14 and 26 June 1773.
43. Ibid., P/49/46, 31 May 1774.
44. Ibid., P/49/46, 23 August 1774.
45. Ibid., 19 August 1774.
46. Ibid., P/49/47, 30 August 1774.
47. Ibid., P/49/58, 27 July 1775 and 3 August 1775 contained in appendix to consultations of 22 December 1775.
48. Ibid., P/50/18, 20 April 1779 and the districts mentioned as having such low prices were those of Burdwan and Bishnupur.
49. See ibid., P/51/17, 20 December 1787 for a statement of such lands in the active delta of eastern Bengal.
50. Ibid., P/49/52, 7 April 1775.
51. *BDR, Midnapur*, vol. 4, no. 91, 7 March 1771.
52. IOR, BRC, P/49/52, 7 April 1775.
53. Ibid., P/49/52, 7 April 1775.
54. M. Alamgir, *Famine in South Asia: Political Economy of Mass Starvation*, Cambridge, Mass., 1980, p. 109; P. J. Marshall, *Bridgehead*, p. 25.
55. *Memoir of a Map of Indostan*, London 1793, p. 349.
56. James Taylor, op. cit., p. 295.
57. IOR, FR, Murshidabad, G/27/1, 28 September 1770 and 25 March 1771.
58. WBSA, CCRM, vol. 6, 29 September 1771.
59. IOR, BRC, P/49/41, 23 September 1773; ibid., P/49/42, 2 November 1773.
60. Ibid., P/49/44, 29 December 1773.
61. Ibid., P/49/46, 24 May 1774.
62. Ibid., P/49/42, 2 November 1773.
63. Ibid., P/49/41, 1 October 1773.
64. The districts of Dinajpur, Mymensing and Chittagong were said to be 'in absolute state of inundation', ibid., P/51/9, 9 August 1787.
65. There is much material on the cyclone of 1787 in IOR, BRC, P/51/9 to P/51/15.
66. Ibid., P/51/22, 25 June 1788. Villages were invariably situated on relatively

higher ground, and this evidence indicates the height to which the flood waters may have attained in low-lying areas.
67. Ibid., P/51/13, 24 November 1787.
68. Ibid., P/51/13, 25 November 1787.
69. Ibid., P/51/22, 25 June 1788.
70. Ibid., P/51/25, 1 October 1788; WBSA, BRFW, vol. 3, part 2, 29 August 1787.
71. Memoir of Sir George Campbell in J.C. Geddes, *Administrative Experience Recorded in Former Famines*, Calcutta, 1874, pp. 426-7.
72. IOR, BRC, P/51/12, 16 October 1787 for damage to crops in Nadia, and ibid., P/51/13, 25 October 1787 for floods in Burdwan.
73. Ibid., P/49/42, 7 December 1773.
74. James Rennell considered the rain during this period ideal for the cultivation of rice as it 'saves them [the peasants] the trouble of watering their lands, and keeping them in the state required for the production of that grain' (IOR, HM, vol. 765, 31 August 1765, p. 147).
75. IOR, HM, vol. 385, 15 July 1789, p. 327.
76. The reference obviously is to the most important rice of the winter harvest.
77. For the Board's Minute, IOR, BRC, P/52/36, 21 October 1791.
78. For a detailed discussion see Rajat Datta, 'Subsistence Crises, Markets and Merchants in Late Eighteenth Century Bengal', *Studies in History* (new series), vol. 10, no. 1, 1994.
79. IOR, BRC, P/52/37, 23 November 1791.
80. Ibid., P/50/58, 22 April 1785.
81. Memoir of Sir George Campbell, in J.C. Geddes *Administrative Experience*, p. 18.
82. WBSA, CCRM, vol. 2, 5 December 1771; IOR, FR, Murshidabad, G/27/4, 20 December 1771.
83. WBSA, Proceedings of the Select Committee, vol. 4, proceedings of July 1770, p. 444.
84. IOR, FR, Murshidabad, G/27/1, 26 September 1770.
85. WBSA, BRFW, vol. 3, part 2, 29 August 1786.
86. IOR, BRC, P/51/2, 20 September 1787; ibid., P/50/67, 11 July 1787.
87. Ibid., P/52/14, petition (*arzi*) of the zamindars of 24-Parganas regarding the impact of the flood of 1787, 7 July 1790.
88. Ibid., P/51/12, 4 September 1787.
89. Ibid., P/51/22, 15 August 1788.
90. Ibid., P/52/36, 21 October 1791.
91. F. Buchanan noted that the 'common fare' of the poor of Rangpur consisted of 'boiled rice, or other grain, which is seasoned with pot ashes and capsicum, and it is only seldom that such persons can procure oil or fish' ('Ronggoppur', vol. 1, book 2, IOL, Ms. Eur. D. 74, fol. 24). The diet of the poor in Calcutta was equally indifferent, being composed of '. . . salt and a little oil,

and one or two other prime necessities; though the vast multitudes . . . obtain only from day to day boiled rice, green pepper pods, and boiled herbs; the step above this is a little oil with rice. The lowest class often want betel and salt, and in place of the latter use the ashes of various plants containing different saline substances' (A. Mitra, *Census of India, 1951: volume 6, part 3: Calcutta City*, Calcutta, 1951, p. 50).

92. IOR, BRC, P/49/47, *arzi* of producers of Burdwan, 30 August 1774.
93. Ibid., P/50/58, 22 April 1785. The reference is to the impact of the partial drought of 1783 in western Bengal.
94. This district was described as the 'heart of the rice country' of eastern Bengal in 1771 (WBSA, CCRM, vol. 2, 31 December 1770).
95. IOR, BRP, P/70/40, 25 April 1788; emphasis in text.
96. See P. J. Marshall, 'The Company and Coolies', p. 28 for the connection between food-prices and wages among the coolies of eighteenth century Calcutta.
97. W.W. Hunter, *Famine Aspects of Bengal Districts*, Calcutta, 1869, pp. 16-17.
98. Rajat Datta, 'Merchants and Peasants', pp. 379-403; also Rajat Datta, 'Agricultural Production, Social Participation and Domination'.
99. See IOR, SCC, P/A/10, 16 February 1770; WBSA, Select Committee Proceedings, vol. 2, Proceedings of July 1770; W.B. Bailey, 'A View of the Population', pp. 560-1.
100. Cf. A. Mitra, *Census of India, 1951*, p. 11.
101. Compare IOR, BRC, P/51/16 and P/51/17, 23 January 1788, 1 February 1788 and 19 March 1788.
102. Price, *Midnapur*, p. 83.
103. IOR, BRC, P/51/17, 1 February 1788
104. M. Alamgir, *Famine in South Asia*, p. 6; emphasis added.
105. Estimate of D. H. MacDowall, Collector of Rangpur in IOR, BRC, P/51/45, 28 June 1789.
106. The receipts from Bihar's revenue have been left out in the present excercise, so have the duties collected on salt and betel-nut.
107. K.N. Chaudhury, *CEHI*, 2, p. 819. The rate seems to have increased during the late eighteenth century Rs. 10 = £1 (see, IOR, HM, vol. 434, p. 648).
108. G.W. Forrest, *State Papers*, 2, p. 265.
109. IOR, SCC, P/A/10, from Richard Becher, Resident at the Darbar, 10 April 1770.
110. Sinha, *Economic History*, vol. 2, p. 56.
111. Seid Gholam Hossein, *The Seir Mutaqherin*, vol. III (original 1783, 1st published 1926), Delhi, 1990, pp. 56-7.
112. K.M. Mohsin, *A Bengal District*, p. 25.
113. IOR, FR, Murshidabad, G/27/1, 26 September 1770.
114. IOR, SCC, P/A/8, 13 November 1768.

DEARTH AND FAMINE 281

115. Ibid., P/A/10, 31 May 1770; Sinha (*Economic History*, vol. 2, p. 56) gives an erroneous sum of Rs. 1,24,860 as *taqavi* during the famine.
116. Sinha, *Economic History*, vol. 2, p. 65; B.B. Chaudhuri, *CEHI 2*, pp. 299-300.
117. For instance, Marshall, *Bridgehead*, p. 18.
118. WBSA, Committee, vol. 2, 10 June-31 December 1771, 24 June 1771.
119. *FR 2*, p. 31.
120. *The Seir Mutaqherin*, vol. III, p. 56.
121. Ibid., p. 26.
122. WBSA, CCRM, vol. 5, 11 May 1771.
123. IOR, FR, G/27/1, 26 September 1770.
124. IOR, FR, Murshidabad, G/27/2, 22 February 1771.
125. IOR, BRC, P/49/50, 27 January 1775.
126. WBSA, Proceedings of the Select Committee, vol. 4, Proceedings of July 1770, p. 444.
127. IOR, SCC, P/A/10, 16 February 1770.
128. WBSA, Circuit at Purnea, 25 January 1773.
129. WBSA, CCRM, vol. 5, 31 May 1771.
130. IOR, FR, Murshidabad, G/27/4, 20 December 1771.
131. IOR, CCR, P/67/53, 20 May 1771.
132. Those of Birbhum, Nadia, Murshidabad, Purnea, Rajshahi, Dinajpur and Rangpur.
133. IOR, FR, Murshidabad, G/27/1, 3 December 1770.
134. Estimates based on the information contained in the relevant volumes of IOR, FR, Murshidabad, IOR, SCC, and WBSA, Circuit. The death toll at Malda pertains only to the loss suffered by the artisanal population under the Company's manufactories (*aurangs*).
135. Even the Black Death, which is said to have decimated 25 million people of a total European population of 80 million, and destroyed 40 or 50 per cent of England's population, raged for nearly three years on the continent after its initial visitation in May 1348. In England, the Death seems to have been most severe in the eighteen months after May 1348. Moreover, these deaths were caused primarily by a pandemic (bubonic plague) on a scale which was never to be repeated on the European continent.
136. IOR, SCC, P/A/10, 28 April 1770.
137. Marshall, *Bridgehead*, p. 19.
138. And the absence of a general epidemic in 1769-70 is all the more convincing because of the fact that a lot more documentation exists for the analysis of this famine than what is available for any of the previous ones. Surely, an epidemic of provincial proportions would have been reported in the numerous, and graphic, descriptions of the misery caused by the famine which are contained in the sources.
139. M.W. Flinn, *The European Demographic System, 1500-1820*, New York, 1981, p. 15.

140. J.C. Sinha, *Economic Annals of Bengal*, London, 1927, p. 105; N.K. Sinha, *Economic History*, vol. 2, p. 65; B.B. Chaudhuri, *CEHI 2*, p. 94.
141. B.B. Chaudhuri, *CEHI, 2*, p. 300
142. Sugata Bose, *Peasant Labour*, p. 19.
143. *FR 2*, p. 510.
144. A.T. Embree, *Charles Grant and British Rule in India*, London, 1962, p. 37.
145. The full text of Cornwallis' remark was: 'I may safely assert, that one-third of the Company's territory in Hindostan, is now a jungle inhabited only by wild beasts. Will a ten years' lease induce any proprietor to clear away that jungle, and encourage the ryots to come and cultivate his lands; when at the end of that lease, he must either submit to be taxed, ad libitum, *for their newly cultivated lands*, or lose all hopes of deriving any benefit from his labour, for which perhaps by that time, he will hardly be repaid' (*FR, 2*, p. 512; emphasis added).
146. 'Final Report of the Aumenee Commission, 25 March 1778', IOR, HM, vol. 206; also see R.B. Ramsbotham, *Studies in the Land Revenue History of Bengal*, Oxford, 1926, p. 132.
147. *FR 2*, p. 31.
148. N.K. Sinha, *Economic History*, vol. 2, p. 62; Ratnalekha Ray, *Change in Bengal*; Nagchoudhury-Zilli, *The Vagrant Peasant*.
149. WBSA, BRFW, vol. 3, part 2, 29 August 1786.
150. IOR, BRC, P/51/13, 25 October 1787.
151. Ibid., P/51/12, 4 September 1787.
152. Ibid., P/51/51, 9 December 1789.
153. Ibid., P/49/52, 7 April 1775.
154. Ibid., P/52/14, 7 July 1790.
155. IOR, BRP, P/70/47, *arzi* of zamindars of Rangpur to D.H. MacDowall, the Collector, 25 June 1788.
156. IOR, BRC, P/51/12, 4 September 1787.
157. Ibid., P/51/12, 20 September 1787 and 16 October 1787.
158. 'Remarks on the Several Collectorships in Bengal in the years 1788 and 1789', IOR, HM, vol. 385, p. 68.
159. John Haldane, Robert Kennaway, Charles Grant and John Bristow, 'Report of the Commercial Occurrences of 1788', in IOR, HM, vol. 393, p. 113.
160. IOR, BRC, P/51/25, 1 October 1788.
161. IOR, HM, vol. 393, pp. 113-14, 120.
162. IOR, BRC, P/51/29, 22 December 1788.
163. Ibid., P/51/29, 22 December 1788 for Dhaka, and ibid., P/51/22, 25 June 1788 for Jessore.
164. Estimate of D.H. MacDowall, collector of Rangpur in IOR, BRC, P/51/45, 28 June 1789.
165. IOR, HM, vol. 206, p. 205.
166. Ibid., vol. 206, pp. 212-13.

DEARTH AND FAMINE 283

167. IOR, BRC, P/49/52, 7 April 1775.
168. J.C. Price, *Midnapur*, p. 82.
169. IOR, BRC, P/49/52, 7 April 1775.
170. Sinha, *Economic History*, vol. 2, p. 54.
171. See WBSA, LCB, vol. 1, 20 August 1770; IOR, FR, Murshidabad, G/27/1, 26 September 1770.
172. WBSA, CCR, vol. 1, 20 May 1771.
173. For estimates of lands under cultivation in different parganas of Birbhum after the famine see IOR, BRC, P/49/42, 25 October 1773, and ibid., P/49/47, 23 April 1774.
174. *BDR, Rangpur*, vol. 1, no. 5, 23 June 1770.
175. WBSA, Circuit at Purnea, 2 February to 9 February 1773, p. 43.
176. IOR, BRC, P/51/13, 25 October 1787.
177. Ibid., P/51/12, 4 September 1787.
178. Ibid., P/51/29, 22 December 1788.
179. See ibid., P/51/12, 4 September 1787 for such migrations in Rangpur; ibid., P/51/13, 9 October 1787 for the situation in Burdwan; and ibid., P/51/20, 18 April 1788 for Rajshahi.
180. Ibid., P/52/14, 7 July 1790.
181. IOR, BRP, P/70/40, 1 April 1788.
182. IOR, BRC, P/51/21, 4 June 1788. Interestingly the flight of peasants from Burdwan to Birbhum became a source of friction between the zamindars of these two districts. The zamindar of Burdwan subsequently demanded the return of these cultivators on grounds that they were the settled resident-cultivators (*khud-kashta raiyat*) of his zamindari and were therefore obliged to cultivate land and pay revenue in Burdwan. The counter-claim proffered by the zamindar of Birbhum was that these migrants had acquired the legal status of non-resident peasants (*pahi-kashta raiyat*) in his lands, and were therefore entitled to stay and cultivate land in his zamindari (see ibid., P/51/34, 8 April 1789).
183. Ibid., P/52/2, part 1, 10 February 1790.
184. For the flood induced havoc in Tipperah and Sylhet see WBSA, BRFW, vol. 3, part 2, 29 August 1786 and IOR, BRC, P/50/67, 11 July 1786 respectively.
185. IOR, BRC, P/51/19, 18 April 1788.
186. Ibid., P/51/29, 22 December 1788.
187. Ibid., P/49/46, 21 June 1774.
188. Ibid., P/50/59, 1 July 1785. It was the fear of a famine, rather than the fact of famine, which seems to have forced people to move out of this district in 1784.
189. *Bangladesh Districts Records*, vol. 1, Dacca, no. 25, p. 77.
190. Ibid., no. 42, 4 October 1794, p. 95.
191. IOR, BRC, P/52/36, 7 October 1791, and ibid., P/53/21, 30 September 1793.

192. Ibid., P/53/36, 21 October 1791. While floods destroyed crops in places like Burdwan and Midnapur, 1791 was a year of an apparently severe drought in 24-Parganas. 'I am sorry', wrote Edward Colebrooke, Collector Calcutta Cutcherry (*kachahri*), 'to be obliged to report to you [Board of Revenue] that the Rice crops have totally failed in the extensive Plains from Bernagore to Russapugla [where] whole maaths [fields] of many thousand Begas (*bighas*) are totally destroyed by the drought!' (IOR, BRP, P/71/45, 15 November 1791).
193. IOR, BRP, P/71/42, 1 August 1791.
194. Ibid., P/72/30, 30 May 1794.
195. WBSA, Circuit at Purnea, p. 43.
196. Marshall, 'Company and Coolies', pp. 33-5.

CHAPTER 6

Subsistence Crises and the Agrarian Order

WHAT INFLUENCES did famines and dearth have upon the state of agricultural production and on the social situation of the peasantry? Numerous attempts have been made to analyse the impact of the famine of 1769-70 on the agrarian economy, but very little exists on the famine of 1788 and on the instances of dearth which were interspersed between these major famines. This lacuna has made it necessary to study the effects of the other events of famines and scarcities, and to position them in relation to the famine of 1769-70 in order to acquire a composite picture of the agrarian consequences of famine and dearth in late eighteenth century Bengal. The following section attempts to reconstruct the consequences of these events on the agrarian economy and to question some of the main postulates of the existing interpretations of 1769-70.

Merchants and Food Supply during
Famine and Dearth

The discussion in Chapter 5 gave rise to two important facts. First, a famine or a dearth, were occasions when the per capita availability of food actually declined. Second, getting supplies by acquiring food from elsewhere, particularly from other provinces, was difficult, if not impossible as a full-scale crisis would never be apparent before late October (i.e. just before the maturation of the *aman* harvest), by which time it was difficult to get food from outside since a 'considerable time must elapse before the ships which may happen to be lying in the river [Hughli] unemployed can be fitted out, as the owners ascertain from what countries the rice can be procured at a price which will be worth their while to procure it'.[1] These facts meant that during a famine or a dearth the province had only its internal resources to depend on for food, and this raises the question of the internal movement of food in bulk by the merchants.

When a scarcity struck, it tended to exacerbate the already high demand for food in the towns and *qasbas*; it also forced an immediate upward spiral in rural demand as the producers now became net consumers of grain.[2] The entire range of issues behind the events of 1769-70 leaves a distinct impression that it was an artificial famine. The artificiality of the famine was hinted at by Warren Hastings when he acknowledged the severity of the Company's revenue collections during the famine months. Hastings was to reiterate this theme subsequently (in 1783) when he wrote:

It appears from the enquiries then made, and from the others which were connected with the famine of 1769, that the first want was artificial, proceeding from the *expectation* of a real want, and from the natural inducement which it afforded to the dealers in grain to withhold it from the market in the hopes of *deriving a larger profit from a more distant sale*, and this, with the compulsive means which were used to force the grain into consumption, and which drove the proprietors into destructive expedients for hoarding and secreting it, [prevailed] as the *principal cause* of the famine which ensued. . . .[3]

Hastings' analysis is important in so far as it points to the actuality of the famine. It is also crucial because his analysis clearly indicates that the famine in 1769-70 can no longer be interpreted as a food shortage occasioned by 'natural' causes alone. His estimation also suggests that the actual misery was caused by an *artificial* failure of supplies, a point which indicates that supplies of food were available but did not, or were not made to reach those places which needed grain most. There is, of course, a clear bias in Hastings' analysis and that lies in his propensity to blame the 'merchants', obviously the indigenous ones. For reasons which are obvious, Hastings did not stress the fact that the Company's own officials were also alleged to have profiteered during the famine.[4]

Unfortunately not much is known about the activities of the indigenous grain merchants during the famine months. But Hastings' suspicion that they did attempt to profiteer was apparently well-founded. As soon as the drought forced prices to rise abnormally, the merchants started buying up from those districts which still had some surpluses from previous harvests. Merchants from Burdwan and Hughli flooded to Midnapur and apparently carried off 'much larger quantities [of grain] than the Country [Midnapur] could afford'.[5] Similarly, the Murshidabad based merchants entered Rangpur, where stocks of food were available and began 'exporting the grain extremely fast, and it will all very soon be taken away except what may be purchased to be kept here'.[6] The net result was a rapid increase in the spot price of grain, and such instances were reported from Burdwan[7] and Midnapur[8] in western Bengal; prices were also pushed up in eastern Bengal though not to the same extent as most districts here had relatively larger stocks of food available.[9]

The chief reason for such mercantile strategies was the need to cater to town demand, but the situation in 1769-70 was different in so far as there was a phenomenal rise in food prices in the *countryside*. In fact the available price data suggest a *higher* price of food in the villages than in the towns during the famine months. For instance, while the average price of rice in Midnapur was Rs. 1.57 a maund between September and December 1769, in Calcutta rice could be purchased at Re. 1 per maund in 1769.[10] In July 1770, coarse rice in Murshidabad was being sold for Rs. 1.14 a maund,[11] but in the villages of Rangpur it was selling at an unheard of price of Rs. 3.07.[12] In November 1770, when the price of rice at Murshidabad was Rs. 1.33 a maund, the price at Natore, 'situated in the heart of the rice country', was Rs. 2.22.[13] The result was a cross-flow of food supplies at very high prices from town to country and from country to town.[14] Therefore the allegation that the merchants *only* hoarded or 'secreted' food during the height of the famine does not seem feasible since an action of that type would go against the very essence of their enterprise of making the maximum profit during times of scarcity.

The role of the East India Company during the famine was apparently unique in the history of the province. In the immediately preceding regime of the Nizamat, the state seldom interfered in grain trading, except during times of scarcity when it sought to regulate prices and prevented merchants from setting up monopolies; the essential idea behind such strategies was to keep supply lines open to towns under threat of scarcity.[15] On the other hand, the Company's political role was inextricably linked to its commercial and revenue interests, which meant that political rule was still intermeshed with variety of trading interests: the interests of its own, those of the private traders and those of its officials who traded for themselves under the guise of their 'native' agents.

With the first symptoms of the drought being apparent, the Company responded by stockpiling to feed its garrisons in Bengal and Bihar 'for six months to come'; the thought of relieving 'the miserable situation' occasioned by the drought was subsidiary to the needs of revenue and feeding the army.[16] Thus in Chittagong 'we have given directions for buying up small quantities of rice for the Calcutta market as the new crop comes in, and have employ'd a Gomasthah in the adjacent parts of the country for the same purpose'.[17] As the drought intensified and the demand and prices increased, the officials and private European traders apparently took it upon themselves to profiteer from the situation. They used their 'native agents' (*gomasthas*) to purchase grain by force[18] which allegedly went to such an extent that the hapless peasants were coerced into selling 'seed requisite for the next harvest'; the rice thus collected was sold at famine prices in the towns of Calcutta and Murshidabad.[19]

Yet, the exact amount of the province's food stocks which the Company or its officials actually cornered is unclear. Estimates of rice monopolized by the state alone range from 60,000 maunds[20] to 1,20,000 maunds.[21] Sinha's estimate appears closer to the real picture because the *dastaks* issued to officials in Bihar and Chittagong during that famine amounted to 57,300 maunds, whereas the total amount of rice actually handled by these officials was 46,000 maunds.[22] The scale of the drought and the entrenched power of indigenous merchants simply did not allow Company officials to corner a major proportion of the market during the famine. Most of Chittagong's grain reserves had already been bought by local merchants before the officials could intervene,[23] and the supplies of Midnapur were snapped up by merchants from Birbhum and Burdwan leading to a steep escalation in local prices for a short time.[24]

Nevertheless it must also be pointed out that some efforts were made by the state to regulate the amount of exports of food from different districts. In Bihar, Thomas Rumbold ordered the suspension of market duties on the importation of rice to Patna and forbade grain to be 'carried out of the province',[25] and in Chittagong: 'Your [the Board's] orders for preventing Monopolys [*sic*] of Grain & to encourage the planting of such Grain and Pulse as can be produced in the dry season have been published thro'out the Province, and we have satisfaction to advise your Honour & ca. that precautions taken by the Chief [at Chittagong] to prevent the exportation on the first appearance of a scarcity has kept the price here very moderate.'[26]

Exports of rice from Calcutta were immediately reduced, partly as a result of insufficient supplies and partly owing to the need to retain available food stocks within the province. The duties collected on exports of rice at Calcutta's custom house fell from Rs. 6,134.68 in 1769 to Rs. 1,876.81 in 1770.[27] However, this effort to alleviate a crisis in the distribution of food was inadequate. Chaudhuri's explanation that this crisis was caused by the interests of the state, private merchants and the Company officials which put immense pressure on the market with the result that prices soared[28] does not take into account the fact that even local officials were also blamed for malpractice. For instance, Muhammad Reza Khan was said to have:

Entirely ruined the country by every oppressive means and methods that he could think of; for he in the very height of the famine, having stopt [sic] the merchants boats loaded with rice and other provisions, bound for the city of Muxadavad [Murshidabad], forcibly purchased from them rice from 25 to 30 seers per rupee, and did retail it out from 3 to 4 seers per rupee, and all other eatables in proportion; yet for all these execrable acts and deeds, he has been winked at by the superior power, who could have at times controul [sic] and compel him from acting in so inhumane and infamous a manner; and by this means several lacks of people have starved and died, after selling every substance of theirs, to procure eatables, at the exorbitant rates aforesaid.[29]

Since Muhammad Reza Khan was later absolved of these charges, this complaint must indicate the hostility of grain merchants to official interference during periods of scarcity. The food-supply situation during the famine of 1769-70 clearly shows that the availability or non-availability of food in particular districts were determined more by the amount of the local harvest actually destroyed by the drought and by the grain merchants whose activities the state could hardly control, than by the profiteering activities of some Company or other officials.

The power of the merchants to determine the flow of food during a subsistence crisis in direct contravention of the state's will is also borne out by their behaviour during periods of dearth. The chief features of a dearth were a sharp decline in the per capita availability of food at localized levels and a escalation in spot prices, but stocks of food were available from the less-stricken areas which meant that the *byapari* could circulate his stocks with greater ease and profit. During the partial drought of 1773 many merchants: 'Have set up private gunges for the reception of all rice and paddy brought to the market by themselves and others, and the more effectually to avoid being detected, in what they are conscious to themselves tends to a monopoly, have resisted and turned out from their golahs and landing places the sircars [officials] and kyalls [weighmen] who are employed by the [Company's] Customs House.'[30]

Establishing 'private gunges' and disposing of the Company's market-officials meant that these merchants tried to avoid paying established duties and they were able to circumvent state authority at the same time. From the Company's point of view, such activities had three major implications: (i) such 'importations and sales admit of collusive compacts which greatly enhance the price', (ii) they freed the merchants 'from the general and wholesome regulations established in publick markets for the humane purposes of preventing an artificial scarcity', and (iii) such actions prevented 'a precise knowledge of the quantity of grain in Town and consequently those timely measures which government upon the apprehension of real scarcity might judge proper to adopt'.[31]

Though the Company was worried about the consequences of mercantile strategies during situations of dearth, there was very little which could actually be done to prevent these consequences. In 1773, these merchants resorted to buying-up of stocks as soon as the first signs of a dearth-panic arose,[32] and the result was a sharp price increase because 'they have leagued together to keep it [the price] up and we [the producers] are perishing with hunger'.[33] Once again we find the spectre of hoarding during a food-crisis emerging as the dominant form of merchant behaviour, but this was only a partial explanation of what happened during the drought of 1773. Of greater importance was the steep rise in the urban prices of food caused by the 'sudden demand at the several *capital marts*' which caused the merchants to export their stocks in that direction rather than sell in the countryside where prices were relatively less.[34] Thus boat-loads of rice were delayed at Bhagwangola by local merchants in order to make the most from a situation of high food-prices.[35] Temporary stock-retention, rather than outright hoarding, seems to have been characteristic of these merchants during bad agricultural years. Thus during the drought of 1775, merchants bringing grain to the city of Murshidabad 'have temporarily ordered their boats to be detained in the Jellingee [river] to see if the weather will hold fair [*sic*] for any time longer'.[36]

Another illustrative example of such strategies comes from the flood of 1784 in eastern Bengal. The flood destroyed the *aus* harvest and severely threatened the *aman*, and a large-scale food shortage was imminent. The merchants withheld stocks from previous harvests, thereby threatening the inhabitants of Dhaka with the prospect of protracted starvation.[37] The initial reaction of the Board of Revenue was to clamp down upon such merchants by instructing the collector to give 'public notice by beat of Tom Tom, that all Traders in Grain who shall refuse to sell or to bring usual supplies to market shall have their Property in Grain seized by Government, and be liable to such other punishment as the Circumstances of their offence appear to dictate'.[38] This seemingly harsh stricture had no effect upon the merchants. The grain dealers simply refused to sell at dictated prices and the impasse continued till October when the collector was forced to concede: 'the dealers were declared to have authority to sell at pleasure' in *bazaars* protected by sepoys. The results were immediate. Rice stocks appeared once again albeit at high prices because the 'general scarcity in the mofussil [countryside] will not allow of its being cheaper'.[39]

Similar tendencies operated during the famine of 1788. In a reply to a questionnaire from John Shore, the principal grain-traders of Calcutta replied that the only way in which merchants all over Bengal could be induced to send supplies to Calcutta and Murshidabad was to free the markets of state control by temporarily suspending 'all gunge duties' and by the abolition of all arbitrary valuation in the prices of food. These would, they said, 'operate greatly to the relief of the inhabitants by enabling the merchants to furnish them with grain so much cheaper'.[40] The traders could lend weight to their claims because they had already purchased the internal surpluses of food available[41] and the demand of the cities could only be met if they decided to bring in this food.[42] Both these aspects are apparent from the following letter written by Bhoj Raj, the principal grain dealer in Murshidabad, to Nawab Mubarak-ud-daulah describing the state of procurement during that famine:

There are three causes for the increasing price of grain in the city of Moorshedabad. First, the beoparries of Calcutta . . . have purchased and keep large quantities of rice in the neighbourhood of the Rara [*Rarh*] region [which is] the name of the districts to the west of Moorshedabad among rice merchants. Second, at Bogwanpoor, where it is known there are 22 Golahs, all the Grain coming from the south [of Bengal] arrives and is sold. This year whatever arrives from the environs and from the south, the Dacca merchants purchase immediately on arrival of the boats and carry it away. It is [therefore] impossible to make any purchases there. Third, what little arrives in the city of Moorshedabad is purchased & carried away by Calcutta merchants.

Under a situation of this kind, it was perhaps not surprising for the state to think that traders 'wholesale & retail have their separate combinations, and have actually created an artificial famine' and caused food prices to rise 'to a scandalous degree'.[43]

The problems faced by the state during a famine, or even a dearth, were two: first, to prevent an over-exportation of food from centres possessing some surpluses, and second, to acquire adequate supplies to feed the towns and the military establishment. As we have seen, forcible embargoes on exports simply did not work. D.H. Curly suggests that an embargo on exports of food and regulation of prices were the two components of traditional famine relief policies followed by the Nawabs of Bengal, and that they were reasonably effective since they could replace 'free' markets with local systems of autarky and regulated distribution in the short-run to relieve the pressures of a famine.[44] This view does not conform to the actual systems of food-distribution during a famine or dearth.[45] Our evidence shows that merchants constantly sought to maintain conditions of 'free' market, and any interference met with immediate retaliation: stocks were deliberately withdrawn from the direction desired by the state. Thus during the famine of 1788, the *amlah* of Rajganj (in Dinajpur) tried: 'To keep the markets in and about Dinagepore supplied with a sufficient quantity of grain for the daily consumption of inhabitants, [for which] I have been obliged to have recourse to the Golahs where the grain is hoarded [but] the Proprietors [i.e. the merchants] have persevered in opposing every mild endeavour I have practised to induce them to comply with my entreaties to relieve the scarcity, which in fact is of their own making. . . .'[46] Food could only be procured by a *negotiated settlement* with the grain-dealers, as was, for instance, done in Malda in 1788. After the initial flexing of state-muscles, the Company was forced to negotiate as the much desired food refused to materialize. Kishan Mangal 'a merchant of that place' negotiated a 'fair' price on behalf of the other local merchants for the available grain stocks, and: 'After [the price] being settled. . . [it] was paid to the dealers who then went away apparently satisfied. The only end purpose was to prevent distress to the Country from its being stripped of provision, and *this required none of the violence which they [the merchants] assert to have been done to them [before].*'[47]

Food-supplies during years of scarcity were therefore a source of constant friction between the state and the local trading communities. Obviously, the state wanted to take the credit for ensuring supplies but the reality of the situation was different. For instance, Hastings described the state's role during the severe dearth of 1783 in the following words:

Few men doubted that the scarcity proceeded from the failure of the preceding harvest, and the consumption of former years. The members who composed the administration chose to put this conclusion to the test. After having applied such measures as were most likely to give a temporary check to the complaints, they appointed a Committee, consisting of the Senior and most Intelligent Servants of the Company, whom they invested with ample powers to collect accounts of the actual quantities of rice existing in the provinces; to compel [*sic*] every proprietor of it to deliver to their agents an exact account of what he possessed; and in the event of want of a due supply in the markets, to contribute to it according to his ability. The threat of confiscation was also proclaimed against any who should attempt to elude the investigation, either by secreting their grain, or by delivering false accounts of it; but if I may trust to my memory, only one instance occurred, in which it was found necessary to inflict that penalty. The result of the measure was that markets were in a short space of time abundantly supplied; the price of grain gradually sank in its level; and from the returns made to the Committee, it appeared that there was a sufficiency to last even a considerable period beyond the next expected harvest.[48]

Hastings' analysis appears an unconvincing apology for the failure of the state to ensure adequate food-supplies during that crisis. Witness for instance, the Bengal government's own admission (in 1788) that: 'The most active temporary interference of Government has been found productive of no other consequence than a *slight alteration* of the pressure of famine on the poor class of natives.'[49]

Later enquiries about the scarcity of 1783 showed that: 'Mr. Hastings allowed the *price to fix itself*, the merchants having represented to him that they were deter'd from bringing it to market because they were obliged to sell it according to an arbitrary valuation. He also *suspended the Calcutta Gunge Duties*, which were five chuttacks per rupee, and ten seers per hundred maunds. . . .'[50]

The mechanisms by which food was made to move from one area to another during a subsistence-crisis does show the organized power of the grain-merchants in the late eighteenth century. The sources of such power were numerous. First, they owned *golahs* and the means of transportation which were crucial not only for acquiring the crucial supplies but for moving them around as well. Second, the grain-merchants were socially cohesive as they came from similar caste groups and, therefore, followed common occupational strategies. They were additionally organized in tightly knit groups (*dala*). People like Thakurdas Nondi (in Rangpur), Bhoj Raj (in Murshidabad) and Kishan Mangal (in Malda) were situated at the head of an organized network of trading activities, hierarchically arranged and spread over a wide catchment area. This enabled effective trading; it also made it possible for traders to disseminate information and to quickly forestall competition, particularly from the state. Third, and perhaps the most crucial, was the intermeshing of trade with agricultural production. In fact one of the major barriers in the way of acquiring supplies during the famine of 1788 was 'the usual and long standing custom of Grain Merchants advancing [money] at the commencement of the season for the crops' which meant that the 'Riauts sell what grain they can spare to the merchants *in preference* of bringing it to publick market to be sold at a fixed rate'.[51] The food surpluses of Birbhum were already hypothecated to the merchants of Calcutta and Murshidabad and as a result they were being provided with grain by the peasants during the famine.[52]

Subsistence Crises and Rural Artisans:
1769-70 and 1788 and Bengal's Textile Producers

One of the most harvest and price sensitive social group in rural Bengal were the textile producers. There is also substantial material in the sources to document the impact which famine and dearth had on them. The following section seeks to reconstruct the effects the events of 1769-70 and 1788 had on their physical and material conditions.

The evidence from Malda and Purnea suggests that between half and one-third of those who died in the famine were the textile producers. The disruption caused by the drought to mulberry and cotton in 1769 and 1770 meant that those who reared silk-worms (*chassars*) and those who grew cotton (*kappas*) were immediately affected. The cultivation of mulberry was an expensive enterprise: 'under the most favourable circumstance mulberry will cost the Husbandman five or six, and often from ten to fifteen rupees per bigha',[53] which meant that once a peasant entered this sector his survival depended on good weather and favourable food-prices. The situation in 1769-70 was precisely the opposite on both counts, and therefore proved disastrous for such producers. There was an 'incredible mortality' among the *chassars* of Rajshahi during the famine[54] and this was for two reasons. First, the high costs involved in the culture of silk-cocoons meant that the *chassars* had no reserves to buy food at famine-point prices. Secondly, the *chassars* belonged to 'only two casts [*sic*] of the Gentoos [Hindus]' who followed this vocation as a specialized occupation since their profession was 'considered as an abomination by the rest of the sects'.[55] These two reasons meant that they were perhaps the most harvest-sensitive of all the affected social strata, and not having enough food-reserves to fall back upon, they died in large numbers. Significantly there appears to have been some mortality among the *chassars* even in Dhaka[56] which was otherwise largely unaffected by the severity of the famine. The deaths among the *chassars* were closely followed by deaths amongst the spinners of silk or cotton thread.

Finally there were the weavers, and it is generally believed that they also died in 'large numbers'[57] but these deaths were especially concentrated in areas like Purnea, Rajshahi, Malda and Murshidabad. The reports of such deaths suggest a higher concentration of excess mortality among the *chassars* followed by deaths among the 'winders and weavers' who 'suffered in proportion' to the *chassars* but not to the same extent.[58] There were two reasons why deaths among the weavers were perhaps fewer in comparison to the deaths of other occupational categories involved in textile production.

Firstly, the weavers functioned in combination with agricultural production: 'the weavers have the double resource of tilling the lands'[59] and would presumably had some food reserves with them. Secondly, they had the relative advantage of having access to the financial advances made by the Company through its commissioned agents (*dalals* and *paikars*) for their finished piece-goods.[60] It is certainly true that the latter system tied the rural weaver to the financial and political dictates of the Company and to the tyranny of the agents; it is nevertheless also true that such advances provided the critically useful and *alternative* financial fund which was used by the weavers to stave off total starvation during a famine. Quite obviously they were unable to fulfil their contractual obligations under such trying circumstances. The sharp increase in the accumulated balances of the *dalals*[61] at Dhaka between 1769 and 1770 (Table 58) appears to have been a result of the failure of the weavers to produce in times of subsistence-crises, but the equally substantial reductions in the amount of outstanding balances between 1771 and 1772 also show that in the case of Dhaka at least the disruption caused by the famine to the textile sector was short-lived.

Elsewhere (in Rajshahi, Murshidabad and Purnea), the plight of the weavers was worse as apparently large numbers died in the famine or in

TABLE 58. BALANCES WITH THE *DALALS* AT DHAKA, 1767-72

Year	Amount of balance (Rs.)
1767	49,103.43
1768	29,763.62
1769	72,234.12
1770	52,433.75
1771	18,636.75
1772	16,758.62

Source: WBSA, PCR, Dacca, vol. 13, 27 November 1776.

its immediate aftermath. Yet, there is some indication that the combination of agriculture with craft production and the Company's advances may have provided, at least some of them, with a temporary way out of the impasse. It was certainly believed by the Company's servants at Qasimbazar that deaths notwithstanding, many weavers and even *chassars* had deserted and 'turned their hands to the cultivation of grain'.[62] Therefore the 50 per cent reduction in the number of looms 'employed' by the Company at Malda may not necessarily have been a result of deaths. Such cross-migration from craft production to agriculture, or even outright desertion by the artisans, certainly occurred in the 24-Parganas where the artisans and peasants were said to have found 'means of supporting themselves during the famine';[63] and in Burdwan and Bishnupur producers had dispersed in large numbers all over southern Bengal[64] which caused a 'diminution of manufacturers'; interestingly, this diminution was not matched by a similar decrease in the number of peasants[65] which perhaps suggests an inter-sectoral movement during the course of the famine rather than an absolute reduction of numbers caused by famine mortality.[66]

The evidence which we have regarding the unrecovered balances from the *dalals* at Dhaka (Table 58) does show that the artisans, at least in that region, did return to work once the disruption was over. Interestingly, the Company's investments for Bengal's piece-goods and silk remained substantially high during the famine as the figures in Table 59 show.

Bengal's piece-goods remained in the forefront of the Company's trade.

TABLE 59. COMPANY'S INVESTMENTS FOR BENGAL PIECE-GOODS AND SILK, 1766-75

Year	Piece-goods (£ sterling)	Silk (£ sterling)
1766	3,29,498	91,602
1767	4,15,774	132,596
1768	5,00,797	137,299
1769	5,76,281	142,328
1770	4,51,152	160,337
1771	5,71,542	170,457
1772	6,97,778	136,270
1773	5,08,622	94,431
1774	4,66,944	160,016
1775	6,59,255	239,514

Source: *Ninth Report of the Select Committee*, 1783, Appendix 6.

Of a total of 8,68,357 piece-goods disposed at the Company's sales in 1771 (from its investment of 1770), Bengal provided 6,04,757 pieces. In the sales of 1772 (made from the investments of 1771) Bengal's share had gone up to 6,26,160 pieces out of a total of 10,67,452 pieces sold in that year[67] and these quantities were higher than the average annual purchases of piece-goods in any period before the famine.[68]

These facts obviously have a significant bearing on the state of textile producers in so far as these data show that the production of cloth did continue in the midst of a famine. The data provided in Tables 58 and 59 show the famine had not made a significant difference to the Company's financial position and this is reflected in the state of its investments for piece-goods before and after the famine. Both these facts imply that the famine had not ruptured the Company's role in Bengal's economy despite the disruption which that event otherwise seems to have caused. But what is not clear is who produced these textiles? Given the scale of famine misery in places like Murshidabad, Qasimbazar, Malda and the death (or flight) of the artisans, the obvious buoyancy of the Company's investment and procurement becomes difficult to explain.

Unfortunately, no direct answers are forthcoming to resolve this obvious paradox; nor have the historians who have examined the context of textile production in the eighteenth century have any explanations to offer.[69] Despite the evidence of weavers dying in the Company's factory at Malda which lay in the heart of the famine-ravaged areas, most artisans in the employment of the Company, especially those who had already received cash advances before the famine, would have had a better chance of survival and continue with their activities. Certainly, advances had a certain inbuilt element of coercion[70] but in critical moments these could make all the difference between survival and mortality. Another probable reason was the geographical asymmetry of the famine which allowed the continuance of artisanal production in the less stricken areas of the province. But both these possible causes do not explain the fairly high investments and *procurement* during the famine years. As Table 59 shows, the investment for piece-goods did decline somewhat between 1769 and 1770 but investment for silk actually increased during these two years. The other significant aspect of this table is the apparently swift recovery immediately after the famine. The only blip was 1773 which seems to have been a bad year for silk investments.[71]

The principal reason for these high-level investments was the gradual recuperation of artisanal production after the famine. By 1774 the *chassars* and *naqads* (silk winders) of Laskarpur, who were quite devastated by the famine,[72] were said to be cultivating 'more lands than they pay rents for'.[73] By 1775 the *naqads* of Rajshahi and those of Laskarpur had fully resumed their activities of moving to adjacent parganas of Rangpur and Dinajpur in search of cheaper priced cocoons from the *chassars* there, and pargana Kaligaon [Colligong in our sources] in Dinajpur had once again 'become the general mart of silk worm seed to their neighbours',[74] and Rangpur's 'mulberry plant is brought to most perfection in the southern districts where of course most silk is manufactured'.[75] Elsewhere, for instance in Burdwan and Midnapur, 'the cultivation of the mulberry plant having been a good deal encouraged, the article of raw silk has rather increased than diminished'.[76] What certainly transpired was an escalation in the price of raw materials both of silk and cotton[77] which also forced up the price of the finished product,[78] but the reason for this, according to the Controlling Committee of Commerce, was the unlawful 'combination among chassars or between them and the Pykars' and by the attempts 'to introduce the Italian [filature] mode of winding silk by Messrs. Wiss and Robinson'.[79]

Did the famine make a material difference to the social situation of the textile producers? Given a drastic reduction in the numbers of the producers and the fact of high investments by the state, one would logically expect the artisans to have been in a better bargaining position after the famine with a possible improvement in their economic positions. Immediately after the famine the *chassars* of Rajshahi were reported to be averse to selling their silk pods to the Company preferring 'to wind off silk themselves'; but this opposition was because of the lower prices being paid to them by the Company than what could be had in the

TABLE 60. PRICES OF RAW SILK IN BENGAL, 1768-71 (*Rs. per seer*)

Name of manufactory	1768	1769	1770	1771
Qasimbazar	7.43	9.18	8.25	8.25
Rajshahi	7.31	7.56	9.87	?
Kumarkhali	5.75	6.25	9.56	·9.75
Rangpur	7.50	7.94	9.50	9.56
Jangipur	10.25	10.81	9.87	10.00
Average price	7.65	8.35	9.41	9.39

Sources: WBSA, CCRM, vol. 6, 25 November 1771; Sinha, *Economic History*, vol. 1, p. 198.

'open' market and because of their immediate opposition to 'winding a large quantity of silk in the European [filature] fashion'.[80] What the *chassars* and the *naqads* essentially wanted were more remunerative prices and the evidence (Table 60) does suggest that some increases were made in the prices paid to them between 1768 and 1771.

But these increases, averaging about 22.74 per cent between 1768 and 1771, were insufficient compared to the nearly 50 per cent rise in the market price of cocoons in the comparable period.[81] This obviously points to an actual deterioration of the living standards of the textile manufacturers and there is evidence to show that this downward trend continued throughout the rest of the century.[82] It is, therefore, reasonable to agree with S. Bhattacharya that the 'Company used its dominant position to fix the terms of exchange, pushing down the share of wages towards the subsistence level'.[83]

The other reason behind the obviously deteriorating social standards of the textile producers was the grip of the Company appointed middlemen (*paikars* and *dalals*) over the silk producers. *Paikars* were: 'A kind of chapmen [who] on the approach of a Bund or Harvest, visit the Houses of the Chassars, and by a small advance of money enable them to cultivate their Mulberry plantations; thereby securing to themselves the whole produce; that is the Harvest becomes *bound for payment*'.[84]

The famine, instead of reducing the pressure of these middlemen upon the producers and winders, actually intensified it, for we are told that in Rajshahi the decline in silk production was not only due to famine mortality, but also because of the 'very unjustifiable influence which the Pykars and Dullols have acquired over the Chassars and Ryotts'.[85] In 1773, the 'oppression' practiced by these people over the artisans was seriously threatening textile production in Jessore as these oppression 'will compel them [the producers] to leave this part of the country if not speedily put a stop to'.[86] In Muhammadshahi, the Company's agents were openly flouting the established terms of contract with the weavers, often forcing the latter to 'deliver cloths at 4 [Rs.] 6 annas [Rs. 4.37] which had cost 6 r[upee]s to make'.[87] In terms of its social consequences, therefore, the famine of 1769-70 actually strengthened the control exercised by the previously established linkages, between the Company's commercial interests and its modes of procurement from the artisans.

The impact of the famine of 1788 is documented in lesser detail, although the scale of artisanal misery in that year appears to have been substantial. There was, however, one major difference in the economic context of the two situations. The famine of 1769-70 occurred in the midst of fairly buoyant investments made by the Company for Bengal's silk and cotton textiles (Table 59) whereas the flood-induced famine of 1788 appeared in the midst of steadily shrinking investments for silk from the beginning of 1780[88] and a corresponding decrease in the price of raw silk being offered to the *chassars* and *naqads*.[89] Production of raw cotton (*kappas*) also declined in some areas of western Bengal[90] while in other centres of production (for instance in Dhaka) there was an increasing dependence on imports 'from the banks of the Jamuna and from the Dakhin' which was much cheaper despite the high costs of transportation and 'undersells cotton of a middle quality in [Bengal] where this article was heretofore abundantly produced', fine cotton was still grown in eastern Bengal, but it was the middling variety which seems to have declined substantially[91] by the time of the famine in 1788.

The constriction in the Company's silk investments and the steady fall in the price of raw silk meant that the income of the *chassars* and *naqads* was already on a downswing in the years prior to the famine. Thus the recovery apparently made by this sector after the famine of 1769-70 (discussed earlier) was short-lived. The case of the cotton weavers was similar to the state of the silk manufacturers. Evidence from some important *aurangs* (manufactories) of Dhaka show a decline in the number of looms and of cotton weavers between 1776 and 1786. This can be seen from Table 61.

The famine of 1788 appeared in the midst of a general contraction in the volume of artisanal production, and therefore its impact on the

TABLE 61. LOOMS AND WEAVERS IN DHAKA'S MANUFACTORIES

Manufactories	Looms and weavers in 1776		Looms and weavers in 1786	
	Looms	Weavers	Looms	Weavers
Dhaka	1,700	5,100	1,542	3,896
Tittabadi	1,200	3,600	741	1,364
Dumroy	1,500	4,500	503	906
Sonargaon	1,600	4,800	902	2,047
Jangalpari	700	2,100	1,213	1,218
Chandpur & Srirampur	300	900	757	1,049
Total	7,000	21,000	5,658	10,480

Sources: For 1776, IOR, BRC, P/49/64, 24 September 1776; for 1786, IOR, HM, vol. 795, p. 20.

artisans was perhaps equally severe as that of the famine of 1769-70. The cyclonic flood in 1787 not only physically destroyed the crops of mulberry and cotton, it also forced an escalation in the prices of basic food grains by as much as 212 per cent (in Dhaka) and 425 per cent (in Mymensing) in the space of a year between 1787 and 1788. The Company made some efforts to relieve the pressures on the textile weavers. Thus, for instance, the weavers of Tittabadi were given a price increase of Re. 1 per piece for finer varieties and Re. 0.50 per piece for the coarser assortments of textiles which proved invaluable for the survival of the artisans there,[92] but such efforts were extremely selective and did not affect the entire range of occupations engaged in the production of cotton textiles. The weavers at Lakhipur and Malda (two premier centres of weaving in Bengal) were not provided with any financial support; the spinners of thread were hardest hit since they 'fell a sacrifice to it [the famine] in large numbers'.[93] The general effects of the famine of 1788 on the cotton producers and on the artisans making cloth were (i) mortality, especially among the spinners, though the exact number of deaths are not known; (ii) inability of the artisans to manufacture cloth because of the destruction caused to raw cotton and the prohibitive prices of thread;[94] (iii) intense physical hardship caused by the massive increase in the price of food; and (iv) desertion of looms and flight from places of production.[95]

The silk industry was also severely hit. The mulberry trees withered and the price of raw silk rose by nearly 100 per cent between March and November 1788,[96] but the price of rice had gone up by more than 200 per cent. The *chassars* were forced to let the silk-cocoons die 'from a conviction of their inability to support the Expense' of rearing them.[97] The silk-winders (*naqads*) were bereft of their specialized vocations, and they reportedly died in large numbers in Malda, Rangpur and Lakhipur.[98] Mortality estimates for the famine of 1788 are not available, but it is nevertheless possible to outline a picture of universal physical and material debilitation for Bengal's silk industry. Three facts provide the background to understand the debilitation caused by the 1788 famine. Firstly, there was the apparently swift reduction in the Company's investments for silk in the province in the years preceding that famine. Secondly, we have evidence that the prices offered to the producers were also proportionately reduced, thereby hitting at their income. Thirdly, one has to keep in mind the rise in the prices of food grains both over a long-term and in the months immediately preceding the famine.

Thus when the famine struck, the *naqads* were the first to die as they were under a double disadvantage, that of a failure in supplies of raw silk (which crippled their enterprises) and the phenomenal rise in agricultural prices (which disabled them from having access to food). In terms of the social spread of famine mortality, it is perhaps possible that deaths in 1788 followed a pattern different to that of 1769-70. In the latter year (discussed earlier) the highest concentration of excess mortality was among the *chassars*; weavers and winders had suffered proportionately, but not to the same extent as the former occupational group. In 1788, on the other hand, the highest deaths were reported from among the *naqads*. Interestingly, the weavers in the east, in areas which were devastated by the flood, 'complained in very strong and moving terms of the distress they suffer from the present high price of the necessaries of life: Rice and salt', but not of any noticeable diminution in their numbers.[99] *Chassars* were similarly distressed since they were forced to abandon their chosen vocation as their resources were reduced drastically,[100] yet they apparently survived, though on entirely miserable levels.

The major impact of the famine of 1788 was the apparently universal immiserization of all those involved in the production of Bengal's textiles. Alamgir's suggestion that famines intensify the 'low-income-low-foodgrain intake equilibrium' amongst the resource-constrained[101] receives support from the case of the textile producers of Rangpur, whose situation in 1788 was described as follows: 'The pay of these people, which is barely adequate to their subsistence in the most favourable seasons, does not now afford necessary food for themselves and their families.'[102] The situation in Rangpur was replicated almost everywhere else. Thus in Lakhipur, the situation in the Company's *aurangs* exhibited: 'Too lamentable a proof of the excessive dearness of provisions in the distressed situation of the Inhabitants, who sell their children to enable them to purchase a few meals to prolong their miserable existence.'[103] And in Malda: 'A destructive famine has swept off a great number of people, and reduced those who remained to the necessity of selling their children for support, a traffic which has been carried on to a great extent.'[104]

The famine of 1788 seems to have exerted a critical influence on the future of Bengal's silk industry. Its influence on the cotton industry was disastrous but not to the same extent. The relative difference between these two sectors (which proved of decisive importance in their post-famine situations) was that cotton cloth, especially the coarser and middling varieties, had an internal market in the province, whereas silk was almost entirely dependent on the Company's world market. It is likely that the steady reduction in the Company's investments for silk after 1780 had already reduced the economic position of this commodity even before the famine; and that famine reduced the potential of any substantial recovery in the remaining years of the century. Unknown numbers of silk artisans were decimated during that famine.[105] The survivors were physically debilitated and materially reduced to dire straits. *Chassars* uprooted their mulberry trees and let the cocoons die because they could not afford to maintain their enterprises in the face of high food prices and a shrunken market for their produce.[106] Those who had agricultural land, presumably the *chassars* and *tantis*, automatically reverted to the cultivation of basic staples: 'every native having land that can be made to produce grain turned to it more and less immediately' was how the surviving silk-producers of Rajshahi coped with the famine of 1788.[107] Whatever little recovery made subsequently in the production centres around Murshidabad[108] was inadequate to revive this industry and to bring it to its old footing.[109]

Compared to silk, the state of cotton production was relatively better despite the terrible famine. Three reasons made it so. Firstly, ordinary types of cotton cloth had a provincial market. Colebrooke's estimate in 1794 computed an annual out-turn of cotton textiles at Rs. 60 million[110] which was about 9.15 times the Company's investment for cotton piece-goods in 1793.[111] Secondly, the Company's involvement in this sphere was maintained at a high profile both before and after the famine. The average annual sale of Bengal's piece-goods in the ten years between 1780 and 1789 was 5,66,412 pieces;[112] for the decade 1792-1801 the average was 7,77,237 pieces.[113] The average share of Bengal's cotton piece-goods was nearly 84.71 per cent of the total sale of piece-goods made by the Company in the eleven years between 1780 and 1790.[114] Thirdly, private traders seem to have made substantial investments in cotton textiles, at least in eastern Bengal, the scale of which is reflected in Table 62 which pertains to the *aurangs* at Dhaka between 1790 and 1799.

The three reasons outlined above prevented the virtual collapse of cotton manufacturing at Dhaka in the aftermath of the famine of 1788, notwithstanding which the productive capacity of this major centre of textile production seems to have declined over the next decade. In fact Table 62 shows a steady fall in output between 1790 and 1794. There was some recovery between 1795 and 1799 but this was at best an indifferent retrieval when compared to the sharp decline of yield in the

TABLE 62. SHARE OF THE COMPANY AND PRIVATE TRADERS IN
PRODUCTION OF COTTON TEXTILES IN DHAKA, 1790-9
(Rs. in 10 thousand)

Year	Total output	For private trade	For company
1790	22.37	14.91	7.46
1791	14.82	10.81	4.01
1792	16.48	10.11	6.37
1793	12.68	6.89	5.78
1794	8.75	4.14	4.61
1795	11.78	6.62	5.16
1796	12.01	6.49	5.52
1797	14.01	8.94	5.07
1798	10.80	6.39	4.41
1799	12.56	7.52	5.04

Source: IOR, HM, vol. 456 F, p. 109.

preceding five years. There is also no reason to believe that there was an improvement in the conditions of the surviving artisans despite the continuing requirement of cloth for overseas trade and the increased demand for labour in a post-famine situation; nor did the famine succeed in easing the pressure of the Company and its agents upon the weavers.[115] Practically the entire output of high-grade textiles was under Company control; advances to the weavers to manufacture textiles often proved inadequate to meet the minimum requirements of working capital; balances accumulated but these were treated as outstanding debts which were to be deducted from the advances of the next season which increased the burden of producer indebtedness.[116] Wages did not keep pace with the rising costs of food which pushed the living standards of these weavers down to the brink of subsistence,[117] a fact which was also acknowledged by James Taylor, the Commercial Resident at the Dhaka factory, in November 1800.[118] All these pressures had 'the unavoidable effect of impoverishing the conditions of a great part of the weavers, and of reducing many to a state of insolvency'.[119] Naturally, there were desertions and a return to agricultural production or to subsidiary occupations like fishmongering; weaving of coarse textiles for the local market seems to have been another way out for such people, but quitting the Company's service was always a difficult task, since all such persons were treated as debtors to the Company, and, as Hossain shows, the Company was remorseless in collecting its dues.[120]

Coping, Recovery and the Peasantry

The climatic origins of dearth had a significant bearing upon the state of agricultural production in the affected areas and of the problems associated with that of coping. Floods meant that the standing crops were washed away and arable land was damaged either because the waters did not recede quickly enough, or the lands became over-silted.[121] In either case, and despite their best efforts, the producers found it extremely difficult to bring these lands back into cultivation after the worst had passed. Droughts, on the other hand, made the crops wither. They also parched the arable, thereby making the act of ploughing a very difficult task.[122] Droughts also meant that water levels in the rivers receded drastically, thereby throwing-up sand-beds (*chars*)[123] to which the cultivators turned in the hope of getting a crop of low-grade rice, and cultivation of this kind seems to have been an important, and recurring, feature of agricultural adaptation during seasons of drought.[124] What is not clear, however, is the extent to which these lands could compensate for the losses suffered in the more productive spots of land, or the degree to which the output from the *chars* could alleviate the prospect of semi-starvation during a drought. On the whole, the possibilities of easing localized food shortages by cultivating *chars* appears remote for a variety of reasons.

Firstly, *chars* were not fixed. They shifted from season to season. These lands were fertile, but production on them was uncertain as it invariably depended upon 'the next change of the rivers'.[125] Secondly, the established arable land was severely damaged by a drought, and the *chars* could hardly have compensated, in terms of output, for the losses suffered.[126] Thirdly, cash crops, like cotton, mulberry and tobacco could not be grown on *char* lands. Therefore, the problems created by droughts (or by floods) for the rural artisans remained unrelenting. Such problems were compounded in those areas which did not produce all three rice harvests. For instance the district of Hijli[127] did not produce any *aus* or *boro* rice; it depended almost exclusively on the *aman* harvest, and upon imports from other rice-producing areas for its domestic consumption.[128] *Boro* rice was not grown in Purnea, Burdwan and Midnapur.[129] Thus, when unfavourable weather curtailed output, alternative strategies of cultivation were inevitably inadequate.

Apart from trying to cultivate the *chars*, droughts also meant that the peasants tended to concentrate on the *jala* (low lying) land as a last ditch effort to procure a harvest.[130] Of course the degree to which such strategies succeeded depended on the intensity of the current drought. During the drought of 1791, the water-courses in Hijli dried up; cultivated lands which depended upon the moisture and sediments (*douk*) left on the *jala* lands by seasonal inundations became parched; thus the cultivators were unable to 'derive any advantage from which their future cultivation can be ameliorated'.[131] A similar situation was reported from the district of 24-Parganas which had largely escaped the horrors of the 1769 famine 'owing entirely to its situation, which being low retained what rainfall and enabled the people to water and preserve their crops'.[132] But in 1791 'in the higher parts which dried up first, the rice plants are burnt up in the same size as they were originally transplanted. In the center parts where the plants have grown to some height, the ears which have formed are perfectly empty husks without any possibility of producing grain. In the lowest parts which [were] the longest moist have been long since entirely dry, the few ears of grain which have filled have a very sickly appearance & cannot on largest calculation realize one-fourth of the usual crop.'[133]

Dearth in late eighteenth century Bengal had severely destabilizing effects. The fact that they occurred periodically after a major famine in 1769 meant that there was very little respite provided to the producers to recoup their losses, or to reclaim and extend production on their own resources. The best spots of land,[134] which produced the most valuable crops and were the most productive, were immediately lost. This meant that the producers were pushed into cultivating marginal lands, which were essentially inadequate to meet current requirements of food and income in the contest of sharply spiralling food prices.

Such situations were loaded with grave economic and social implications. Firstly, producers were invariably forced into working their lands under a state of persistent uncertainty. According to Philip Francis, 'The proportionate value of lands fluctuates in all countries according to the industry or ability of the owners. In this country [Bengal] more particularly it depends on accidents of drought, inundation or favourable season, of which no general calculation can be made.'[135]

Secondly, such conditions forced lands under cultivation to fluctuate sharply from season to season, 'both in extent and quality'.[136] This meant that there was a constant movement from the more productive lands to relatively marginal lands with each cycle of uncertain weather, or there was a large scale lateral shift of an entire village to seek out lands which were relatively better placed: 'a village in Bengal', noted Buchanan, 'is removed 4 or 5 miles with very little inconvenience indeed' at the slightest sign of adverse weather, 'such as an inconvenient shower'.[137] In Rangamati, where the Brahmaputra forced cultivated lands to fluctuate every year, the cultivators had no choice but to 'fix on any unoccupied land whenever they may be injured or destroyed by the periodical inundations', which meant that they, and the landed-proprietors, 'cannot depend upon sure income from particular places'.[138]

But such shifts did not necessarily provide relief. Apart from the physical inconvenience of leaving behind their settlements, movements of this sort succeeded in placing the cultivators in a number of difficulties. The land which they previously cultivated was laid waste (*pateet*), and they had to start afresh on other lands which placed additional pressures on their already meagre resources. In these new lands the migrants were given access to *pateet* lands of varying productivity,[139] usually on concessional terms of revenue; but there was very little done in the way of reducing the financial burden of bringing such lands into cultivation as reclamation was 'attended with great expence and more labour'.[140] They were, therefore, placed in the unenviable position of having to incur debts from the landed-proprietor or from the grain merchant. Quite often the zamindars or merchants provided these people with the initial costs, but these were once again loaded with high interest rates.

Dearth, therefore, were situations of intense short-term crises of subsistence and production, but the fact that they recurred periodically in Bengal meant that these short-run crises tended to fuse into one another. The natural rhythms of agriculture were disrupted, so were the economic foundations of agricultural production. Such instances placed crippling burdens on the productive resources of the cultivators thereby making them physically and economically dependent on other sections of agrarian society.

The famine of 1788 was another major catastrophe. Though, not as intense as the famine of 1769-70, it nevertheless, had severe consequences for the peasant-economy. The flood and the cyclone took an immediate toll of lives as people drowned. The flood also killed large quantities of cattle. In its immediate aftermath, therefore, there was a steep rise in the price of draught animals which meant that the cultivators were unable to plough their lands.[141] The damage done to the established arable land was also enormous. Thus in Jessore:

The effects of these calamities are not confined to any one present injury. They prevented the cultivation of those lands on which the late or dry crops are produced as the Ryotts were *obliged to sow them on lands infinitely less productive* than those they had usually sown them on, which were either covered with water, or so wet as to be unfit for use. That ground on which Cullye [lentils] is grown was lost. . . . Some crops approaching to ripeness, as til [oil-seed], kudgoor [dates], the first crop of cotton & ca., others in a less forward state, as mustard, barley, tobacco, peas of several kinds, and in some places the seed only sown [were] entirely destroyed.[142]

Even those districts which had not suffered from the cyclone, but had been extensively ravaged by floods, the scale of agricultural devastation was vast. Thus, Purnea, still reeling from the famine of 1769-70,[143] was severely flooded in 1787 when even its *colla* (high) lands were submerged. The district of 24-Parganas was also severely affected by a combination of a flood and a small-pox epidemic. In both districts, therefore, there were sharp reductions of lands under cultivation, high food prices, agricultural disruption, desertion and mortality.[144]

The additional problem during the famine of 1788 was the inability of the peasants to revert to the *jala* or *char* lands. The *jala* lands were totally submerged, and the soil so waterlogged was difficult to drain, since 'if the land is too moist it will not vegetate at all'.[145] The *char* lands were similarly rendered uncultivable. Thus in eastern Bengal, 'The lands in the churs [*chars*] are in a great measure overgrown with reeds and canes, and the small quantity that is sown, the Ryets cannot take care of from the dread of Tygers, [*sic*] Buffaloes, Hogs & ca. that swarm in the surrounding jungles.'[146]

It is, nevertheless, significant that even amidst such difficult circumstances the peasants tried to salvage a crop of rice at the slightest available opportunity. In Rangpur, there was a patch of fine weather for a fortnight before the cyclone of November 1787 when 'the waters had in great measure subsided and the Ryotts in all quarters were diligently employed in the important business of transplanting rice'. The cultivators of Rajshahi were similarly engaged in trying to retrieve a rice harvest when the cyclone struck, leading to fresh floods which totally crippled all potential of a self-induced recovery.[147] The surviving population was reduced to abject debilitation, both physical and material. Thus, in Dhaka:

Those who remain have no cattle to plough their ground, and are much reduced in point of bodily strength that their end is fast approaching, and those with whom a little of the necessaries of life are left are from their distressed situation altogether unable to cultivate their former possessions; in consequence of which the lands cultivated for the last two years remain fallow.[148]

The circumstances leading to peasant immiserization, shaped over the preceding years, were strengthened by the famine of 1788. Desperate for subsistence, the cultivators were forced to part with their 'implements of tillage', eat into their seed-stocks, if any, and even to sell their children[149] as a last-ditch attempt to survive in a situation so heavily biased against them, as the following description shows: 'To procure cattle, seed and the requisite implements of cultivation, and at the same time subsist their families till the reward of their labours be reaped, is attended with more difficulty than most of the husbandmen are able to accomplish.'[150]

Three concerns central to our comprehension of dearth and famines are: (i) that these occurrences were primarily critical crises of subsistence, (ii) they generated conditions of 'progressive aggravation'[151] in the living conditions of a whole range of producers and labourers because of their acute shortage of resources, and (iii) they created profound dislocations in the rhythms of economic production which could not be rectified by taking recourse to a 'moral economy'. Therefore, for the producers the problems of coping with, or recovering from, a dearth or a famine could only be resolved by looking for sources located *outside* their traditional boundaries. The questions are, who were these agencies, and what was the price exacted by them for such help?

After the famine of 1788, the cultivators of Dhaka:

> Have borrowed sums at extravagant interest to enable them to cultivate a part of their lands; which will absorb in a greater degree, if not wholly their expected profits, so that at the commencement of the next season their situation is consequently little improved, and the same necessity for borrowing money every year will remain. Hence it follows they are likely to proceed in the shackles of vexatious and oppressive usurers, and consequently [*sic*] under difficulties, that must ever prevent those exertions necessary perhaps even to their existence.[152]

This is an extremely important statement which shows the mechanisms and extent of peasant impoverishment after a famine. It also points to the constraint of resources as a critical factor in hampering the pace of self-generated recovery after crises of such magnitude.

There was practically no institutional financial support for the afflicted either to cope with or to recover from such disasters. There was, furthermore, very little support from the village community as all traditional ties of dependence and mutual help within it were shattered under the impact of such events. Was there a 'moral economy' of survival? Essential to the 'moral-economy' argument are the conceptions of shared-poverty and mutual assistance in a society living on the edge of frequent crises of subsistence. Such threats are seen as the main motivating force in the creation of elaborate patron-client relationships and village hierarchies which are therefore mutually sustaining.[153]

The evidence from eighteenth century Bengal provides very little in the way of supporting such conceptions. Customary expectations of succour from local landed-proprietors were the first to disappear. Each cycle of adverse weather—harvest failure—subsistence crisis created a fresh run of demand for *taqavi* and remission of revenue from the producers which the local zamindar or talluqdar was unable to provide. Incidents such as a dearth or a famine immediately reduced their incomes and affected their capacity to help the distressed producers. The state's financial burdens (i.e. its revenue demand) remained largely unrelenting thereby hitting at their ability of granting remissions of revenue.

Some efforts to lessen the burdens on the producers were, nevertheless, made by the landed-proprietors from time to time. Thus after the famine of 1769-70 the talluqdars in the revenue-paying circle (*chakla*) of Murshidabad were said to be spending 'a great effort in improving their lands & in extending cultivation at great expense, in advancing money to the ryotts to clear the lands for cultivation and devoting their time and attention to superintend & direct the improvement of them'.[154] Zamindars of Malda had 'zealously' advanced '10 or 15000 ru[pee]s for Tuccavy' in order to reduce the sufferings of the peasants owing to the 'Calamity inflicted by the Heavens' in 1769.[155] But such efforts were too few and too little to achieve any substantial recovery in a post-crisis period.

Dearth or famine posed financial problems for such people as these events jeopardized their incomes from agricultural production. The following petition from the zamindars of Rangpur during the flood of 1787 reveals the likely conditions of small landed-proprietors during natural disasters:

We are poor zemindars, and you are not ignorant of the least circumstance of our condition for the last eighteen months. We have no other income than our private lands . . . and even that though managed with the greatest frugality is hardly sufficient to defray the necessary expenses of cloths and sustenance in so much that we can only afford to wear the meanest dresses even when our duty calls us to attend upon you [the collector]. We have this year been under the *necessity of mortgaging to the bankers the small quantities of private lands which were allowed to us*. . . . We are in the utmost distress, and have no prospect of supporting ourselves.[156]

Larger zamindars (like the Rajas of Burdwan, Nadia, Birbhum, Dinajpur and Rajshahi) were not so crippled, but they were certainly constrained which meant that the peasants in such lands were unable to retrieve their economies in proper time. Thus the peasants in the northern parganas of Nadia (which had been badly hit by the drought and famine of 1769-70) could not commence agricultural operations 'from the suspense they are in to know from whom they are to receive advances to enable them to go through the business of cultivation'.[157] Large tracts of intrinsically good quality land were still lying *pateet* in this district in 1776 because of the financial inability of the zamindar to provide *taqavi*.[158] During the disaster of 1788 the cultivators in Rajshahi had no traditional sources of help to turn to since this district 'is peculiarly circumstanced, having at present in fact no zemindar to any useful purpose, nor any inferior landholders in a condition to do much to assist the Ryotts'.[159]

However, such financial help provided by these landed-proprietors should not be seen as an extension of traditional ties of patronage and dependence. Bengal was a highly commercialized economy, and the available evidence suggests that the assistance given to the peasants to settle their ruined affairs was designed by the zamindars to tie the cultivators in the grip of extended indebtedness, the ultimate purpose of which was to have unrestricted access to their produce or to prevent subsequent desertions. It is perhaps extremely significant that all seed-advances were to be returned 'two-fold to the person who furnished it'[160] and each money-advance was invariably taxed 'with a heavy interest', while repayments were 'fraudulently devalued'.[161] Therefore, the facts that villagers were forced to abandon their habitations, their fields and a whole set of established relations at the slightest sign of dearth, or that they could actually die of starvation, or were forced to sell their last belongings (even their children, and perhaps their wives) are significant for they show a total collapse of the existence of a so-called 'moral-economy'.

The problems created by the famine of 1769-70 for Bengal's agrarian economy were compounded by the subsequent instances of dearth and by the famine of 1788. Dearth were short-run crises of subsistence caused by localized or partial harvest failures, and interspersed, as these were, between two major famines, dearth created unstable conditions in the productive enterprise of the province's small-peasant economy. Such conditions were re-enforced because of the virtual absence of state help in nursing a crippled economy or to ameliorate the conditions of the peasantry by providing some form of financial assistance. The failure of the state either to design a cogent famine-relief policy or to create state-operated channels of agricultural credit was accepted by the Company after the famine of 1788 in the following words: 'The most active temporary interference of Government in the famines has been found productive of no other consequence than a slight alteration of the pressure on the poorer class of natives.'[162]

The absence of institutional assistance, the unrelenting influence of weather upon agricultural production and the almost chronic shortage of resources, meant that the average peasants in Bengal found it impossible to make a self-induced recovery on their own assets, even after one harvest failure. The price situation during and after such harvest-disasters also worked against the producers. Bad agricultural seasons were also times of high food-prices but these prices, caused by a shrunken produce, posed problems of survival for the peasantry because of the inexorable revenue squeeze. The fact that cultivators had to buy their food in such times meant that they were under the double pressure of meeting the state's financial claims and finding money for their survival under extremely adverse conditions and drastically curtailed incomes. In the countryside, starving people ate into their seed reserves. The result was their subsequent inability to sustain production on their own resources in the next agricultural season.

One important aspect which emerged while discussing the causation of dearth and famine is that most peasants were thrown into the market to purchase food from a specialized community of grain-merchants who were the only major agency for the bulk distribution of food during such situations. The second important aspect, central to the entire matrix of agricultural production, was that these peasants were dependent on the merchants for a wide variety of loans of consumption and capital. The economic constraints which engendered these two aspects assumed critical proportions during a subsistence crisis, and the result was that all recovery made subsequently, or the manner in which producers coped, was made to depend on the dictates of these grain-traders-cum-usurers. If the producer sought the market to survive during a crisis, the market found the producer in its aftermath.

The description of the plight of the peasant's of Dhaka during the famine of 1788 (mentioned earlier) is an excellent representation of their consumption needs during a famine and their subsequent productive requirements coalescing to create the *sine qua non* of their dependence on such traders and moneylenders. A similar situation was recorded for the cultivators of western Bengal after the famine of 1769-70, to survive and recover from which they were forced to borrow 'from different merchants at a most exorbitant and unheard of premium'.[163] Such loans were not limited to famines alone; they were contracted with equally crippling consequences in times of dearth. Peasants were forced to desert because the merchants in Burdwan stopped providing rice to them. They were forced into petitioning the state to 'oblige the merchants to turn the grain in [their] hands to us, setting what profit they deem just as that was the only way in which we can purchase our subsistence and preserve our lives'.[164] 'Without the assistance of merchants the Ryotts suffer the greatest distress'[165] seems a good summary of the structure of merchant-peasant linkages effected through the medium of peasant consumption during times of scarcity.

Equally, if not more, significant were the loans which the merchants provided to the peasants to *commence* their agricultural operations. These were loans of working capital and as such had the closest bearing on the productive enterprise of these peasants. B.B. Chaudhuri suggests that problems in 1769-70 were aggravated because the merchants had 'abruptly reduced' these loans during the drought of 1769 'causing thereby a sudden withdrawal of a large supply from the market which was already under severe pressure';[166] but there is no evidence to show that this actually happened. Circumstantially also this seems largely improbable. The monsoon of 1769 failed, and it is not at all clear why advances were 'abruptly reduced' when these were invariably made six months *before* the lands were actually cultivated for any particular crop. What presumably happened during that famine was that merchants who had made such advances may have faced problems of bad debts and losses, but the system of advances was not stopped since we find these merchants 'oblig[ing] the Ryotts to purchase grain for seed at a very high price' almost immediately (in 1771) after that famine.[167] It is, perhaps, true that the threat to stop, or discontinue, such advances was a potent weapon used by the merchants to force the peasants to capitulate in cases of dispute, since such threats usually forced the 'Ryotts to dispose of their grain on any terms, for one third, often for half less than the customary market price'.[168] Briefly stated these loans were the first steps in what was subsequently to become a vicious cycle of debt and debt-servicing, the full implications of which have already been discussed in Chapter 4.

Ray's argument that capital for agricultural recovery was provided by the rich peasants in the aftermath of the famine of 1769-70[169] does not appear to have any validity either in the context of that famine or in the larger framework of subsequent instances of dearth and famines. The main reason, why 'rich' peasants were unable to employ their capital in agricultural developments, lies in the atmosphere of economic uncertainty created by the frequent bouts of dearth after 1769 and the famine of 1788. It is perhaps feasible that some peasants were able to fight-off the intrusive pressures of these events and the merchants who came in their wake. But these would, at best, only be isolated instances and need not reflect the creation of a rich peasant 'class' in any generic sense. Our evidence suggests that all nascent possibilities for a class of this type to emerge in the eighteenth century were rudely shattered by two major famines and a succession of dearth years.

It must also be remembered that the major problem after a severe crisis of subsistence was that of regressive indebtedness which affected *all* categories of peasants in varying degrees. Under such a situation, the gains, if any, made by the surviving peasants were bound to be short-lived as the next crisis would invariably throw up a new crop of dispossessed migrants into the ranks of the rural poor. The substantial proliferation of sharecroppers (variously called *adhiars, bargadars* and *projakisans*) in the late eighteenth century (Chapter 2), despite the favourable land-labour ratio and demand for labour, can be mainly explained by the protracted impoverishment of the small-peasant by dearth and famine. The distinction between the rich and the poor in peasant society was extremely small and often shattered by even the slightest variation in agricultural output or price.

Yet, years of crisis were not without their victors and these were primarily the grain-merchants for whom such years were extremely advantageous for a number of reasons. Firstly, the bulk distribution of food was entirely in their hands, a weapon which they effectively used to manipulate prices and to circumvent state-directed embargoes on the export of grain from deficit areas. Certainly, there were occasions when individual merchants incurred losses from speculation and hoarding but, as the Board of Revenue accepted (in 1791), these 'pecuniary losses were no indemnification to the State for the consequences which resulted from grain having been withheld from the market during the height of public distress'.[170] Thus individual losses do not detract from the fact that the grain-merchants, as a distinct social group, benefited most from all types of scarcities. Secondly, grain-trade was generally combined with money-lending and the provision of capital and consumption loans, the effective power of which has been discussed earlier. Once again some merchants may have faced problems of bad or unrecovered debts in case of mortality; but these were occupational hazards they had to risk. These risks in no way weakened the framework of their social domination over the peasantry. The fact that their refusal to furnish additional loans or call in outstanding debts could bring even the slightly recalcitrant peasant to capitulate indicates the extent to which they had actually managed to succeed in their designs.

On the surface, the balance between human and natural conditions of existence in eighteenth century Bengal was precariously weighed against the former. The slightest adversity could destabilize the rhythms of agricultural production. Yet, every effort was made to adapt to adverse conditions in a set of short-run and long-run responses. While the short-run responses were largely the peasants own strategy to cope with an immediate crisis of subsistence, the long-term response was more complex and involved non-peasant groups in a substantial way and also reinforced their domination over rural society. This emerges as a major way of coping with ecological uncertainties, the most important mode of which was the lateral expansion of the agrarian boundaries of the region. As it has been discussed in Chapter 1, this expansion was the systematic push into those areas which were less prone to such crises (the 24-Parganas, Jessore, Bakarganj and Chittagong) and to Midnapur which was the last major opening in what had hitherto been a frontier of the province.

Whether this reclamatory thrust was sufficient to alleviate distress remains a moot point. Since society was stratified, it is natural that a substantial section of Bengal's population would remain harvest sensitive. This probably explains the famine-related deaths suffered in the province in the year 1788. However, the suggestion that there was a strong demographic growth in the period between 1770 and 1860, caused in the main by a high birth rate outdistancing a high death rate,[171] may have been an outcome of this process of agrarian expansion. It is certainly significant that, notwithstanding the exaggerated estimates of mortality for the famine of 1769-70, Bengal did not suffer another devastation of that magnitude till the middle of the nineteenth century, if not till the major famine of 1943.[172]

NOTES AND REFERENCES

1. IOR, BRC, P/52/36, 21 October 1791. Even in the mid-nineteenth century a price rise after the arrival of the winter harvest was fraught with serious implications of a severe scarcity during the next year, but even as late as that

period the colonial state had been unable to collect enough evidence 'by which to gauge its full and terrible significance' (Hunter, *Famine Aspects of Bengal Districts*, Calcutta, 1869, p. ix).
2. This is not to say that under relatively normal circumstances they were self-sufficient. Rural artisans invariably had to buy food and the peasants, barring a few with relatively larger resources, had to take recourse to the local *haat* to purchase a number of items of food (like lentils, vegetables, salt) which they needed. Even basic staples were often purchased. Their dependence on the market was intensified during seasons of scarcity.
3. Minute of Warren Hastings to the Council, 20 November 1783, incorporated in consultations of IOR, BRC, P/50/68, 8 September 1786; emphasis added.
4. BM, Add. Mss. 29076, fol. 75, and 29132, fols. 380, 470.
5. *BDR, Midnapur*, vol. 4, From E. Baber to C. Russell, 7 March 1771, p. 53.
6. WBSA, LCB, vol. 1, 17 August 1770, p. 40.
7. BPC, IOR, P/1/44, 20 November 1769; IOR, BRC, P/49/42, 7 December 1773.
8. *BDR, Midnapur*, vol. 4, 7 March 1771.
9. This fact was noted by the Board of Revenue in their Minute (October 1791) on famines in Bengal, cf. IOR, BRC, P/52/36, 21 October 1791.
10. W.B. Bailey, 'Statistical View', p. 516.
11. WBSA, CCR, vol. 1, Proceedings of July 1770, p. 444.
12. WBSA, Letter Copy Book of the Resident at the Darbar, vol. 1, 7 August 1770.
13. WBSA, CCRM, vol. 2, 31 December 1770.
14. See ibid., for rice trade from Murshidabad to the villages of Rajshahi; WBSA, Letter Copy Book, vol. 1, 7 August 1770 for trade in Rangpur between Govindganj, its principal grain mart, and the countryside; also Hunter (*Annals of Rural Bengal*, p. 410), for trade between Calcutta and Nadia during the famine.
15. See Chapter 4.
16. IOR, Letters Received, E/4/20, 23 November 1769.
17. *BDR, Chittagong*, vol. 1, 30 November 1769, p. 100.
18. Anderson Papers, BM, Add. Ms. 45430, Richard Becher to David Anderson, 3 January 1773, fols. 256-7.
19. BM, Add. Mss. 29076, fol. 75, and 29132, fol. 380.
20. Sinha, *Economic History*, vol. 2, p. 57.
21. B.B. Chaudhuri, in *CEHI 2*, p. 299.
22. Public Consultations, vol. 38 (National Archives of India), pp. 315-16, 413, 416.
23. BPC, IOR, P/1/51, 6 February 1772.
24. See Chapter 5.
25. SCC, IOR, P/A/9, 1 August 1769.

26. *BDR, Chittagong*, vol. 1, from J. Reed and T. Lane to H. Verelst, 30 November 1769, pp. 100-1.
27. IOR, BRC, P/49/52, 16 May 1775.
28. B.B. Chaudhuri, in *CEHI*, 2, p. 299.
29. Letter of Huzurimal to Company, 10 October 1770, BM, Add. Ms. 29132, fol. 380.
30. IOR, BRC, P/49/38, 19 February 1773.
31. Ibid.
32. Ibid., P/49/42, 2 December 1773.
33. Ibid., P/49/47, 30 August 1774.
34. Ibid., P/49/42, 7 December 1773; emphasis added.
35. Ibid., P/49/51, 13 February 1775.
36. Ibid., P/49/58, 22 December 1775.
37. *Bangladesh District Records, Dacca District*, 10 July 1784, p. 81.
38. Ibid., 15 July 1784, p. 83.
39. Ibid., 5 October 1784, pp. 97-8.
40. IOR, BRC, P/51/17, 1 February 1788.
41. Ibid., P/51/19, 17 April 1788 and 18 April 1788; *BDR, Dinajpur*, vol. 1, 19 January 1788, p. 89.
42. 5,000 maunds of rice per day were required to feed the population of Calcutta and Murshidabad alone (ibid., P/51/16, 1 February 1788, and P/51/17, 14 March 1788); ibid., P/15/17, 1 February 1788.
43. Ibid., P/51/16, 1 February 1788.
44. 'Rulers and Merchants', unpublished Ph.D. thesis, pp. 49ff.
45. Additionally the entire evidence of price regulations and restrictions on exports provided by Curly pertains to food shortages after the famine of 1769-70, which he interprets (without explaining why) as a continuation of 'traditional' famine-relief policies.
46. *BDR, Dinajpur*, vol. 2, 14 June 1788, p. 231.
47. Ibid., vol. 1, 19 January 1788, p. 89.
48. 'Copy of a Memorial on the High Prices of Wheat, February, 1800', BM, Add. Ms. 29233, fols. 132-132a.
49. IOR, BRC, P/51/22, 15 August 1788; emphasis added.
50. John Shore to Council, ibid., P/51/17, 1 February 1788.
51. Ibid., vol. 2, 14 June 1788, p. 231; emphasis added.
52. IOR, BRC, P/51/19, 17 April 1788.
53. Whereas the cultivation cost of rice was 'not above one, two or at best three rupees a bigha' (WBSA, CCRM, vol. 6, 19 November 1771).
54. Ibid., vol. 6, 11 November 1771.
55. Ibid.
56. WBSA, Controlling Committee of Commerce, vol. 1, 15 October 1771; also IOR, BRC, P/49/52, from Richard Barwell to Council, 7 April 1775.
57. Sinha, *Economic History*, vol. 1, pp. 160-1.

58. WBSA, Controlling Committee of Commerce, vol. 1, 19 April 1771.
59. WBSA, CCRM, vol. 8, 5 December 1771.
60. Ibid.
61. Middlemen used by the Company to procure raw silk or textiles from the *chassars* and weavers.
62. WBSA, Controlling Committee of Commerce, vol. 1, 19 April 1771.
63. IOR, BRC, P/49/52, 7 April 1775.
64. J.C. Price, *Midnapur*, p. 82.
65. IOR, BRC, P/49/50, 20 January 1775.
66. This point receives substantiation from the history of famines in pre-industrial Europe where it has been noted that famines and pandemic immediately diminished the numbers of artisans and wage-earners, partly owing to deaths but equally by the spontaneous movement of people from the former sectors in order to replenish the diminished numbers of peasants and employers (J.D. Chambers, *Population, Economy and Society in Pre-Industrial England*, Oxford, 1972, p. 14).
67. 'Account Bengal Piece Goods sold the Hon'ble Company's Sales, 1771 to 1790', IOR, HM, vol. 401, p. 138.
68. See Chaudhuri, *Trading World of Asia*, Table C 22, pp. 544-5.
69. For instance Hamida Hossain notices the large investments and sales made by the Company in the famine period (*The Company Weavers*, p. 67) but does not provide an explanation as to why this was possible under such critical circumstances when she herself notes (ibid., pp. 9-15) that famines created insurmountable barriers to producers and to production.
70. See ibid., pp. 20ff. for a good account of the element of coercion in the Company's procurement of textiles in the late eighteenth century.
71. The reason for this perhaps lies in the fact that 1773 was a year when agricultural prices had slumped drastically all over Bengal.
72. IOR, SCC, P/A/10, 28 April 1770.
73. WBSA, PCR, Murshidabad, vol. 2, 23 May 1774, p. 417.
74. Ibid., vol. 7, 4 December 1775.
75. WBSA, PCR, Dinajpur, vol. 3, 26 June 1775.
76. IOR, BRC, P/49/50, 27 January 1775.
77. Controlling Committee of Commerce, 22 August 1771, IOR, HM, vol. 769, p. 361.
78. Hossain, *The Company Weavers*, pp. 52-5.
79. IOR, HM, vol. 769, p. 361.
80. WBSA, CCRM, vol. 6, 11 November 1771, letter from Boughton-Rouse to Committee.
81. IOR, HM, vol. 769, 22 August 1771, p. 361.
82. See S. Bhattacharya in *CEHI 2*, p. 288.
83. Ibid., p. 289.
84. Controlling Committee of Commerce, IOR, HM, vol. 769, p. 359; emphasis added.

85. WBSA, CCRM, vol. 6, 11 November 1771.
86. IOR, BRC, P/49/41. 10 August 1773.
87. Ibid., P/49/42, 5 November 1773.
88. The reduction in the Company's investment for silk from 1780 was adduced as one of the principal causes for the decline in the breeding of silk-cocoons and of silk-weaving in Rajshahi (IOR, BRP, P/71/10, 25 June 1789), and in Rangpur (ibid., 25 June 1789) and in Midnapur (ibid., P/71/11, 25 June 1789). The declining state of silk production in Rangpur was poignantly described by D.H. MacDowall (the Collector there) as follows:

> Formerly a large quantity of silk was purchased by the Company & greater advances being received by the Ryotts, great attention was paid to the cultivation of mulberry. Afterwards the Company declining to purchase any more silk, left off making advances; the Ryotts are now therefore afraid that should they once more cultivate the mulberry grounds as formerly & the Company once again put a stop to their silk investment a great loss would then immediately fall upon them. The Ryotts have therefore in consequence almost entirely abandoned it again.

89. The figures provided by Sinha (*Economic History*, vol. 1, p. 198, n. 7) show a steady fall in the prices of Bengal wound silk from Rs. 8.75 per *seer* in 1781 to Rs. 7.75 in 1783 to Rs. 6.96 per *seer* in 1788.
90. Thus, for instance in Purnea, 'the cultivation of cotton was formerly more abundant, but declined in consequence of the imports of this article from Patna and Mirzapore which undersell the cotton [grown in] Purnea' (IOR, BRP, P/71/10, 15 June 1789).
91. Colebrooke, *Remarks*, pp. 31-2.
92. Sinha, *Economic History*, vol. 1, pp. 168-9.
93. 'Report of Commercial Occurrences, 6 August 1789', IOR, HM, vol. 393, pp. 255-8.
94. In Malda, the price of thread had gone up by 50 per cent in one season (ibid., p. 257).
95. Ibid., pp. 259-60 for flight of weavers of the Sonamukhi *aurang*.
96. 'Report of Commercial Occurrences of 1788', IOR, HM, vol. 393, pp. 112, 115.
97. Ibid., p. 112, also p. 115.
98. Ibid., pp. 113-14, 118.
99. Ibid., 21 January 1788, p. 113.
100. Ibid., p. 115.
101. *Famines in South Asia*, p. 47.
102. IOR, HM, vol. 393, 28 July 1788, p. 118.
103. Ibid., 23 January 1788, p. 113.
104. Ibid., 25 January 1788, p. 114.
105. Hossain (*The Company Weavers*, p. 12) suggests that the scale of mortality among the textile producers 'presumably . . . was not dissimilar in scale to the previous experience [in 1769-70]'.

106. IOR, HM, vol. 393, 6 November 1788, p. 112, and 26 April 1788, p. 115.
107. IOR, BRP, P/71/10, 25 June 1789. The response of the *chassars* in 1788 was similar to that of the weavers in 1769-70 when it was noted by the factory at Qasimbazar that many of them 'have turned their hands to the cultivation of grain' (WBSA, Controlling Committee of Commerce, vol. 1, 28 March 1771, p. 23).
108. See Sinha (*Economic History*, vol. 1, pp. 196-7) for a statement that silk production and trade in Bengal was able to make only a partial recovery in the rest of the century.
109. See Figure 1; also Marshall (*Bridgehead*, p. 159) for a suggestion that silk exports from Bengal to markets in northern India had also sub-stantially declined in this period.
110. *Remarks*, p. 130.
111. The Company's investment for 1793 was Rs. 65,53,763, vide., Sinha, *Economic History*, vol. 1, pp. 178-9.
112. Figures in IOR, HM, vol. 401, p. 138.
113. Marshall, *Bridgehead*, p. 106.
114. IOR, HM, vol. 401, pp. 138-41. But the impact of the famine of 1788 was clearly visible in these sales. The share of Bengal's cotton in the Company's sales were as follows (ibid., p. 138):

Year	% share of Bengal's cotton in total sales
1785	93.71
1786	92.88
1787	91.58
1788	81.34
1789	76.71
1790	84.74

115. See Hossain, *The Company Weavers*, pp. 140ff. for a discussion of the depressed state of cotton weavers in the Dhaka *aurangs* after 1790.
116. Ibid., pp. 150, 153.
117. Ibid., p. 162.
118. IOR, HM, vol. 456 F, pp. 201-5.
119. Ibid., p. 211.
120. *The Company Weavers*, pp. 165-8.
121. Thus a partial flood in eastern Bengal (in 1773) caused by the Padma and Burril rivers: 'not only . . . destroy[ed] the harvest on the ground and swe[pt] away whole villages [thereby] ruining the Ryotts & occasioning great desertion; but when the waters subsided, they left large tracts of land which was before cultivated, entirely choked up with sand & so impoverished as to preclude all hopes of bringing it again cultivation'

(IOR, BRC, P/49/44, 29 December 1772). The flood of 1783 caused similar damage to crops and lands in Dhaka (see S. Islam, ed., *Bangladesh District Records*, vol. 1, 10 July 1784, pp. 77-81).
122. Droughts invariably meant that 'the sowing lands are waste, the arable not ploughed' (WBSA, PCR Murshidabad, vol. 7, 4 December 1775). The drought of 1791 had rendered 'the whole soil harder & more cracked than it generally is' (IOR, BRP, P/71/45, 16 November 1791).
123. See IOR, BRC, P/51/40, 15 July 1789 for a description of the types of *chars* in Bengal.
124. This feature is revealed in the fact that such lands were invariably the cause of intense disputes between the zamindars as they 'claim the property thereof as annexed to their respective lands' (IOR, BRC, P/51/40, 15 July 1789).
125. F. Buchanan, 'Survey of Ronggoppur', IOL, Ms. Eur. D. 75, vol. 1, book 1, fol. 109.
126. *Chars* were also caused by erosion, when rivers 'sweep large tracts of country in some places and deposit it in others' (IOR, BRP, P/70/26, 16 March 1787) so 'one person's property is carried away and another's enlarged' (F. Buchanan, 'Survey of Ronggoppur', IOL, Ms. Eur. D. 75, fol. 109). However, the net results of such shifts were fluctuating and often reduced the lands for arable purposes, reasons which became instrumental in making landed-proprietors clash with each other (ibid., fol. 110; also IOR, BRP, P/72/6, 22 June 1792; IOR, BRC, P/53/69, 14 December 1793).
127. Hijli was a major salt producing area and therefore had a sizeable concentration of artisans.
128. IOR, BRC, P/52/38, 2 December 1791.
129. See IOR, BRP, P/70/40, Appendix to Proceedings of April 1788 for a detailed list of types of rice produced in different districts of Bengal.
130. Turning to *jala* lands during floods was impossible as these lands were the first to be submerged.
131. IOR, BRC, P/52/38, 2 December 1791.
132. Ibid., P/49/52, 7 April 1775.
133. IOR, BRP, P/71/45, 16 November 1791.
134. These lands were so denoted 'owing both to the quality and their relative situation . . . in regard to rivers, to the facility of obtaining water when wanted for its cultivation & at other times of preventing inundation' (IOR, BRC, P/53/8, 15 November 1793).
135. Ibid., P/49/65, 5 November 1776.
136. IOR, BRC, P/51/22, 23 July 1788.
137. 'Survey of Ronggoppur', IOL, Ms. Eur. D. 75, fol. 110. This statement additionally suggests the insufficiency of resources in the hands of the peasants.
138. IOR, BRP, P/72/30, 13 May 1794.

139. This was in keeping with the strategies framed by zamindars to improve the state of cultivation in their lands. See WBSA, PCR, Burdwan, vol. 15, 22 June 1776; IOR, CCR, P/68/5, 21 September 1780; IOR, BRP, P/70/26, 10 April 1787; ibid., P/70/41, 18 January 1788; IOR, BRC, P/51/25, 1 February 1788; ibid., P/53/8, 2 October 1793, for instances of such strategies in different zamindaris.
140. IOR, BRC, P/51/25, 1 February 1788. Peasants were required to cultivate waste lands in these areas. The lands given was variously called *porroah* or *mahsulat pateet*, which denoted previously cultivated land now lying waste.
141. Ibid.; also IOR, BRP, P/70/40, 25 April 1788. The toll on the lives of draught animals was another major difference between the devastation caused by floods and droughts. During droughts, when crops withered, the parched grain stalks became a major source for the survival of these animals as they were let into these fields to graze (see, for instance, IOR, SCC, P/A/10 proceedings of April 1770, and IOR, BRC, P/49/47, 30 August 1774). During a massive flood, even this meagre basis of sustenance was cut-off for these animals. Consequently cattle-mortality was higher in a flood.
142. IOR, BRC, P/51/22, 23 July 1788; emphasis added.
143. In 1783 Purnea was still suffering from the consequences of 1769-70 whereby there was 'a deficiency in the population and that above one half the lands [were] lying waste merely from the want of ryotts to cultivate them' (IOR, BRC, P/50/46, 10 June 1783).
144. See for Purnea, IOR, BRP, P/70/18, 31 August 1786 and IOR, BRC, P/51/17, 19 March 1788; for 24-Parganas, IOR, BRC, P/52/14, 7 July 1790.
145. 'Remarks on the Several Collectorships in the Years 1788 and 1789', IOR, HM, vol. 385, p. 68.
146. IOR, BRC, P/51/29, 22 December 1788.
147. See IOR, BRC, P/51/12, 20 September 1787 and ibid., P/51/13, 20 November 1787 for Rangpur; ibid., P/51/29, 22 December 1788 for Jessore.
148. IOR, BRC, P/51/29, 22 December 1788.
149. Ibid., P/51/51, 9 December 1789; IOR, HM, vol. 393, pp. 113-14, 120.
150. IOR, BRC, P/51/51, 9 December 1789.
151. For a discussion of famines leading to 'progressive aggravation' of the poor in early modern France see, Aymard and Mukhia, *French Studies in History*, vol. 1, p. 15.
152. IOR, BRC, P/51/51, part 1, 9 December 1789.
153. See David Arnold, *Famine*, pp. 80-6 for a discussion of the 'moral-economy' thesis in relation to subsistence crises.
154. WBSA, CCRM, vol. 8, 30 December 1770.
155. Ibid., vol. 2, 10 December 1770.

156. The collector of Rangpur vouched that this petition 'contains a true and faithful description of the state of the district', IOR, BRC, P/51/12, 4 September 1787.
157. WBSA, CCRM, vol. 5, 23 May 1771.
158. IOR, CCR, P/67/62, 13 June 1776.
159. IOR, BRC, P/51/20, 7 May 1788.
160. IOR, BRP, P/71/26, 26 April 1790.
161. IOR, SCC, P/A/9, 16 August 1769.
162. IOR, BRC, P/51/22, 15 August 1788.
163. WBSA, CCRM, vol. 5, 18 May 1771.
164. IOR, BRC, P/49/42, 22 August 1774 and 7 December 1773.
165. IOR, BRP, P/71/30, 11 October 1790.
166. *CEHI 2*, p. 299.
167. WBSA, CCRM, vol. 6, 29 September 1771.
168. IOR, CCR, P/68/7, 4 May 1781.
169. *Change in Bengal*, pp. 56-7.
170. IOR, BRC, P/52/26, 21 October 1791.
171. Sugata Bose, *Peasant Labour*, p. 23.
172. See Leela Visaria and Praveen Visaria, 'Population, 1757-1947', in *CEHI 2*, pp. 528-31; also Paul Greenough, *Prosperity and Misery*, pp. 281-5.

Conclusion

THE DISCUSSION so far has tried to trace the multiple trajectories of agrarian commercialism in rural Bengal in order to show a process by which both production and circulation of commodities became linked to a market in the late eighteenth century. There was an irreversible integration of people's lives with the market and with the agencies who controlled the access to such institutions, and this fact is one of the principal features of a commercialized society. In fact, the survival of a whole set of people in such a commercialized society comes to be entirely linked to the vicissitudes of markets and prices. Quantities and modes of sale (whether enforced or spontaneous) are therefore not the only phenomena in a commercialized economy. Equally crucial are matters relating to the amount of commodities which people are able to buy back from such markets, and the procedures involved therein. Even a temporary withdrawal from the market (caused by high prices, inadequate supplies or mercantile strategies), therefore, invariably lead to severe 'entitlement' crisis for a whole range of consumers and producers.

A recent re-examination of the cultivation of opium and indigo in colonial Bengal by B.B. Chaudhuri[1] has drawn attention to the fact that the production or the sale of cash crops through the advance-payment system, so-called 'forced commercialization', did not prevent 'market conjunctions', especially prices, in determining the *autonomous* response of the cultivators towards these crops; nor could the monopolistic policies of the state obstruct alternative (that is indigenous) sources of credit from percolating into these sectors of production.

Therefore, the postulate that only a spontaneous sale of the producers surplus can be construed as a normal process of commercialization[2] also labours under the assumption that only stable conditions are conducive to such a process. Such a hypothesis overlooks the power of hazards and uncertainty as the other important factors in commercial growth. The influence of risk is particularly vital in understanding economic decisions in pre-industrial societies because of the influence of weather on production and the absence of a firm infrastructure. Risk breeds uncertainty, which in turn leads to the creation of networks and strategies to cope with them and to circumvent them as well. But these strategies are not inimical to a commercialized economy; they are on the contrary vital to its existence and development. The influence of dearth and famine on the economy of eighteenth century Bengal provides an important empirical canvas to test such hypotheses.

One important conclusion which can be reached from the analysis of dearth and famine is that these subjected *all* social groups to the vicissitudes of the market. A bad agricultural year was only a proximate cause of a famine or dearth, the actual deprivation was caused by a per capita food availability decline because people could not afford to buy food at prevailing prices and there were practically no other 'traditional' avenues open to them to get sustained relief. Secondly, situations of dearth resulted in a flight of grain to areas with higher food prices and this flight was a *conscious* mercantile strategy to reap maximum profits from a situation of acute distress. The flight of grain was not always directed towards the cities. It is true that in general price differentials were higher in the towns but a dearth or a famine forced prices in the countryside to rise at par with those prevailing in towns. There was therefore a cross-flow of grain with the merchant as its pivot. Thirdly, merchants invariably disregarded attempts made by the state to impose embargoes on the export of food during crisis years.[3] This fact is crucial insofar as it shows that mercantile accumulation occurred under conditions of great uncertainty, and that it was done not in subordination to the state but in direct *contravention* of its authority. A famine (even a dearth) in eighteenth century Bengal can therefore no longer be visualized as a mere food shortage in a subsistence economy. These episodes reveal a highly commercialized economy in the throes of massive crises of subsistence. It was precisely this commercialization which influenced the availability or absence of food for the harvest-sensitive strata in the province.

The need to circumvent or cope with such crises furthered the pace of commercialization. For the producers the essential questions were those of survival and economic reproduction. Our evidence shows that both requirements were linked to the larger issue of productive resources which were continuously hit by such crises. In sum it was impossible for the majority of the peasants in Bengal to cope and recover from the effects of such episodes on their own assets. This, in turn, led them to look to external agencies of support. Their problems were worsened by the absence of any substantial state help. This led to the creation of coping strategies which were loaded with a number of significant economic implications. Firstly, the market became the central determinant of agricultural production, and secondly access to the market was henceforth mediated by the grain merchants. Insecurity was one crucial factor in commercializing Bengal's rice production in the eighteenth century. The very act of coping, thus, became an act of commercial production and exploitation.

The element of risk was also responsible for the creation of strategies of circumvention. For the grain merchants, such tactics were vital if they had to take advantage of the rising demand for food in the province and to avoid the obstacles posed by frequent harvest failures. The devices shaped by the merchants had two components. Firstly, efficient storehouses (*golahs*) were established to stock sizeable amounts of grain. These *golahs* had to be cost-effective yet possess the ability to store perishables in them without damage for long periods of time. The most common device was a storehouse made of baked mud on a bamboo frame which was thatched with mats. A *golah* 40 feet long, 10 feet broad and 20 feet high, with a storage capacity of 1,500 maunds of rice could be constructed at total cost of Rs. 40. This was certainly cost-effective considering that rice could be retained in it for five years and paddy for about 10 without any damage.[4] The fact that one of the appellations of a principal graintrader was *golahdar* (owner of *golahs*) is a significant indication of the importance of these storehouses. Ownership of *golahs* added to the capacity of a merchant to ride the crests of any food shortage. The post-1770 period helped in furthering the importance of these places. In the aftermath of the famine of 1769-70, the Company decided to encourage free movement of grain by abolishing all duties levied upon it 'in its transportation from the country'. This proved immensely beneficial to the merchants as it significantly reduced the costs of transport. The net outcome of this decision was a proliferation of *golahs* and 'private gunges for the reception of all rice and paddy brought to market by themselves and others'.[5]

Ownership of *golahs* had other advantages. These meant that merchants could not only stock supplies for long periods, but they could also keep their goods in constant circulation by selling their old stock at regular intervals and replenishing the portion sold by fresh purchases during the harvest. In the case of rice, which could be safely stored in a *golah* for five years, the merchants adopted the following strategy: in the fifth year they would sell, or lend, one-third of the portion in store from their first year's purchase, and replenish by purchasing fresh rice from the market 'at the height of the harvest when rice is always cheapest'. This, incidentally, was profitable venture on both counts. When sold, undamaged old rice fetched a higher price than that of new rice, thereby providing a substantial margin of profit, which was also increased by making these sales coincide with 'annually at the season when rice usually rises in price'. When loaned to others, the old rice was given at the relatively higher price prevailing before the fields were sown, whereas repayments were taken at the time of harvesting the crop and according to the rates established at the time of making such advances. The peasants invariably had to part with a larger proportion of their crop because of this seasonal variation in price.[6]

The second important component of mercantile strategies of circumvention was the subordination of the small-peasant. Our evidence has shown a resource constrained peasantry being repeatedly made to face periodic bouts of economic dislocation by famine and dearth. Our evidence has also revealed that the only viable coping strategy open to these producers was through a continuous dependence on merchants for the provision of seed and consumption advances. The way such advances worked, clearly led to the subjection of a whole range of producers to the financial dictates of the local merchant. This kind of subordination was not limited to the production of rice alone, though the fact that even this so-called subsistence sector was brought within the ambit of mercantile domination must speak volumes for the commercialization of agricultural production in the eighteenth century. Sugarcane produced in Birbhum was based on such arrangements with merchants who then sent the entire amount of refined sugar to Calcutta, often at the expense of domestic consumption within the district; tobacco grown in Nadia and Rangpur was partly financed through advances by the merchants of Calcutta, Murshidabad and Dhaka, and the rest was purchased by their agents on the spot; ginger produced in Rangpur was sold immediately by the farmers to merchants since 'the whole produce is paid for in advance'.[7]

A recent study suggests 'dependent commercialization' as a typological categorization of Bengal's economy during the late eighteenth and early nineteenth centuries. Basing itself on the model of dependent commercialization developed by Philip C. Huang for northern China,[8] this study characterizes dependent commercialization as the control exercised by the intrusive 'foreign' merchant capital which brings the 'agricultural production process firmly under its sway but stops short of capitalist accumulation and consolidation of land', which in the case of Bengal was brought about by the Company engineered integration of Bengal's textiles with the world market.[9]

There is, undoubtedly, some justification in seeing a degree of dependency in the case of Bengal's textiles in the eighteenth century because of its connections with the world market, but one must also remember that the Company's investment for cotton seldom cornered more than 10 per cent of the output from Bengal. The remaining 90 per cent was traded and consumed within the country and the Company had no control over the structure of local trade.[10] The evidence about cotton also shows that there was a decline in its cultivation within the province throughout the eighteenth century, and that districts producing cotton cloth were depending on a very 'large importation from the banks of the Jumna and the Dakhin';[11] for instance, the cotton industry of Nadia required 16,800 maunds of de-seeded cotton per year, of this local procurement amounted to 4,000 maunds while the rest (16,400 maunds) was imported *via* Patna, Mirzapur and Surat.[12]

The other reason why the typology of dependent commercialization would have a limited applicability for Bengal lies in the nature of its landed gentry. Of course, the entire range of zamindari interface with the rural economy cannot be encapsulated in terms of market choices alone. Peter Robb's assertion that a 'wide range of controls related mainly to the extraction of rent' were operative in rural society, provides a note of caution.[13] Instead, we should see these landed proprietors as a variegated lot having a symbiotic relationship with the production and marketing of agricultural produce, a relationship underpinned in the last instance by extraction of the surplus. Even charity, altruism or benevolence were other forms of social investment legitimizing their status as the gentry. Even if their basic motivation was 'increased rental income',[14] the participation of the landed gentry in production was neither sporadic nor accidental. A substantial portions of zamindari incomes (at times as high as 35 per cent) were derived from the sale of produce grown in their *khamar* lands. On these lands, surplus was appropriated on the basis of a relationship between investment and labour use, going beyond the maximization of rental income, and implying a notional wage for labour which the zamindars tried to depress. Sharecropping on the *khamar* was more than the attempted maximization of rental income.

Zamindars were the dominant but not the only landed proprietors in Bengal. The role of talluqdars and *la-kharaji* holders also shows the intermeshing of landed property with production. One has also to consider landed property in the context of social stratification. People holding zamindari, talluqdari and *la-kharaji* rights represented different grades of property owners. They constituted a composite rural elite positioned over a small-holding, and often under-differentiated, peasantry. Their position was often bestowed by birth or descent, but could also easily be acquired by ready money. They were without doubt *the* rural rich and performed two critical productive functions. They were central players in the extension of the arable land and they established markets, thereby sustaining a high form of commercial networking in rural society.

Therefore, as suggested earlier, the dynamics of agrarian commercialism, at least, in the eighteenth century can become explicable by looking at the entire process from the standpoint of a triadic integration—of agricultural production, trade and the peasant-household—with a regionally integrated market. The evidence of cotton also suggests that the integration of the market went much further than the provincial level, and the existence of a comprehensive interregional nexus in agricultural produce was a symptom of this integration.

The developments in the domain of rice production in Bengal were extremely significant for four reasons. First, rice alone amounted for more than 40 per cent of the total agricultural output, and developments in that sector would naturally have the major bearing on the agrarian economy. Second, the prices of rice escalated at a faster rate than the prices of sugar (a cash crop) throughout the eighteenth century. This had a significant effect on the inter-sectoral distribution of crops in favour of rice. Third, the response of the cultivators was in tune with the state of relative prices. Greater effort seems to have been expended by cultivators to the cultivation of rice both in absolute terms (that is by reclaiming new land), and in relation to cotton and sugarcane. Fourth, new tenurial relations, like sharecropping, emerged in the eighteenth century as a response to expansion in output and an increase in the prices of agricultural produce as did the *jotedar* on the other end of the tenurial scale.

The regime of the East India Company and the political-economy of Company rule intensified the formation of a regionally integrated market in the province, not only for high-value commodities like cotton or silk, but for basic food grains. Appropriation of revenue was integrally linked to the expansion of Bengal's internal market. But this expansion was more than just an incidental outcome of the state's political will being manifested through the extraction of the surplus in cash. It was more a product of a rising demand for food in the province at prices which were now linked on a provincial scale. The instances of dearth and famine (including the massive famine of 1769-70) were integrally linked to this prevalent demand for food in an integrated market. Though these phenomena were caused by partial or general crop failures, their significance and intensity can be fathomed only from looking at these events as severe dislocations in the food market and the price mechanism. They also forced society to make structural adjustments in their aftermath. These adjustments were necessary for one stratum to cope with its problems and for another to circumvent unforeseen difficulties which resulted in the creation of different economic strategies by all social groups in the countryside. The net result of these strategies was to intensify the interconnections between agricultural production and the market to an unprecedented degree.

Therefore, a reinterpretation of the existing notions of the divide between the commercial and the subsistence, or that between the notions of forced and spontaneous linkages with the market, is clearly necessary in the light of such developments. Dependence on the cultivation of rice does not interfere with the growth of petty commodity production and spread of 'commercial' farming in a small peasant economy. The face of rural Bengal in the eighteenth century was transformed when agricultural production, the peasant household and trade became combined in a triadic relationship with the market. The intervention of the state, especially in the second half of the century, intensified these interlinkages.

Questions nevertheless remain. How are we to contextualize these commercial tendencies? In what manner is the market-expansion, both quantitatively and in terms of their networks, an indicator of the nature of commercialization? Since coercive methods were also used to acquire control over markets and labour, did they militate against a structural integration of production with the market? Lastly, is it at all possible to assign a typological label to this process?

An interesting suggestion made by S. Bhattacharya is that of 'the domination effect'.[15] Bhattacharya uses this concept to study the unfolding of colonial control over the textile industry in the late eighteenth century. The three elements necessary for this 'effect' to become operational are: (1) the position of a pre-emptive buyer, (2) 'extra-economic' compulsions, especially that of power and the sanction of law to control producers, and (3) and the ability to displace or marginalize competition from indigenous capital. Therefore a combination of market and non-market 'operations' and 'institutions' is necessary for this 'effect' to control production. In the context of colonial control, 'domination', especially its third component, was vital for the furtherance of metropolitan trade, and it is in this context that the concept of 'domination effect' would be applicable most effectively.

It is true that domination was a vital constitutive element of Bengal's agricultural production, and that it had permeated into every corner of rural society in this period. It is also true that domination in rural Bengal entailed a combination of economic and non-economic methods of coercion. Yet, it would be difficult to apply the concept of 'domination effect', as developed by Bhattacharya, into rural Bengal because domination here did not entail the displacement of 'native' capital. On the contrary, the use of a variety of market and non-market strategies to ensure steady returns on investments and to acquire a grip over producers were devices shaped by merchants to cover the risks of investing among peasants with practically no resources or reserves, and in an economy often destabilized by nature calamities.

Yet, the entire gamut of mercantile participation in agricultural production cannot be explained away in terms of hedging their investments against increasing uncertainties. Pressing as these circumstances were, mercantile calculations were equally influenced by the facts that there existed a palpable demand for food in the province, and especially in the countryside, at prices which were deemed profitable for a long-term investment in what had become an integrated regional market.

The agrarian situation in the eighteenth century was characterized by clusters of small-commercial regimes—formations in which the producers' surplus was appropriated by overwhelmingly 'mercantile' mechanisms of interest-bearing capital without introducing any changes in the existing labour process or land use. It would, nevertheless be foolhardy to dismiss such regimes as 'unproductive' on grounds that they operated on an elaborate system of loans given at usurious rates of interest. Quite often such advances were the only major source of rural credit available to the petty producer and the high interest rates which accompanied them were, to use Bhardwaj's words, a system of 'interlinked exploitation', denoting not only the returns on cash lent but also the 'hypothecation of produce and labour services' of borrowers with very little security to offer.[16] The province's rice producing economy was subsumed under indigenous merchant-capital, thereby bringing about a silent transformation of petty production in agriculture.

NOTES AND REFERENCES

1. 'Agricultural Commercialization in Eastern India during British Rule: A Reconsideration of the Notions of "Forced Commercialization" and "Dependent Peasantry", in Peter Robb (ed.), *Meanings of Agriculture,* pp. 35-70.
2. Developed for instance by Amit Bhaduri, 'Class Relations and Commercialization in Indian Agriculture', in K.N. Raj et al. (eds.), *Essays on the Commercialization of Indian Agriculture,* Delhi, 1985.
3. See, 'Abstract of an Examination of Several Grain Merchants of Calcutta' by John Shore, in IOR, BRC, P/51/17, 1 February 1788; also see, S. Islam (ed.), *Bangladesh District Records,* vol. 1, 5 October 1784, p. 97.
4. WBSA, Grain, vol. 1, 31 October 1794; ibid., 17 October 1794.
5. See IOR, HM, vol. 217, pp. 44-9; IOR, BRC, P/49/38, 19 February 1773 and 23 March 1773.
6. See WBSA, Grain, vol. 1, 30 October 1794, 15 November 1794, 18 November 1794.
7. IOR, BRP, P/71/10, 1 June 1789; ibid., P/72/10, 11 November 1792; F. Buchanan, 'Survey of Ronggoppur', IOL, Ms. Eur. D. 75, vol. 2, book 2, fol. 16.
8. *The Peasant Economy and Social Change in Northern China,* Stanford, 1985.
9. Sugata Bose, *Peasant Labour and Colonial Capital,* pp. 42-4, quotation from p. 44.
10. Rajat Datta, 'Rural Bengal', pp. 254-5; also Datta, 'Merchants, Markets and Subsistence Crises', pp. 88-9 for the Company's failure to control internal trade.
11. H.T. Colebrooke, *Remarks,* p. 84.
12. IOR, BRP, P/72/10, 15 June 1789.
13. Peter Robb, 'Peasants' choices? Indian agriculture, and the limits of commercialization in nineteenth century Bihar', *Economic History Review,* xlv, 1 (1992); and also see Peter Robb, 'Variant Meanings of Agriculture: Implications for Research and Policy', in Peter Robb (ed.), *Meanings of Agriculture,* pp. 9-10.
14. B.B. Chaudhuri, 'Rural Power Structure', pp. 143-4.
15. S. Bhattacharya, in *CEHI 2,* pp. 287-8. Bhattacharya derives this model from Francois Peroux, 'The Domination Effect and Modern Economic Theory', *Social Research,* XVII, 1950.
16. Krishna Bhardwaj, 'A Note on Commercialization in Agriculture', in K.N. Raj et al. (eds.), *Essays on the Commercialization of Indian Agriculture,* p. 343.

APPENDIX 1

The Company's Tax Burden and the Economy

Table 63 provides the data of assessed revenue (*jama*) of Bengal for various years between 1700 and 1790. It also constructs an index of that *jama* with the year 1755 as the base. Year 1755 has been chosen as it marks the closing years, at least formally, of the Nizamat and for the fact that the *jama* is said to have been inflated by conjectural additions of *abwab* and *mathot* (cesses and imposts) which, for James Grant and John Shore, were the salient features of the Nizamat's revenue administration after Murshid Quli Khan.[1]

TABLE 63. BENGAL'S REVENUE ASSESSMENT, 1700-90

Year	*Jama* (Rs. in million)	Index (1755=100)
1	2	3
1700	11.81	71.056
1701	12.05	72.503
1702	12.50	75.210
1703	12.54	75.451
1704	12.65	76.113
1705	12.67	76.233
1706	12.67	76.233
1707	12.68	76.294
1708	12.68	76.294
1709	12.68	76.294
1710	12.68	76.294
1711	13.4	80.626
1712	13.43	80.806
1713	13.57	81.649
1714	13.57	81.649

1	2	3
1715	13.88	88.514
1716	13.94	83.875
1717	14.03	84.416
1718	14.03	84.416
1719	14.03	84.416
1720	14.09	84.777
1721	14.11	84.897
1722	14.29	85.981
1728	14.24	85.679
1755	16.62	100.000
1763	20.47	123.165
1764	17.70	106.498
1765	17.67	106.318
1766	16.03	96.450
1767	16.65	100.180
1768	22.16	133.333
1769	22.15	133.273
1770	22.57	135.800
1771	22.57	135.800
1772	22.49	135.319
1773	22.78	137.064
1774	22.49	141.336
1775	23.07	138.809
1776	22.85	137.485
1777	22.60	135.981
1778	25.76	154.994
1779	25.38	152.708
1780	25.26	151.986
1781	25.51	153.489
1782	27.90	167.870
1783	28.02	168.592
1784	27.27	164.079
1790	25.93	156.017

Sources: IOR, BRC, P/51/18, 2 April 1788, Appendix no. 7; IOR, HM, vol. 122, p. 753; Verelst Papers, IOL, Ms. Eur. F. 218/15, fols. 30, 31; Sinha, *Economic History*, vol. 2, p. 105; BM. Add. Ms. 29205, fol. 258.

These figures show that though there was an increase in the assessment revenue of the province after 1765, the increase was comparatively moderate considering the fact that the *jama* did not even double between 1755 and 1790. A modest increase in the assessed revenue surely vindicates James Grant's argument that Bengal continued to be under-assessed both under the Nizamat and under the Company's administration.

What about actual collections (*hasil*)? Figure 18, showing *jama* and *hasil* figures between 1766 and 1784 becomes significant. What strikes in Figure 18 is the steadily rising disparity between assessment and actual collections despite the consistent increase in both. Whether the Company's collections were of an unprecedented nature would depend on the state of collections under the Nizamat, of which we know very little. Alivardi Khan was credited with collecting nearly Rs. 15 million from his assessment of Rs. 16.62 million in 1755. In other words he succeeded in collecting 90.25 per cent of the gross assessment. Mir Qasim raised the *jama* to nearly Rs. 20.47 million in 1763 (Table 63). The collections he made, were substantial, though considerably less than those of Alivardi Khan and also somewhat uneven. These can be seen from figures of his collections from the districts of Dinajpur and Rangpur:[3]

District	Jama	Hasil	% Collected
Dinajpur	29,10,885	1,8,22,156	62.56
Rangpur	11,29,324	6,68,692	59.21

Therefore, despite all the efforts in that direction, the Company was unable to break through the typical Mughal revenue administration's bottleneck, that of the continuing discrepancy between *jama* and *hasil* and the problems arising from that.

Nevertheless, there were some major differences between the Company's revenue administration and that of its predecessor. For instance, under the Nizamat, the abiding principle of revenue administration was to adhere to the *assul jama bandobust* (original assessment) established by Akbar. Subsequent increases were made from time to time by *abwab* and *mathot* (additional imposts) which were then added to this *assul* to provide new *jama* figures.[4] The Company's administration was different because, not only did it retain the previous policy of using *abwab* and *mathot* to revise assessments, it introduced the previously unheard policy of taxing districts in relation to their actual productive resources. The detailed investigation of the 'Aumeen Commission' in 1778 bears adequate testimony of the Company's desire to get at the productive capacities of different districts. The papers of that Commission conclusively show that after 1765, the Company's revenue increments were based on four sources:

Talaash Beshee: Increases made by searching for secreted land.
Beroopher Beshee: Increases made from identifying lands sown with high quality crops.
Comar Utpan Beshee: Increases made from grain producing lands.
Abwab and Mathot: Additional taxes and imposts levied on the *assul*.³

Nevertheless, loopholes remained. Charity grants continued to pose tremendous problems for the administrators throughout this period, but the central thrust of the Company's policies made it a qualitatively different regime than its predecessor.

These considerations are vital for understanding the economic role of the Company and to contextualize Bengal's revenue history in this period. I have already drawn attention to the state's intervention in the market as a device to free it of its major internal restrictions imposed by zamindari control. The available evidence indicates that the steps taken to this effect were substantially forceful and met with very little *effective* opposition from the local landed proprietors. A proliferation in marketplaces was its most immediate outcome. The increase in their numbers, or their establishment in previously deficient areas enabled the peasantry to relate more easily to wider commercial networks and facilitated the creation of a pervasive gamut of mercantile functions. It, therefore, appears that a greater degree of appropriation was accompanied by increasing the facilities for circulation of commodities.

Also, the Company's rule occurred at a time when the collection of revenue in cash had already become a 'general' phenomenon and it took this process a step further. This meant that nearly 40 per cent of Bengal's agricultural output taken as revenue⁶ was annually converted into cash. The chief indicator of the connection between revenue and cash is the influence of agricultural prices (Chapter 1) on the province's peasant economy. The state also persistently refused all suggestions to revert to revenue in kind during periods of price slumps in order to reduce the pressure on the peasants,⁷ which implies a major departure from one of the abiding principles of Mughal land-revenue administration. Thus by 1769, even sharecroppers in an extensive zamindari of Burdwan were being made to sell the crop and then pay the zamindar in *cash*, a process which seems to have intensified in subsequent years.⁸

The other important aspect of revenue was the manner of its assessment and collection. Irfan Habib has shown the existence of a fairly extensive system of collecting revenue in cash in pre-colonial India; but revenue-in-cash was *one* form in which the surplus was appropriated.

Myriad forms of non-monetized appropriation (like *batai, kankut, ghalla-bakshi*) existed along with the Mughal *zabt*, and the state allowed the cultivator the option of paying revenue in kind even in the *zabti* regions.[9] There was, undoubtedly, a fairly extensive cash-nexus in Medieval India, but a multitude of non-cash forms of taxation indicate that such a nexus was perhaps *unevenly developed* in the major portion of the empire. For instance Bengal was not a *zabt* area during Mughal rule.[10]

The distinctive feature of revenue in the eighteenth century is that there was a continuous tendency to appropriate revenue *only* in cash. While various types of crop-sharing persisted in Bihar,[11] the Nawabs of Bengal collected their revenue in cash, which meant that the portion of the crop earmarked as revenue was sold in the *bazaar* by the producers at prevailing prices. This is borne out by a number of later inquiries made by the Company on the revenue administration during the reign of the 'soubahs' (Nawabs) of Bengal. According to the Board of Revenue's estimate (made in 1788), the total amount of silver currency in the economy during the first half of the century amounted to more than Rs. 12 crore,[12] of which Rs. 6 crore seems to have been the money in annual circulation (each *sicca* rupee 'cannot be issued and received more than once in two years').[13] This estimate and the *jama* figures available suggest that the gross assessment levied by Murshid Quli Khan (in 1722) of Rs. 14.29 million[14] amounted to 23.82 per cent of the circulating medium, whereas Alivardi Khan's *jama* (in 1755) of Rs. 16.62 million (Table 63) represented 27.7 per cent of the currency in circulation. This circulating amount had increased substantially during the later years. The amount in annual circulation in 13 districts in the early years of the nineteenth century was firmly estimated at Rs. 5.74 crore.[15] Thus the fact that there was hard currency available facilitated the collection of revenue in cash. It is, therefore, no accident that the silver *sicca* had become the 'general currency' of Bengal, which was '13 oz. 15 devts or 13 devts better than the English standard', and whose purity was maintained by 'recoinage after every three years'; it is also not surprising that this *sicca* had 'become the standard of weight throughout Bengal; the seer is composed of so many siccas, and the maund at forty seers'.[16]

The connection between revenue and commercialized agricultural production can be seen from the connection between revenue, crops and agricultural seasons. Peasants in Bengal paid revenue in three instalments (*kists*) coinciding with the three harvest seasons: *aman* (winter or *kharif*), *aus* (spring or *rabi*) and *boro* (an intermediate harvest). In general, *aman* rice was commercially the most valuable, being produced mainly for sale. 'The Khureef [*kharif*] crop is undoubtedly more productive than that of Rubee' were the words used by the Board of Revenue to describe the value of the former harvest in Bengal.[17] It was this harvest which determined the major amount of revenue collected because most of the important 'cash' crops like cotton, mulberry, tobacco and sugarcane were either sown or harvested during this principal agricultural season.[18] *Aus* rice was relatively inferior and was essentially consumed by the 'lowest and poorer classes of people'. *Boro* produced the coarsest quality rice and was grown in 'low marshy ground after the waters have subsided'.[19] The proportion which each bore to revenue in different districts is outlined in Figure 19 which is based on evidence available from the districts of Dhaka, Rangpur, Jessore and Midnapur.

The connection between revenue and commercial agriculture as shown in Figure 19 is perhaps self-explanatory.

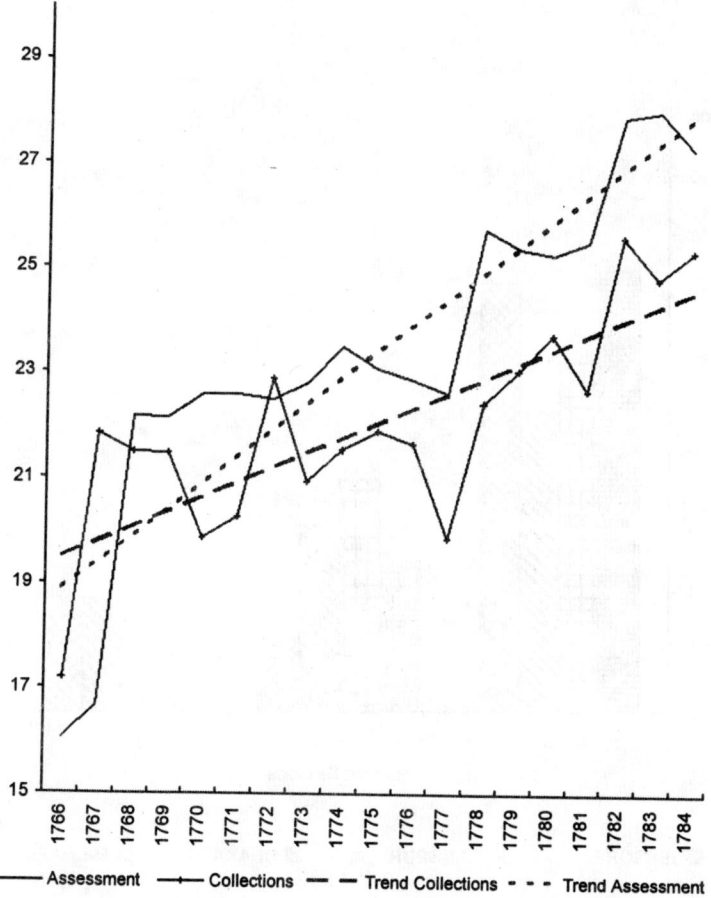

Sources: IOR, BRC, P/51/18, 2 April 1788, Appendix no. 7; IOR, HM, vol. 122, p. 753; Verelst Papers, IOL, Ms. Eur, F. 218/15, fols. 30, 31; Sinha, *Economic History*, vol. 2, p. 105; BM, Add. Ms. 29205, f. 258.

FIGURE 18. BENGAL'S REVENUE, 1766-84 (*Rs. in 10 thousand*).

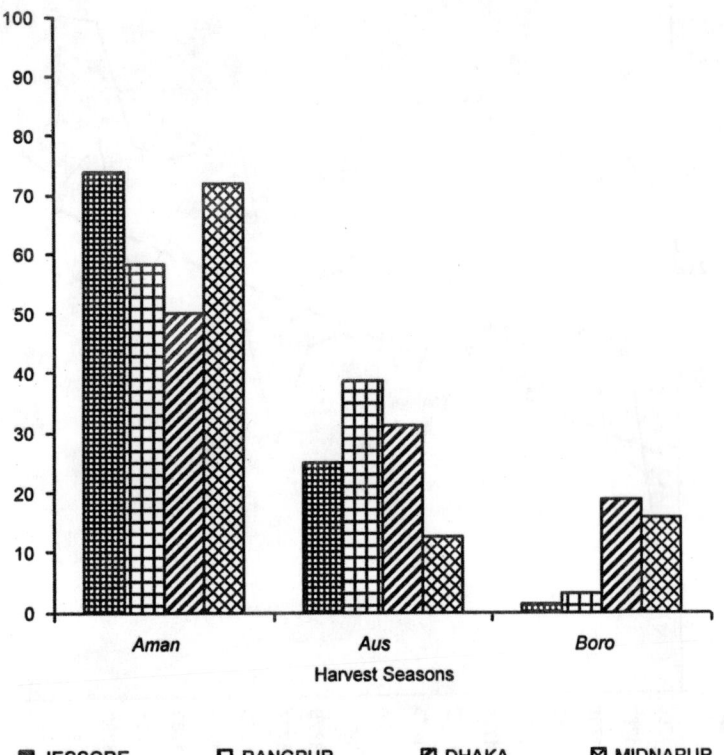

Sources: Jessore: IOR, BRC, P/51/13, part 2, 10 February 1788; Rangpur: ibid., P/71/27, 8 May 1787; Dhaka: ibid., P/70/38, 4 February 1788; Midnapur: ibid., P/70/37, 11 February 1787.

FIGURE 19. REVENUE AND HARVESTS AS PER CENT OF *JAMA*.

NOTES AND REFERENCES

1. See the Minutes by James Grant and John Shore in *FR 2*.
2. Compare, IOL, Verelst Manuscripts, Ms. Eur. F. 218/15, fol. 30 and ibid., F. 218/9, fol. 23.
3. IOR, SCC, P/A/9,13 September 1769; Sinha, *Economic History*, vol. 2, p. 29.
4. IOR, HM, vol. 122, pp. 771-3; IOL, Ms. Eur. F. 218/15, fols. 30-1.
5. See BM, Add. Mss. 28086 to 29088, esp. 29088 for figures.
6. John Shore's Minute of 18 June 1789, in *FR 2*, p. 192.
7. See the suggestion of George Bogle to collect revenue in grain during the massive price slump of 1771 and 1772 in D.C. Ganguly, *Select Documents of the British Period of Indian History*, Calcutta, 1958, p. 90.
8. Compare, IOR, SCC, P/A/9, 16 August 1769 and IOR, BRP, P/72/4, 6 July 1792.
9. Irfan Habib, *Agrarian System of Mughal India*, Bombay, 1963.
10. See the information on Mughal and Nizamat land-revenue assessment in Bengal in IOR, BRC, P/49/70, 4 April 1777. For information on how these worked in different districts, IOR, BRP, P/70/16, 7 June 1786 for Birbhum; ibid., P/70/37, 8 May 1787 for Rangpur and IOR, BRC, P/51/22, 23 July 1788 for Jessore.
11. IOR, BRC, P/52/40, 13 January 1792.
12. IOR, BRP, P/70/40, 25 April 1788.
13. Ibid., 18 April 1788.
14. IOR, BRC, P/51/18, 2 April 1788.
15. Figures in D.B. Mitra, *Monetary System in the Bengal Presidency, 1757-1813*, Calcutta, 1991, pp. 76-8.
16. John Shore's Minute on the Coinage of Bengal, 28 October 1789, in IOR, BRC, P/51/50, 28 October 1789.
17. IOR, BRP, P/70/27, 8 May 1787.
18. See Chapter 1.
19. WBSA, CCRM, vol. 5, 30 April 1771; IOR, BRP, P/70/38, 4 February 1788; WBSA, Grain, vol. 1, 17 October 1794.

APPENDIX 2

The Problem of Bullion and the English East India Company in Bengal

Some historians see a vital connection between the economies of precolonial Asia and the injections into them of regular supplies of bullion, particularly silver, during much of the seventeenth and early eighteenth centuries. In arguments heavily tinged with idioms from the 'price-revolution' in Europe, the flow of bullion into India (and into China) is said to have engendered a sustained rise in prices during this period.[1] The supposed cessation of bullion after 1765, and the reversal of its flow, is seen as a deflator not only of prices but also of the short-lived economic vibrancy of the preceding century or so.[2]

A discussion of the problem of bullion in this period must address three issues. First, there is the still unresolved problem regarding the component of bullion in the Company's shipments to Asia and the pattern of its supposed contraction, especially in Bengal after 1765. Second, the share of Bengal in the bullion being shipped to Asia has to be calculated, if not for the entire period (for there is a constraint of evidence) then for a good number of years in order to get a representative trend. Third, the effects of the trend have to be analysed by looking at the behaviour of prices, assessments of revenue and fluctuations in the rates of exchange (*batta*) between the different currencies in circulation. The latter is a particularly good exercise for Bengal's currency was silver-based and fluctuations in the amount and quality of silver were bound to be reflected in the rates of *batta*.[3]

That the Company bought goods from Asia by paying for them in silver, and that this was a cause of serious heartburn in Britain are facts which are commonplace. There was, however, another component in the outgoing cargoes from England and these were the 'goods & stores'—items like woollens, copper, iron and lead—which they endeavoured to sell abroad for, apart from reasons of trade, the Company was bound by charter to export British manufactures worth at least a tenth of its trading capital every trading season. A comparison of the relative shares occupied by these two components of exports between 1732-3 and 1804-5 are established in Figure 20.

Figure 20 shows that there was clearly a marked shift in the pattern of Company investments in Asia from the middle of the eighteenth century. Though treasure continued to arrive for trade, there was a noticeable attempt to increase the merchandise component of this trade. This was to be the dominant feature till 1784 when exports of treasure began to pick up once again. In 1792-3 the value of bullion exported to India and China was £ 3,69,033 while in 1793-4 it slid to an abysmal £ 10,290. On the other hand, 'goods & stores' exported amounted to £ 12,55,746 and £ 11,37,538 in those two years.[5] *A reduction in the flow of bullion had started prior to the acquisition of the Diwani in 1765*, and this fact modifies notions of the sudden stoppage of bullion for trade after acquisiton of the Diwani in 1765.[6] Nevertheless, bullion on private accounts continued to arrive in Bengal in substantial quantities. The total values of bullion and merchandise imported by private traders from London in the 10 trading seasons between 1796-7 and 1805-6 were *sicca* Rs. 4,57,84,281 and *sicca* Rs. 3,01,58,339 respectively (see Table 66).

The details of the Company's imports in these years are not available, but the impression one gathers is one of relative restriction rather than a cessation of bullion as a component of its trade. Bullion comprised £ 33,94,093 (37.48%) of the total imports of £ 90,57,203 made by the Company into Bengal between 1800-1 and 1804-5.[7] The other aspect of the problem of bullion in Bengal's economy is the regional distribution of investments in treasure and the share of Bengal.

With regard to the share of this bullion actually received by the province of Bengal, there are two difficulties in reconstructing the regional profile. First, the figures available pertain to exports of bullion to Asia and they do not always mention Bengal's share. Second, the evidence is not continuous. Nevertheless, the data available for a number of years between 1707-8 and 1804-5 are given in Table 64. The geographical distribution of the Company's bullion in India is provided in Table 65.

Another source from which bullion continued to arrive in Bengal in substantial quantities towards the end of our period was private trade which received a substantial boost after the passing of the Commutation Act of 1784. Table 66 gives the figures in the 10 trading seasons between 1796-7 and 1805-6.

The evidence regarding flows of bullion indicates quite emphatically

TABLE 64. BENGAL'S SHARE OF COMPANY'S EXPORTS OF BULLION FOR TRADE WITH ASIA (in £ sterling)

Season	Bullion into Asia	Bullion into Bengal	% of Total
1	2	3	4
1707-8	3,71,911	70,153	18.86
1708-9	5,52,152	41,856	9.58
1709-10	5,08,905	47,475	9.28
1710-11	5,27,832	46,044	8.72
1711-12	4,66,274	42,704	9.16
1712-13	3,00,418	65,044	21.65
1713-14	3,34,318	57,126	17.09
1714-15	4,20,351	1,20,917	28.76
1715-16	3,98,662	1,39,520	34.99
1716-17	5,20,089	1,40,441	27
1717-18	6,03,762	2,12,359	35.17
1718-19	6,11,283	71,319	11.67
1719-20	6,90,057	61,524	8.92
1721-32	No Information*	No Information*	
1732-3	3,93,377	1,27,528	32.42
1734-5	4,90,992	1,41,918	28.9
1735-6	4,82,985	1,56,270	32.35
1737-8	4,92,270	1,03,314	15.06
1738-9	4,74,525	1,09,472	20.97
1739-40	4,27,901	1,56,669	23.07
1740-1	4,84,927	1,43,328	36.61
1741-2	4,37,550	1,67,475	29.56
1742-3	5,80,878	93,796	24.56
1744-5	4,58,544	1,12,342	16.15
1745-6	4,76,853	89,340	18.73
1746-7	5,60,020	1,88,362	33.63
1747-8	3,37,289	1,33,026	39.44
1748-9	6,68,719	2,73,372	40.93
1749-50	6,20,305	3,68,528	59.41
1750-1	9,36,182	3,31,269	35.38
1751-2	8,32,438	2,01,809	24.24
1752-3	9,44,254	1,71,384	18.15
1753-4	6,68,890	1,57,656	23.57
1754-5	6,20,376	40,820	8.58
1755-6	7,95,005	1,57,415	19.8
1757-97	No Information*	No Information*	
1797-8	7,19,235	5,18,415	72.08
1798-9	7,03,078	1,00,512	14.29
1799-1800	3,31,862	2,31,666	69.81
1800-1	5,81,370	2,24,254	38.57

APPENDIX 2

1	2	3	4
1801-2	11,23,717	10,05,549	39.48
1802-3	11,42,590	10,11,486	88.53
1803-4	17,23,263	11,31,219	65.64
1804-5	17,50,370	11,46,632	65.51

Note: *The gaps for the years 1721 to 1732 and 1757 to 1797 are because of the absence of comparative data of bullion imported into Bengal.

Sources: K.N. Chaudhuri, 'Treasure and Trade, 1660-1720', *Economic History Review*, vol. 21, no. 3, 1968, p. 497; *Reports, Third Report, 1773*, p. 75; *Reports, Ninth Report, 1783*, pp. 110-11; IOR, L/AG/10/2/2, fols. 178, 211-13; IOR, L/A/G/ 10/2/4, fol. 148; D.B. Mitra, *Monetary System in the Bengal Presidency, 1757-1813*, Calcutta, 1991, pp. 11-12.

TABLE 65. DISTRIBUTION OF COMPANY'S BULLION
IN INDIA (*in £ sterling*)

Year	Total for India	Bengal	Bombay	Madras
1747-8	2,28,027	1,33,016	95,011	
1748-9	4,49,077	2,73,733	1,75,344	
1749-50	6,06,054	3,68,526	85,510	
1750-1	7,27,957	3,31,269	1,13,793	2,82,895
1751-2	5,46,859	2,01,809	81,767	2,63,283
1752-3	6,48,665	1,71,384	77,026	4,00,255
1753-4	4,30,434	1,57,657	1,21,691	1,51,086
1754-5	4,59,039	40,820	1,08,854	3,09,365
1755-6	5,30,704	1,57,415	1,80,735	1,92,554
1756-7	1,88,045	35,423	1,52,622	
1797-8	7,19,235	5,18,415	418,38	1,58,982
1798-9	7,03,078	5,02,054	1,00,512	1,00,512
1799-1800	3,31,862	2,31,666	1,00,196	
1800-1	5,81,370	2,24,254	1,30,713	2,26,403
1801-2	11,23,717	10,05,549	1,18,168	
1802-3	11,42,590	10,11,486	1,00,974	30,130
1803-4	17,23,263	11,31,219		
1804-5	17,50,370	11,46,632	2,00,793	4,02,945

Sources: *Reports, Ninth Report, 1783*, pp. 110-11; IOR, L/AG/10/2/4, p. 148.

that Bengal received more of the Company's bullion during the closing years of the eighteenth century than it had ever acquired before. This influx can partly be explained by the rising costs of maintaining the Company's military and civil establishments over the course of the

TABLE 66. MERCHANDISE AND BULLION IMPORTED INTO BENGAL BY PRIVATE TRADERS, 1796-7 TO 1806-7 (*in sicca and £ sterling*)

Year	Merchandise (*in siccas*)	£ sterling	Bullion (*in sicca*)	£ sterling
1796-7	64,77,819	7,36,115.7	48,71,052	5,53,528.6
1797-8	53,47,032	6,07,617.2	27,27,729	3,09,969:2
1798-9	39,17,779	4,45,202.1	29,13,594	3,31,090.2
1799-1800	61,12,355	6,94,585.7	33,82,063	3,84,325.3
1800-1	42,34,493	4,81,192.3	47,32,088	5,37,737.2
1801-2	2,49,507	28,353.06	42,63,133	4,84,446.9
1802-3	4,31,136	48,992.72	34,92,181	3,96,838.7
1803-4	10,89,904	1,23,852.7	49,78,806	5,65,773.4
1804-5	16,07,642	1,82,686.5	92,85,328	10,55,150
1805-6	6,90,672	78,485.45	51,38,307	5,83,898.5
Total	301,58,339	34,27,083.43	4,57,84,281	52,02,758.0
Average	30,15,833		45,78,428	

Sources: IOR, HM, vol. 68, pp. 898, 901; IOL, Ms. Eur. G. 33, fol. 139.

century.[8] The conclusion of the wars in Mysore which saw specie moving from Madras back into Bengal because of the payment of war indemnities to the Company[9] and the massive surpluses of Bengal's trade with China and London[10] were the other reasons for this influx on Company account at the turn of the century. Bullion was also necessary for the rapidly increasing investments for Bengal goods: £ 1,38,000 in 1757-8, £ 10,36,667 in 1797-8, rising to £ 16,19,650 in 1812-13.[11] The specie imported by private traders was entirely consumed by Bengal's economy in the form of investments.[12]

While it is difficult to demonstrate the differential apportionment of this bullion between the many sectors of the Company's interests in Bengal, there is one aspect which emerges quite clearly and significantly. Since Bengal received bullion exclusively in silver,[13] and since the major currency of transaction, the *sicca* rupee, was also silver-based, the increasing arrival of bullion on Company and private accounts largely stabilised the exchange rate (*batta*) of the *sicca* in relation to the gold *mohur*, and this increased the amount of silver currency in circulation as can be seen from Figure 21 which shows the relative shares (in percentage) of gold and silver in the revenue received from Bengal between 1796-7 and 1808-9.

If the evidence, as it exists, points to an increase in the quantities of bullion reaching Bengal during the close of the eighteenth and the early years of the nineteenth centuries, it still leaves one unresolved issue. This is of a monetary nature and has an important position in the historiography of the early colonization of Bengal's economy. This pertains to the complaints of bullion 'famines' made by observers in the second half of the eighteenth century which is seen by some modern historians as evidence of the disruption of the established patterns of trade (i.e. bullion for commodities) which was established during the seventeenth and early eighteenth centuries.

The evidence of bullion would give us very strong reasons to believe that there was more cash available in Bengal in the closing years of the eighteenth century than ever before. Strong administrative interventions to free markets from various local constraints were designed to facilitate the circulation of both cash and commodities. In other words, the institutions of exchange were widened in Bengal in this period, and this was crucial in sustaining a commercial economy of an extremely high order.

The Problem of the Disappearance of Bullion after 1765: Some Tentative Explanations

The disappearance of silver from the Bengal market for varying lengths of time was not a sudden development of the post-1765 period alone. In 1722 the Company's silk investments in Qasimbazar came to a grinding halt for 'their having no money & none being procurable in the place'. Even the provincial financial system was stymied at this acute shortage, and serious delays in the payment of tribute to the Mughal emperor resulted in the process. Thus, 'Jaffer [Murshid Quli] Cawn is under very great affliction that money is so scarce, he having had several positive orders to send away the Bengal treasure, which is upwards of 35 laack [3.5 million] rupees behind . . .'.[14] A similar bullion panic surfaced in 1729 when even the 'minimum 20 chests to supply the Patna factory' were not available.[15] In both instances the Company bailed itself by borrowing heavily from the Dutch in 1722 and from the French in 1729.

Yet, there was a general feeling that there was an absolute decline in the monetary position of Bengal in the second half of the century. Thus, for James Steuart writing in 1772: 'the complaints of a scarcity of coin in Bengal, once so famous for its wealth, are so general that the fact can hardly be called into question'.¹⁶ A recent analysis of banking in early colonial Bengal also speaks of the 'economic distress caused by the shortage of currency as a result of reduced import of silver bullion' after 1765.¹⁷

How realistic were such assessments? Given the fact that there was hardly any attempt made to calculate the quantity of currency in circulation in different periods of the eighteenth century, all statements by contemporary observers and officials about shortages were impressionistic; at best they were inspired guesses made by people extremely worried about the prospects of finding adequate funds for increasing commercial investments in the province from the middle of the century.

Nevertheless, the facts that Company investment for Bengal goods had increased substantially after the grant of the Diwani in 1765, and that there was a substantial reduction and errancy in the export of treasure for trade with Asia between 1761 and 1769 are indisputable.¹⁸ These would imply that the territorial revenues of Bengal ceded to the Company after 1765 were not proving adequate in meeting the civil, military and commercial expenditures of the Company, leading thereby, to the locking up of silver (the only bullion acceptable for commercial transactions in the province) in the economy, and this was particularly troublesome for the Company's investments. The problem, therefore, was caused by demand for goods exceeding the supply of cash. This difficulty came to a head in 1768 when the bloated demand, evident from the statement that 'your [the Company's] demands, and those of other Nations are beyond what the country is able to supply', sat heavily on a constricted supply side whose implications were stated as follows: 'Gold is not current at the Aurungs [manufactories], and we shall with Difficulty be enabled to raise a sufficient quantity of Silver for the Provision of the ensuing year's Investment'.¹⁹

Since Bengal's currency followed a silver standard, and the standard unit of transaction was the silver *sicca* rupee, it was therefore naturally sensitive to the minutest ebb and flow of bullion. Given this fact, the fluctuations in the rates of *batta* (shroffage) between the *sicca* and other currencies would also indicate the variations in the quantities of bullion (especially silver) in circulation in the economy. A sudden or sharp increase in the *batta* of the *sicca* rupee relative to other currencies, would indicate a shortage of silver. On the other hand, a certain degree of steadiness in both would clearly indicate an economy with an adequate, if not abundant, supply of silver in circulation.

Table 67 gives the monthly rates of *batta* on the Arcot rupee from

APPENDIX 2

TABLE 67. MONTHLY RATES OF *BATTA* ON ARCOT RUPEES
IN THE *BAZAARS* OF DHAKA, 1769-73

	For *sicca* Rs. (%)	For *sonaut* Rs. (%)
September 1769	7.19	5.31
October	7.63	5.17
November	7.49	4.08
December	6.45	2.53
January 1770	3.9	1.08
February	3.41	1.28
March	5.2	1.1
April	3.2	1.56
May	2.06	—
June	3.4	—
July	4.96	—
August	7.8	3.2
September	7.8	3.2
May 1771	4.36	3.72
June	7.46	1.12
July	7.6	2.56
August-September	6.9	2.88
October	7.38	3.06
November	7.31	3.37
December	9.04	3.48
January 1772	10.56	5.2
February	10.58	4.56
March	7.35	2.35
April	—	—
May	7.64	2.45
June	8.07	3.06
July	8.72	3.84
August	8.37	3.84
September	8.22	3.84
October	9.06	4.98
November	9.53	5.39
December	9.07	4.16
January 1773	10.15	4.58
February	9.95	5.3
March	9.36	4.24
April	6.59	1.86
May	7.6	2.22
June	7.6	1.46

Source: WBSA, Proceedings of the Provincial Council of Revenue, Dacca, vol. 6, 18 May 1775.

south India, *vis-a-vis* the *sicca* and *sonaut* rupees, in the markets of Dhaka between 1769 and 1773.

Notions of a great scarcity of silver and silver currency are difficult to sustain in a situation where the *batta* on *sicca* rupees averaged 7.06 per cent in the 38 documented months between 1769 and 1773, and the average *batta* on *sonaut* rupees for 35 documented months was 3.2 per cent. Given the fact that the *batta* charged on *sicca* rupees was 12.5 per cent during the silver 'crisis' of 1729 (cited earlier),[20] the rates in Dacca appear strikingly lower by comparison.

Yet, Company officials often complained about a shortage of specie affecting investments. Bullion was imported, but quite obviously these imports were not commensurate with the increasing requirements for investment (discussed earlier). There was, henceforth, greater pressure on the territorial revenues after 1765 since the money received as revenue was locked up for varying lengths of time in investment for goods, leading thereby temporary cycles of monetary dislocation. As Verelst wrote in 1769: 'the whole amount of the Lands [Territorial Revenue] is swallowed up in one gulp . . . nor does any part of it return into the circulation, except the sum issued for our investment, so that there ensues an annual loss to the currency equal to the difference between the aggregate of the investments and disbursements and the total of the Revenues. This, if continued, must in time draw in all specie.'[21]

The main reason for the temporary disappearance of money from the market during the initial years of the Company's direct administration of Bengal's finances was the gap between financial requirements and the revenue surpluses of Bengal brought about by the abysmal failure of the new state to effect a viable system of fiscal administration. In these years investments were often out of tune with production cycles and these resulted in temporary, though severe, disruptions in its supply, while the needs of war-finance and administrative overheads resulted in the diversion of resources from one head of account to another in order to meet short-term expediencies.[22] Moreover, the initial attempts by Clive and Verelst to enforce a uniform currency based on a gold standard in place of the one established on silver, which undervalued silver by 17.5 per cent (under Clive) and 5.5 per cent (under Verelst), had resulted in the disappearance of silver from the market in 1766 and in 1769.[23] Therefore, the temporary disruptions in Bengal's money market were the direct results of the Company's interference in the established monetary system of the province and do not reflect an absolute reduction or shortage of currency.

The Reversed Flow of Bullion after 1765 and the China Connection: A Re-examination

The supposed reversal in the direction of bullion after 1765 is unconvincing as an argument for it is based on inadequate and questionable evidence. Both Warren Hastings and John Shore, who hinted at these reverse flows, were surprisingly uncertain about the quantities involved and of the sources from which such conclusions could be drawn.[24] Moreover, the various reports commissioned by the select committees of the British parliament on Asian affairs, particularly those submitted in 1772-3 (the *Third Report*) and 1782-3 (the *Ninth Report*) make no mention of the export of bullion from Asia to Britain. The emphasis instead is on the procurement and export of merchandize and bills of exchange. The *Ninth Report* also makes a strong argument against silver as an item of import: 'to send silver into Europe would be to send it from the best to the worst market',[25] an argument which is perfectly compatible with changing British perceptions of the role of Asian trade as a means of acquiring raw materials for British manufactures and re-exports during the closing years of the eighteenth century and later.[26] In the detailed accounts and estimates that we have of the Company's trade after 1765, bullion is conspicuous by its absence as a major item of export from Bengal on Company account as can be seen from Table 68 pertaining to exports in the years 1802 and 1803.[27]

Another component of the reverse flow of bullion argument centres around the financing of the trade with China. The logic of this trade was summed up in 1768 in these words: 'The enlargement of the Trade to China to its utmost Extent, is an object we have greatly at heart, not only from the Advantages in prospect, by gaining a superiority and thereby discouraging Foreign Europeans from resorting to that Market; but also from a National Concern, wherein the Revenue is materially

TABLE 68. COMPANY'S EXPORTS FROM BENGAL,
1802 AND 1803 (*in sicca Rs.*)

	A Total Exports	B Merchandise	C Bullion	C as % of A
1802	68,62,147	68,18,897	43,250	0.63
1803	67,60,058	66,70,800	89,258	1.32

Source: IOL, Ms. Eur. G. 33, fol. 139.

invested therefore to prevent all disappointments from the want of a sufficient stock in China for providing Cargoes for the Ships now bound thither.'[28]

It is, therefore, true that Bengal's revenue was seen as an investment for China trade, and as early as 1772, James Steuart was quick to point to 'the specie carried out by the Company for the China market' amounting to £ 7,20,000 between 1769 and 1771 as 'one article of drain' from Bengal.[29] However, the actual siphonage of bullion was much less if we see this 'drain' from Bengal in relation to the inflow of bullion into Asia during the same period for which the figures are shown in Table 69.

These figures suggest that a total of £ 2,23,684 in silver may have moved from Bengal to China between 1768-9 and 1770-1, £ 4,96,316 less than Steuart's estimate of £ 7,20,000. In any case £ 2,23,684 was a mere 2.35 per cent of £ 95,22,389 which the Company earned in Bengal in those three years.[30]

However, by 1773, the Bengal Council had decided that the surplus of Bengal's revenue could not be used to finance trade with China and that monies to be remitted to Canton must emerge from private individuals.[31] Therefore, an acceptance of Steuart's estimate (which was no more than an inspired guess on his part) would obviously lead to an error of over-stressing the bullion component in the 'drain' from Bengal. Moreover, the Company's trade with China was not entirely bullion based. 'Goods and Stores' exported from England were not insignificant: £ 1,79,245 in 1768-9, £ 1,52,797 in 1769-70 and £ 1,47,082 in 1770-1;[32] and if these are added to the quantities of bullion sent directly from Britain into China, then the amount to be met by remittances of treasure from Bengal would have been of no major consequence.

TABLE 69. BULLION INTO AND FROM CHINA,
1768-9 TO 1770-1 (*in £ sterling*)

Seasons	Bullion imported	China investment	Deficit on China trade
1768-9	1,62,583	2,33,045	70,462
1769-70	2,42,998	2,93,210	50,212
1770-1	3,02,625	1,99,615	1,03,010
Total	7,08,206	7,25,870	2,23,684
Average	2,36,068.66	2,41,956.66	74,561.33

Sources: *Reports, Third Report, 1772-1773*, p. 75; IOR, L/AG/10/2/2, pp. 235-6.

TABLE 70. BALANCE OF THE COMPANY'S CHINA TRADE, 1761-2 TO 1770-1 (*in £ sterling*)

Seasons	Receipts	Disbursements	Surplus(+) or Deficit (-)
1761-2	2,37,801	2,71,629	- 33,828
1762-3	3,07,468	3,23,479	- 16,011
1763-4	3,93,797	4,05,974	- 12,177
1764-5	8,25,442	5,96,332	+2,29,110
1765-6	8,23,522	5,44,269	+2,79,253
1766-7	3,04,512	4,52,466	-1,47,954
1767-8	4,22,493	6,17,248	-1,94,755
1768-9	4,48,973	5,47,330	-98,357
1769-70	6,19,271	4,80,279	+1,38,992
1770-1	8,21,624	7,66,049	+ 55,575
Total	52,04,903	50,05,055	

Source: *Reports, Third Report, vol. 4, 1773*, p. 69.

Unfortunately, the evidence regarding the Bengal-China-bullion connection is so scanty that no firm calculations are possible, but it could not have been in the range of £ 2,40,000 a year, and this is obvious from the balance sheet of Company finances in China between 1761-2 and 1770-1 (see Table 70).

It is clear that the Company's overall surplus in China in the decade between 1761-2 and 1771-2 was £ 1,99,848, an enviable record indeed considering the fact that it had failed to sell 15 million lbs. of Chinese tea as late as 1767.[33] It is quite apparent that Bengal's money was not officially necessary in China trade, and by the time of the Commutation Act (1784) the Company had already began to depend on direct exports from Europe to Canton to finance the galloping trade in Chinese tea.[34]

Though the Company did not exhaust Bengal's revenue as a fund for its China investments, private traders used their profits from country trade and individuals their personal fortunes in this channel. Their modus operandi was either to ship merchandise (cotton, tin or opium) in return for tea or to invest in bills of exchange issued at Canton (for money received in India) on London.[35] This was undoubtedly the spout through which much wealth was drained from India. The total amount of such bills on all accounts (that is remittances and those drawn at Canton between 1757 and 1784) was £ 1,18,81,025[37] while another £ 57,92,643 worth of bills were issued between 1785 and 1796.[37] The implications of these figures for the problem of estimating the amount of wealth drained from Bengal will be discussed below.

TABLE 71. TRANSFER OF BULLION TO OTHER PRESIDENCIES, 1761-2 TO 1770-1

Year	Remittances of bullion to other presidencies (in £ sterling)	Remittances of bullion from other presidencies (in £ sterling)
1761-2	1,12,502	63,281
1762-3	56,250	Nil
1763-4	22,500	Nil
1764-5	1,16,437	Nil
1765-6	2,92,302	Nil
1766-7	2,93,445	Nil
1767-8	3,49,635	38,838
1768-9	3,08,575	88,791
1769-70	40,050	34,513
1770-1	Nil	Nil
Total	1,591,696	2,25,423

Source: Reports, Third Report, vol. 4, 1773, pp. 60-1.

Bullion from Bengal was used to meet deficiencies in other presidencies, and the figures we have for the years (Table 71) between 1761-2 and 1770-1 indicate that such transfers were indeed substantial.

The above figures show that there was an average net transfer of £ 1,51,841.4 worth of bullion per year from Bengal to meet exigencies in Madras and Bombay. However, it would be an error to take this as an example of drain of bullion from India since the purpose of these transfers was to meet demands in other Indian provinces and did not entail the physical removal of these resources from India. Additionally, whether the transfer of resources from Bengal to other parts of India was actually more burdensome after 1765 has to be seen in relation to the transfer of imperial tribute from this province to Delhi during the first half of the eighteenth century. The mechanisms of such transfers were described by James Steuart in these words: 'Before the existence of an internal Revenue, in favour of European Nations who traded in the commodities of the country [i.e. before 1765], an equivalent of silver was constantly brought from Europe for all the goods exported from Indostan.' But the silver 'coined or uncoined' which came into Bengal also tended to disappear frequently in the form of tribute to Delhi so that quite often there was 'hardly any currency left in Bengal to carry on trade or even go to market for provisions and necessaries of life, till the next shipping arrives to bring a fresh supply of silver'.

Since Steuart quite unequivocally believed that the grant of the Diwani had resulted in a 'drain' of wealth from Bengal,[38] the disappearance of silver in the pre-Diwani period must point to the internal drain from Bengal in the guise of the imperial tribute to Delhi, and evidence indicates that this was indeed substantial, averaging about Rs. 50,00,000 worth per year in the years between 1721 and 1758.[39] In other words, there was a siphonage of at least Rs.18,50,00,000 worth of bullion (about £ 2,10,00,000) in the 37 years between 1721 and 1758 from Bengal, for which the province received (as in the case of the Company's bills of exchange after 1765) no adequate recompense. Moreover, the sapping away of resources in the first half of the eighteenth century also entailed (along with the transfer through *hundis*, or indigenous bills of exchange)[40] the *physical* removal of bullion from Bengal. This is implicitly the impression one gets from Steuart's description cited earlier, and explicitly too for Murshid Quli Khan despatched the tribute from Bengal (stated to be Rs. 1,35,00,000 in his time) '*in bags*' which were 'put into two hundred carts, and escorted by a guard of 300 cavalry and 500 infantry'.[41] Of this, the amounts remitted through *hundis* were small. In 1719, remittance through *hundis* amounted to Rs. 2,50,000 whereas in 1720 it was a meagre Rs. 1,00,000.[42] In other words, north India was extracting about 98 per cent of Bengal's tribute in the form *khazana*, that is, precious metals (mainly silver), during the tenure of Murshid Quli Khan. After its acquisition of the Diwani the Company continued this tradition of sending tribute to the Emperor in Delhi though on a much reduced scale and for a very short period after which this practice was discontinued. Between 1765-6 and 1770-1, a total of £ 15,53,781 (roughly Rs. 1,36,73,272) was sent out of Bengal as imperial tribute,[43] just Rs. 1,73,272 more than what Murshid Quli Khan was reportedly *dispatching every year* to Delhi during the early years of his rule.

Therefore, the chief problem associated with the drain of wealth, that is, the unilateral and unrequited transfer of resources to meet the political and commercial requirements of the Company, which is seen as a specific feature only of the second half of the eighteenth century needs to be reformulated. At best, the two phases represent two stages in the utilization of Bengal's surpluses. The first stage (in the first half of the eighteenth century) represents the inter-regional transfer of wealth from Bengal while the second (the classic drain in the second half of the century) marked its internationalization through the complex circuits of trade and remittances made under the aegis of the Company and other private traders. Though it is indisputable that the second stage marks a qualitatively different, an early-colonial, mode of utilization of Bengal's surplus, there is very little ground to believe that its economic impact at this stage was anything more than marginal.

How much was actually taken out from Bengal during the second half of the eighteenth century? An estimate puts the net drain from Bengal between 1757 and 1780 at £ 3,80,00,000.[44] This gives us a siphonage of £ 1.58 million per year. This estimate is inflated for it includes two highly inflated and implausible figures as part of that drain (i) £ 23,58,298[46] as transfer of bullion to other presidencies between 1761-2 and 1770-1 (the correct figures are given in Table 71), and (ii) £ 24,00,000 in 'exports of silver to China' between 1766 and 1780.

The intention is not to deny the fact of the drain but to suggest that the data on which some estimates are based and accepted are extremely inflated and some of the indices used (like the supply of bullion to other presidencies and to China for most of this period) are unsustainable. Keeping these objections and qualifications in mind, the following figures are given in Table 72 as the estimates of the *probable* drain from Bengal for the 37 years between 1757-8 and 1793-4.

TABLE 72. ESTIMATES OF THE DRAIN FROM BENGAL, 1757-8 TO 1793-4

Period	Span	Total	Average	
1757-8 to 1765-6	9 years	£ 30,00,000	£ 3,33,333	(Rs. 29,33,333.33)
1766-7 to 1770-1	5 years	£ 47,05,935	£ 9,41,187	(Rs. 82,82,445.6)
1771-2 to 1778-9	8 years	£ 99,45,424	£ 12,43,178	(Rs. 1,09,39,966)
1779-80 to 1783-4	5 years	£ 56,82,915	£ 11,36,583	(Rs. 1,00,01,930)
1784-5 to 1787-8	4 years	£ 85,59,824	£ 21,39,956	(Rs. 1,88,31,612)
1788-9 to 1790-1	3 years	£ 35,10,675	£ 11,70,225	(Rs. 1,02,97,980)
1791-2 to 1793-4	3 years	£ 20,28,475	£ 6,76,158.33	(Rs. 54,09,266.6)
Total		£ 3,74,33,248		
Deduct		£ 2,25,423*		
Net Drain		£ 3,72,07,825		
Net Annual Average		£ 10,05,616.8		

Note: *On account of bullion received from other presidencies between 1761 and 1771, see Table 71.

Sources: Calculated from *Reports, Third Report, vol. 4, 1773*, pp. 311-12; P.J. Marshall, *East Indian Fortunes, passim* and esp. p. 255 for 1757 to 1766; IOL, Ms. Eur. D. 281, fol. 34 for 1767 to 1791; and IOR, L/AG/10/2/3, p. 210 for 1792 to 1794.

This revised estimate of about £ 1 million being drained from Bengal every year between 1757 and 1794 would still be a sizeable amount of unrequited extraction. Yet, the connection between the drain and the economic performance of Bengal remains problematic. Bagchi uses the data of the gross output in 1794, provided by Colebrooke,[46] in making his estimates about the extent of the drain in the late eighteenth century, but this exercise has an ingrained methodological defect, for it treats data for one year as a representative trend for half a century. Since estimates as detailed as given by Colebrooke are not available for other years, an alternative methodological procedure would be to use *jama* figures available for different years as indicators of the gross output from time to time. This would be so for two reasons: (i) revenue estimates invariably express a certain proportion of the gross *agricultural* output, and (ii) the drain was seen as the transfer of the surplus of Bengal's land revenue through the mechanisms of trade, taxation and remittances.[47]

We know that the *jama* in the first half of the eighteenth century corresponded to about 33 per cent of the gross agricultural produce[48] whereas in the second half it had risen to 40 per cent.[49] On this basis, Table 73 is an attempt to show how much of Bengal's gross agricultural

TABLE 73. THE DRAIN AS A PORTION OF THE GROSS PRODUCE, (SELECTED YEARS) (*Rs. in lakhs*)

Year	A *Jama* of Bengal	B Estimated agricultural product	C Amount drained	C as % of B
1722	14.29	42.87	13.5[a]	31.49
1728	14.24	42.72	10[a]	23.41
1756	18.684	56.052	10[a]	17.84
1768	22.165	55.4125	8.28[b]	14.95
1772	24.493	61.235	10.94[b]	17.86
1784	27.265	68.1625	10[b]	14.67
1790	21.743	54.357	10.3[b]	18.95
1793	26.801	67.002	5.41[b]	8.07

Note: [a] these figures are for specific years; [b] these figures are the averages of the series of years given in Table 72.
Sources: Rajat Datta, 'Rural Bengal', p. 58; Sirajul Islam, *Murshid Quli Khan*, p. 229; Verelst Papers, IOL, Ms. Eur. F. 218/5, fol. 30, F. 218/9, fol. 23, and F. 218/15, fol. 31; Marshall, *Bridgehead*, pp. 51-2; N.K. Sinha, *Economic History*, vol. 2, pp. 105, 157, 181; BM, Add. Ms. 29205, fol. 258.

output was actually eaten into by the drain. It should also be emphasized that this exercise should be made, data permitting, for both the halves of the century for any kind of comprehensibility to emerge because, as has been argued earlier, the transfer of imperial tribute from Bengal to Delhi by the Nazims was also unilateral and unrequited.

In fact the impression that one gets from Table 73 is that the rate of exploitation of Bengal's resources through the drain of wealth tended to stabilize at a high equilibrium as the century progressed with a marked decline towards the end, but the data are too fragmentary to allow any further elaboration.

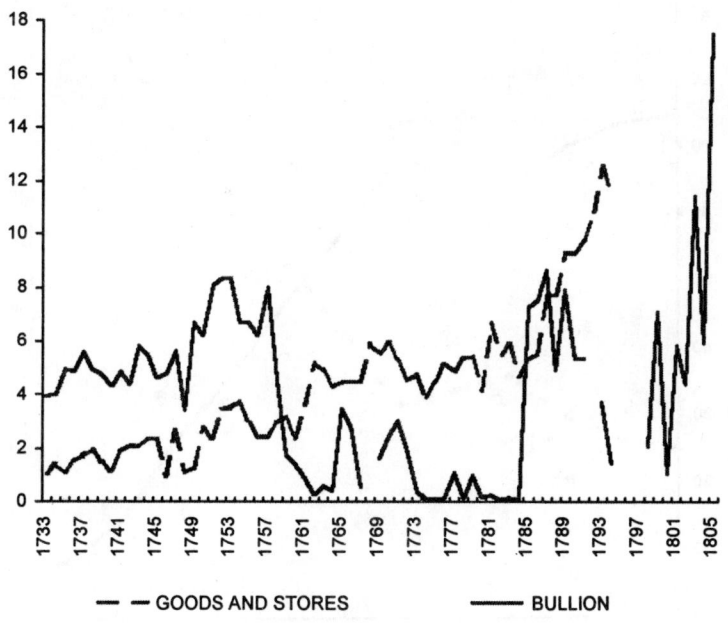

Note: Goods and Stores up to 1794 only.

Sources: *Reports, Third Report, 1773*, p. 75; IOR, L/AG/10/2/2, fols. 178, 211-13, 220; ibid., L/AG/10/2/4, fol. 148.

FIGURE 20. COMPANY'S GOODS AND BULLION EXPORTS TO ASIA, 1733-1805 (£ *in 10 thousand*).

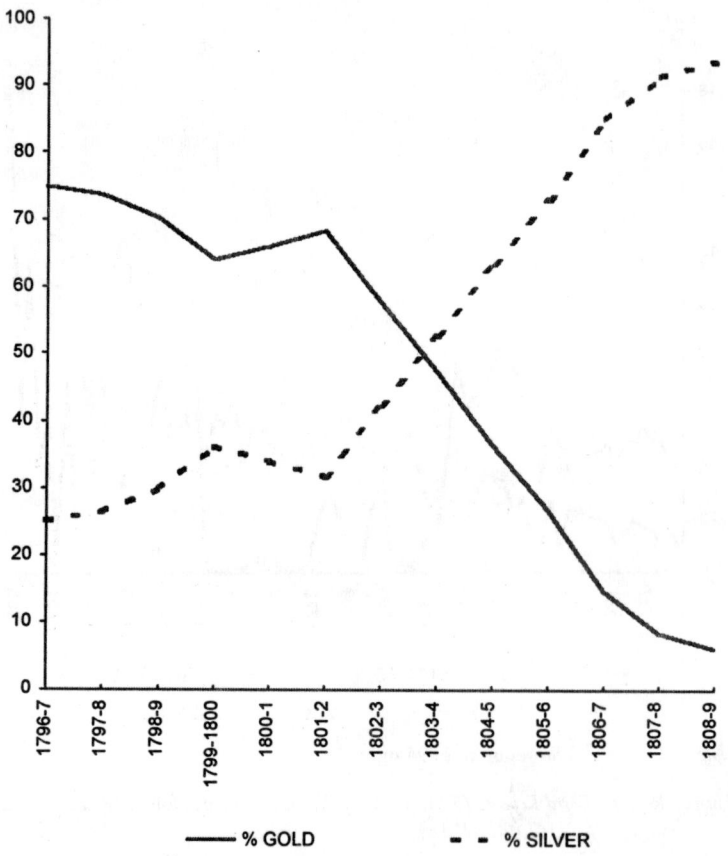

Source: D.B. Mitra, *Monetary System in the Bengal Presidency, 1757-1813*, Calcutta, 1991, pp. 76-8.

FIGURE 21. GOLD AND SILVER IN BENGAL'S REVENUE 1796-7 TO 1808-9 (*as % of total*).

APPENDIX 2

NOTES AND REFERENCES

1. Irfan Habib, 'Monetary System and Prices', in *CEHI, 2*; Irfan Habib, *Caste and Money in Indian History*, Bombay, 1985; for China W.S. Atwell, 'Notes on Silver, Foreign Trade and the Late Ming Economy', *Ch'ing-shih Wen-t'i*, vol. 3, no. 8, December 1977; also Atwell, 'International Bullion Flows and the Chinese Economy, *circa* 1530-1650', *Past & Present*, no. 95, 1982.
2. See particularly, Habib, 'Caste and Money', pp. 35ff.
3. See John Shore's Minute on currency and coinage in Bengal, 28 October 1789, IOR, BRC, P/51/50, 28 October 1789.
4. P. J. Marshall, *Problems of Empire: Britain and India, 1757-1813*, London, 1968, p. 79.
5. IOR, L/AG/10/2/2, p. 236.
6. One would perhaps be wrong to seek a connection between the reduction in the export of bullion to Asia from 1757 with the wealth taken privately by the British in Bengal after their victory at Plassey. This wealth, estimated at £ 21,69,665 between 1757 and 1765 (*Reports, Third Report, 1773*, pp. 311-12) has to be seen in relation to £ 24,09,260 in bullion exported to Asia during these same years (ibid., p. 75).
7. Compare, IOR, L/AG/10/2/4, p. 148 and L/AG/10/2/5, pp. 92-3.
8. See *Reports, Third Report, 1773*, p. 61 and IOL, Ms. Eur. D. 282, fol. 13 for figures.
9. H. Furber, *John Company at Work*, p. 253.
10. Between 1802-3 and 1805-6 Bengal's surplus of trade with China and London was *sicca* Rs. 74,87,678 and 1,98,50,526 respectively (A. Tripathi, *Trade and Finance in the Bengal Presidency*, Calcutta, 1956, pp. 88-9 for figures).
11. IOL, Ms. Eur. D. 282, fol. 13; IOR, L/AG/10/2/5, p. 86.
12. Milburn, *Oriental Commerce*, vol. 2, London, 1813, p. 137.
13. *Reports, Ninth Report, 1783*, pp. 110-11.
14. IOR, BPC, P/1/15, 26 March 1722.
15. Ibid., P/1/18, 17 July 1729.
16. 'Memoirs of Coinage in Bengal', 1772, IOR, HM, vol. 62, p. 163.
17. Shubhra Chakrabarti, 'Intransigent *shroffs* and the English East India Company's currency reforms in Bengal, 1757-1800', *IESHR*, vol. 34, no. 1, 1997, p. 79.
18.

Year	Company Investments in Bengal (£ sterling)	Bullion to Asia (£ sterling)
1760-1	3,56,850	92,136
1761-2	3,95,550	27,089
1762-3	3,20,077	56,857
1763-4	2,76,772	40,017

1	2	3
1764-5	4,37,511	3,45,404
1765-6	5,65,461	2,81,875
1766-7	6,58,341	54,968
1767-8	7,42,288	Nil

    Compare, IOL, Ms. Eur. D. 282, f. 13 and *Reports, Third Report, 1773*, p. 75.
19. Verelst Papers, IOL, Ms. Eur. F. 218/5, fol. 29-30.
20. BPC, IOR, P/1/18, 17 July 1729.
21. Letter to Court, 5 April 1769, para 8, *Fort William-India House Correspondence*, vol. 5, *1767-1769* (ed. N.K. Sinha), Delhi, 1959.
22. D.B. Mitra, *Monetary System*, pp. 181-2.
23. Ibid., pp. 32-5.
24. See J.C. Sinha, *Economic Annals*, p. 43.
25. *Reports, 1782-83*, p. 15.
26. See P. J. Marshall, *Problems of Empire*, pp. 78ff.
27. IOL, Ms. Eur. G. 33, fol. 139. The amounts are in *sicca* rupees.
28. *Fort William-India House Correspondence*, vol. 5, *1767-1769*, Delhi, 1959, 11 November 1768, p. 136.
29. 'Memoirs of Coinage of Bengal', IOR, HM, vol. 62, p. 164.
30. See ibid., pp. 60-1 for a detailed statement of the Company's income in Bengal between 1761-2 and 1770-1.
31. P. J. Marshall, *East Indian Fortunes*, p. 98.
32. IOR, L/AG/10/2/2, p. 235.
33. P. J. Marshall, *Problems of Empire*, p. 89.
34. Ibid., p. 90.
35. P. J. Marshall, *East Indian Fortunes*, p. 98.
36. P. J. Marshall, ibid., p. 255. J.C. Sinha's estimate of £ 2,40,00,000 worth of bills between 1757 and 1780 is a highly inflated amount (*Economic Annals of Bengal*, p. 54).
37. IOR, L/AG/10/2/3, pp. 52 and 210.
38. James Steuart, 'Memoirs of Coinage of Bengal', IOR, HM, vol. 62, pp. 163, 165; Steuart, *The Principles of Money Applied to the Present State of Coin of Bengal*, London, 1772, pp. 62-3.
39. J.N. Sarkar, *Bengal Nawabs*, Calcutta, 1952, p. 14; IOL, Verelst Manuscripts, Ms. Eur. F. 218/5, fol. 42; IOL, Orme Mss. 'India', IV, fol. 875 cited in Marshall, *Bridgehead*, op. cit., pp. 51-2.
40. Balmukund Mehta, *Balmukund Nama*, Letter to Murshid Quli Khan, October 1719 and April 1720 (ed. and trans. Satish Chandra), Bombay, 1972.
41. IOL, *Tracts*, vol. 120, pp. 63-4.
42. *Balmukund-Nama*, pp. 44, 80.

43. *Reports, Third Report,* 1773, p. 535.
44. J.C. Sinha, *Economic Annals of Bengal,* p. 54.
45. Ibid., p. 49. Siphonage of Bengal's resources (the drain of wealth through remittances of private fortunes to England and through bills of exchange) should not be conflated with the question of bullion being sent to other presidencies for in the latter case wealth sent from Bengal was utilized within India albeit for military purposes. It would be very difficult to prove that the Mughals were using the funds from Bengal for purposes other than the strengthening of the state.
46. H.T. Colebrooke calculated the gross output from Bengal *c.* 1793 as Rs. 32,91,30,000 (*Remarks,* pp. 15-16).
47. For the latter point see James Steuart's cogent exposition in IOR, HM, vol. 62, pp. 163-5.
48. Sirajul Islam, *Murshid Quli Khan,* p. 90.
59. John Shore's Minute of 18 June 1789, in W.K. Firminger (ed.), *FR 2,* p. 192.

Bibliography

Only primary sources—manuscripts and other unpublished records, and other contemporary and near contemporary published works—which have been used in this book are being listed here. References to secondary sources have been limited to the text.

Manuscripts

AT THE BRITISH MUSEUM

Add. Ms. 19286	F. Buchanan, 'A Journey Through the Provinces of Chittagong and Tippera, 1798'.
19288	Sir John Anstruther, 'A Report on the Commerce of the East Indies, 1811'.

HASTINGS PAPERS

Add. Mss. 29086 to 29088	'Papers Relative to The Aumeen [Amini] Commission'.
29089	General Account Books, 1778-1785.
29021	Papers Relating to Ganga Govind Sing, Dewan of the Calcutta Committee of Revenue.
29202	Particulars of the Canongoes of Bengal, 1786.
29205	Papers Relating to the Revenues of the East India Company in India.
29209	Memoirs and papers relating to the History, Geography and Trade of India and the Adjacent Countries.
29120	History of Kissen Gur [Krisnanagar], the Zemindarry of Raja Kissen Chund.
29120	'A History of Jungle-Terry [Tarai] Disticts' by James Brown, 1776.
29120	'Remarks on the Rivers and Coasts of Bengal' by James Ritchie, 1775.
29131	Original Letters of Robert Clive.
29132	Correspondence with Harry Verelst and Luke Scrafton.
29132	Correspondence with Henry Vansittart.

29233   'Copy of a Memorial on the High Prices of Grain, 1800'.

OTHER MANUSCRIPTS

Add. Ms. 34123   Register of Papers Relating to the English and Dutch East Indies.
44061   Letters and Papers of Robert Clive.
45429 to 45435   Anderson Papers: Correspondence of Anderson with Richard Becher.

AT THE INDIA OFFICE LIBRARY

Ms. Eur. D. 281   Historical Review of the External Commerce of Calcutta, 1750-1850.
D. 282/G. 33   Tables of External Commerce of Bengal, 1758 to 1830.
D. 283   Historical Sketch of the Taxes on the English Commerce in Bengal, 1633 to 1820.
D. 284   Memoranda on the Trade of India.
D. 75   F. Buchanan, 'Survey of Ronggoppur'.
G. 11   'Statistical Tables of Ronggoppur'.
Francis Papers   Mss. Eur. D. 22 and 23.
                      Mss. Eur. E. 12 to 14, 27, 28, 32, 37 to 39.
                      Mss. Eur. F. 7 and 8.
                      Mss. Eur. G. 6, 8 and 14.
Ms. Eur. F. 95   Robert Kyd, 'Some Remarks on the Soil and Cultivation on the Western Side of the River Houghly'.
Orme Mss   O.V. 28 and O.V. 134.
                      'India', VII and XVII.
Verelst Papers   Mss. Eur. F. 218/2 to 33.

Unpublished Records

AT THE BRITISH MUSEUM

Add. Mss. 29016 to 29036   Proceedings of the Council at Fort William, 1774-1777.
29071 to 29073   Minutes, Council at Fort William, Revenue Department, 1775-1778.
29074 to 29075   Extracts of Revenue Minutes, 1775-76 and 1779-80.

29706   Proceedings of the Committee of Circuit at Nadia.
29077 and 29078 Reports from the Provincial Councils of Revenue.
29082 and 29083 Councils Revenue Transactions.
29090   Abstracts of Revenue, 1772-1785.
29094   General Statements of Salt Accounts of Bengal, 1770-1785.

AT THE INDIA OFFICE RECORDS

Accountant General's Department, General Ledgers and Accounts Books.
Bengal Public Consultations.
Bengal Revenue Consultations.
Bengal Board of Revenue Proceedings.
Bengal Board of Revenue, Miscellaneous Proceedings.
Board of Control & Parliamentary Accounts, L/AG/10/2/2.
Factory Records at Dacca, Qasimbazar and Murshidabad.
Home Miscellaneous Series.
Letters from Coast and Bay.
Letters Received from Bengal.
Proceedings of the Calcutta Committee of Revenue.
Proceedings of the Committee of New Lands.
Proceedings of the Committees of Circuit.
Select Committee Consultations.

AT THE WEST BENGAL STATE ARCHIVES, CALCUTTA

Board of Revenue, Extracts from Miscellaneous Revenue Proceedings.
Chittagong Records, Letters Received, 1771-1785.
Controlling Council of Revenue at Murshidabad.
Controlling Council of Revenue at Murshidabad, Appendix to Proceedings.
Letter Copy Book of the Resident at the Darbar.
Proceedings of the Board of Revenue at Fort William.
Proceedings of the Controlling Committee of Revenue.
Proceedings of the Board of Revenue, Khalsa Branch.
Proceedings of the Committees of Circuit.
Proceedings of the Provincial Councils of Revenue:
    Dacca   1774 to 1779
    Murshidabad 1773 to 1780
    Burdwan   1774 to 1779
    Dinajpur   1774 to 1780
Proceedings of the Select Committee.
Proceedings of the Board of Revenue, Judicial Branch.
Revenue Department, Proceedings of the Governor-General in Council, Grain Branch.

Revenue Department, Proceedings of the Governor-General in Council, Sayar Branch.

AT THE NATIONAL ARCHIVES, NEW DELHI

Proceedings of the Public Consultations, Secret and Separate (1768-1773).

Persian Treatises in Translation

Anonymous, *The Risala-i-Zir'at* (tr. Harbans Mukhia), *Perspectives in Medieval History*, Delhi, 1993.
Hossein, Seid Gholam, *The Seir Mutaqherin*, vol. III (original 1783, 1st published 1926), Delhi, 1990.
Mehta, Balmukund, *Balmukund Nama* (ed. and trans. Satish Chandra), Bombay, 1972.

Contemporary or Near Contemporary Printed Works

Bolts, W., *Considerations on Indian Affairs*, London, 1772.
Boughton-Rouse, C.W. B., *Dissertation Concerning Landed Property in Bengal*, London, 1791.
Clive, Robert, *An Address to the Proprietors of the East India Company Stock*, IOL, *Tracts*, vol. 113
Colebrooke, H.T., *Remarks on the Husbandry and Internal Commerce of Bengal*, Calcutta, 1793 (Calcutta, 1884).
Dalrymple, A., *Plan for Extending the Commerce of this Kingdom and the East India Company*, London, 1799, IOL, *Tracts*, vol. 84.
Gladwin, F., *A Narrative of the Transactions in Bengal*, Calcutta, 1788.
Hastings, W., *Memoirs Relative to the State of India*, London, 1786.
Holwell, J.Z., *India Tracts*, London, 1774.
——, *Interesting Historical Events Relative to Bengal and the Empire of Indostan* (in 2 parts), London, 1765 and 1767.
Johnstone, J., *Letter to the Proprietors of the East India Stock*, London, 1776, IOL, *Tracts*, vol. 50.
Orme, R., *Historical Fragments of the Mogol Empire*, London, 1782.
Pattulo, H., *An Essay on the Cultivation of Lands and Improvements of the Revenue of Bengal*, London, 1772.
Price, J., *Five Letters From a Free Merchant in Bengal to Warren Hastings*, London, 1783, IOL, *Tracts*, vol. 53.
Prinsep, J., *Bengal Sugar*, London, 1794, IOL, *Tracts*, vol. 436 A.
Rennell, J., *Bengal Atlas: Containing Maps of the Theatre of War and Commerce on that Side of Hindostan*, London, 1781.
——, *Memoir of a Map of Indostan*, London, 1793.

Scrafton, L., *Reflections on the Government & ca of Indostan*, London, 1763.
Smith, N., *Observations on the Present State of the East India Company*, London, 1771, IOL, *Tracts*, vol. 84.
Tennant, Reverend W. , *Indian Recreations*, 3 vols., Calcutta, 1804-1808.
Vansittart, H., *Narrative of the Transactions of Bengal*, London, 1766.
Verelst, H., *A View of the Rise and Progress of the Present State of the English Government in Bengal*, London, 1772.

Index

abadkari process 71
agrarian commercialism 28–30, 112–13, 324
agrarian economy 164–6, 185–200, 216–20, 328–9
agricultural price movement 66–8, 194–200
agricultural production
  agricultural price movement impact on 66–8
  commercialisation of 21–30, 108, 112–13
  cultivation cost 62–3
  dynamics of 66–8
  global economic network impact on 28
  inefficiency of 54–6
  land use and labour involvement in 38–54
  merchants' participation in 26, 217–19, 331
  organization of 38
  output 38–54
  pattern and processes of 37–75
  political developments impact on 28–30
  rationality of 56–66
agricultural reclamation 68–75
al-tamgha land 160
Alamgir 256, 302
aman rice 39
Amini Commission (1778) report 74, 99–100, 102, 121, 159, 163, 264
Asiatick Researches 66, 196
aus rice 39

Bailey, W.B. 66, 196–7
Bajekal 26

Bakarganj, socio-economic development of 68, 70–1
Banerjee, Kumkum 26
bargadari system 105–6, 109, 112
Bayly, C.A. 22, 107
baz-i-zamin (revenue free) lands 102, 164–6
Becher, Richard 162–3
Bengal
  absence of fallowing 55
  agrarian economy of 161–6, 185–200, 216–20, 328–9
  agricultural prices during dearth and famine 254–6
  agricultural prices movement 194–200
  agricultural reclamation 68–75
  British conquest of 28–9
  cash crops cropping pattern 56, 241, 324
  choice of crops by peasants in 46–54, 61
  circumvention strategies 325–7
  commercial transactions 185–220
  Company's role in economy of 296–304
  crop rotation in 55–6
  cultivation pattern in 56–7, 241, 265, 324
  dearth and famine in 238–71, 285–94, 325
  demand and consumption pattern in 185–94
  dependent commercialization 327–8
  ecological crises and vulnerability 248–54
  ecology and crises of subsistence 241–8
  economy and factor of demand 185–94
  food market during dearth 188–9, 194–200

food shortages in 239–40
horticultural techniques introduced in 58–9
integration in global economic network 28
jotedari system 102–7
la-kharaj 164–6
landed proprietors and economy 161–4
local marketing networks 200–6
manure use in agriculture in 54–5
markets proliferation in 29–30
merchants of 213–20, 285–94, 325, 327
migration from during famine 268–71
occupations in rural society 189–91
peasant production and agrarian commercialism in 21–30, 108
peasant stratification in 113–23
peasantry and peasant satisfaction 88–99
political dimensions 28–9
population of 265–6
price control and profits by merchants 213–16
price differential between town and country 199–200
reconstruction of peasant society 119–21
regional food markets 194–200
revenue collection and relief in famine 256–60
rich peasants in 122–3
river system in 68
rural elite and local economies of 166–76
sharecropping in 107–13
social development in 70–1
soil fertility 37
state attitude towards markets 200–6
subsistence crises 241–8, 285–315, 326
talluqdari in 161–4
villages in 187–8
zamindari official and influenced ryotts collusion in 119–20
zamindars source of income 157–61
Bernstein 91
betel-leaf
commercial importance 48–9
cost of planting 48
cultivation of 47–9
merchants 201
monopolies in trade in 49

betel-nut cultivation 62
Bhardwaj, Krishna 331
Bhattacharya, S. 299, 330–1
Bhoj Raj 291, 293
boro rice 39–40
Bose, Sugata 91, 121
Buchanan, Francis 44, 48–9, 54, 56, 72, 111–16, 191, 194, 210, 306
Buchanan-Hamilton survey 38, 57, 90, 113–14
byapari, social profile of 208–11

caddurie land 39
cash advances system, for rice cultivation 25
cash cropping 25, 56, 241, 324
chakeran-zamin 165
Chaudhuri, B.B. 91–3, 104, 262, 289, 313, 324
circumvention strategies 325–7
Colebrooke, H.T. 26, 47, 49, 51, 55–6, 60, 63, 90, 120, 198, 218, 303
colla lands 246
comar *pahikashta* 98
commercial transactions
agrarian economy and 185–220
byapari social profile 208–10
factor of demand and economy 185–94
intermediaries role 211–13
local marketing networks 206–8
mercantile strategies 211–13
merchants and 213–20
price control and profits by merchants 213–16
price movements 194–200
regional food market and 194–200
state control circumventing 211–13
state's attitude towards markets 200–6
consumption loans to peasants 217–18, 252, 314
Cooper, A. 113
Cornwallis 175, 264
Cotton
commercial importance of 53
cost of cultivation 63
cultivation of 51–3
impact of famine on production of 302–4, 328
industry 302–4, 328

INDEX

investments in 51
output 534
Curly, D.H. 210, 292

dalals, balances with 295–6
damages due to inundation 247
dearth and famines
   agricultural prices during 254–6
   causes of 238
   desertions in 262–6
   East India Company role in 287–90
   ecological crises and vulnerability 248–54
   ecology and subsistence crises 241–8
   food shortages and 239–40
   food supply during 285–94
   harvest damaged in 250
   institutional relief 259–60
   merchants strategies in 285–94, 325, 327
   migration in 268–71
   mortality 260–8
   revenue collection and relief during 256–60
   state's role during 292–3
dewanaya system 116
Dharm Narain 67
domination effect 331
dowrah land 39
Dreze, Jean 239
droughts 243–6

East India Company, role in economy of Bengal 287–90, 329–30
Eaton, Richard 22, 70
ecological crises and vulnerability 248–54
ecology and subsistence crises 241–8

famines in Bengal 24, 68, 73, 91–4, 97, 100–1, 118–19, 238–71, 285–94, 325
   agrarian consequences of 92–4
   mortality 94, 260–8
   see also dearth and famine in Bengal
floods 246–7, 250–2
food grains
   market 188–9, 194–200
   shortage 239–40

supply 285–94
trade in 207
Francis, Philip 306

gacchoya paan 47, 49
ganthidar 102–3, 105
Geertz, Clifford 24
Ghoshal, Gocul 73
Ghoshal, Jaynarain 72
Gibson, A.J.S. 196
Gleanings in Science 196
golahs, proliferation and advantages of 326–7
Greenough, Paul 242
Grose, John 259
gutchdari tenure 102–4
   zamindars' role in formation of 102–3

Hamilton, W. 187
Hari Das 58
Harrington, J.H. 114–15
harvest seasons 38
hasil land 164–5
Hastings, Warren 37, 93, 213, 244, 257, 260–1, 286, 292–3
Herklotts, G. 196–7
Huang, Philip C. 327
Hunter, W.W. 73, 92–3, 254
Husken, F. 107
Hussain, Akhtar 197
Hussain, Ghulam 261

Irish solution, to problem of famine and dearth in Bengal 59

jala (low-lying) lands 39–40, 246
jalpai land 74
jama on *khudakashta* and *pahikashta* peasants 94–5
Johnstone, John 187
jotedari system 90–1, 102–7
   bargadar relationship 106
   factors influencing shape and size of 103–5
   labour deployment and control in 105
   trends in 102–7
   zamindars' interest in evolution of 103, 106

jotedars 90–1
jute cultivation 2

kasama system, Phlippine 107
kedokan system, Java 107
Keydar, Caglar 23
khamar lands 110, 157–61
Khan, Alivardi 200
Khan, Jung Sayyad Ali 168
Khan, Mohammad Reza 259, 289
Khan, Murshid Quli 200
khas pateet land 73
khas talluqas of Murshidabad 162
khudakashta peasants 92–9, 104
Kishan Mangal 292–3
Kyd, Robert 39–40, 42, 54, 58–9, 68, 190

la-kharaji and agrarian economy 164–6, 328–9
Labrouse, C.E. 195
landed proprietors 161–4
leguminous plant, output of 40–1
lentils and vetches production 40–2
local marketing networks, commercial profile of 206–8
Ludden, David 23

MacDowall, D.H. 212
mahsulat pateet land 73
markets
  commercial profile of networks of 206–8
  East India Company's regulation 203–4
  integration 195–6, 198, 202
  Nizamat's attitude towards 200
  proliferation of 205–6
  state attitude towards 200–6
  zamindars and talluqdars control over 204
Marshall, P.J. 22, 271
Marx, Karl 23
mazkuri talluqa 161–2
mercantile intermediaries 27
merchants
  agrarian economy and 216–20
  and food supply in famine 285–94
  price control and profits by 213–16
Mubarak-ud-daulah 202, 259, 291
mulberry, cultivation cost of 62–3, 65

muqarrari tenure 99, 104

najai tax 93–4
nij-jote land 157, 159
no-abad land 72
Nondi, Thakurdas 293
Norfolk system, of crops rotation 56

oil seeds, cultivation and production of 42–3

pahikashta peasants 93–9, 106
Paterson, J.H. 89, 110, 191, 193
pateet (waste) land 164–5
peasant economy
  commercialised agriculture and 24
  in China 23–4, 26
  in Japan 23–5
  in pre-colonial Asia 23–6
  la-kharaji and 164–6
  price of rice impact on 25
  rural elite and 166–76
  talluqdars and 161–4
peasant household, labour-consumer balance in 23
peasant production
  agrarian commercialism and 21–30
  antipathy 59–60
  cash crops 56, 241, 324
  constraint to cultivation of different crops 64
  crops selection 46–54, 61
  cultivation methods and land use 58
  cultivation cost 62–3
  food grains 56, 63
  impact of land revenue on 64–5
  oil seeds 56
  perception of risks 60
  potatoes cultivation 59–60
  pulses 56
  rationality of 56–66
peasants of Bengal
  categories of 121–3
  consumption loans 217–18, 314
  coping and recovery from subsistence crises 305–15
  famine and 91–3
  jotedari system 102–7

jotedars 90–1
khudkashta 92–3
land holdings 88–9, 113–17
loans to 217–20, 311–14
pahikashta 92–9
production loans 218–20, 252
residents 93
rich peasants 122–3
satisfaction 88–99
sharecroppers 89–90
sharecropping by 107–13
society 99–102, 113–23
stratification of 113–23
village *mandal* and 99–102
Postan, M.M. 194
Pott, R.P. 210–11
Pre-colonial peasant economy, of Asia 23–4
price movement 194–200
production loan 218–20, 252

Ray, Ratnalekha 92–3, 113, 121, 313
reclamation of agricultural land 68–75
regional food market 194–200
Rennell, James 37–8, 187, 246
resident peasants 93
rice
 cultivation cost 63
 cultivation of 38–40
 land use for cultivation of 39
 manure use for 54
 preparation of land for 39
 price movement 196–8
 prices under dearth 255
 production of 25–6, 329
Robb, Peter 328
Roy, Shitab 259, 261
Roy, Durlabh 259
Rumbold, Thomas 259, 288
rural elites, of Bengal 166–76

Sanyal, H. 118, 210
Sen, Amartya 239
sharecroppers 89–90, 105
sharecropping 90, 107–13
 agrarian commercialism and 112–13
 appropriation of relations in 108–10
 crop division in 108–9
 factors responsible to take to 110–11

 systematic use of 110
Shore, John 90, 118–19, 166, 204, 214, 261, 265, 291
sih-fasli (three harvest) land 56
silk industry, impact of famine on 298, 301–4
Sinha, N.K. 91–3, 262, 268, 288
Skinner 206
slave labour 160
Smout 196
Stuart, James 218
subsistence crises
 coping with and recovery from 305–15, 326
 ecology and 241–8, 285
 food supply in 285–94
 merchant's strategies in 285–94, 325, 327
 rural textile artisans in 294–305
sugarcane
 commercial advantage of production of 61–2
 cultivation of 43–7
 labour utilization in cultivation of 45–6

talluqdars 161–4, 328–9
talluqdari-raiyat relations 163–4
Taniguchi, S. 113–18, 123
taqavi fund 163
Taylor, James 39, 247, 304
Tennant, William 37–8, 55
textile producers, impact of famine on 296–305, 328
thika pahikashta 98
Tiwari, Ganesh Das 49
Tobacco
 cultivation of 50
 output 49, 51
trading intermediaries 211–13

vagrant peasants 92–3, 95
Vansittart, George 73
Vansittart, Henry 201, 203
village *mandal*
 duties of 100
 election of 100, 102
 land holdings 100
 nature of 99

376 INDEX

peasant society and 99–102
power of 101
voroj (betel-leaf) garden 47

zamindars
  cash disbursements 172
  cash salaries and wages of officials of 173–5
  East India Company interest in 159
  expenditure patters of official serving 174–5
  expenditures of 171–4
  income source of 157–61
  interest in evolution of *jotedari* 103, 106
  khamar lands 157–61
  landed proprietors and economy 161–4
  personal expenses 172
  production in *khamar* or *nij-jot* lands 159–6
  slave labour use 160
  talluqdars 161–4, 328–9
  unpaid utilization of labour by 159–60